BASIC GUIDE TO
Industrial Hygiene

BASIC GUIDE TO Industrial Hygiene

Jeffrey W. Vincoli, CSP

JOHN WILEY & SONS, INC.

New York • Chichester • Weinheim • Brisbane • Singapore • Toronto

Originally published as ISBN 0-442-01960-2

Published simultaneously in Canada.

Library of Congress Cataloging-in-Publication Data

Vincoli, Jeffrey W.,
    Basic guide to industrial hygiene / Jeffrey Vincoli.
        p.   cm.
    Includes bibliographic references and index.
    ISBN 0-471-28678-8
    1. Industrial hygiene. I. Title. II. Series
RC967.V56 1995
363.11—dc20
                                94-38553
                                CIP

10 9 8 7 6 5 4 3 2

To my brother,
Joseph A. Vincoli,
and my sister,
Cheryl Ann Parker,

in appreciation for their lifelong support and understanding;
for their many personal, selfless sacrifices;
for their outspoken confidence in me;
and for the valuable lessons of life that I have learned from them.

Contents

Preface

The *Basic Guide to Industrial Hygiene* is the fourth in Van Nostrand Reinhold's series of Basic Guides. As is the case in the other books of this series, each of which focuses on topics of concern and general interest to today's practicing safety and health professionals, this text has been designed to provide the reader with a fundamental understanding of the subject. It is hoped the reader will also have a desire to pursue further, more detailed study.

This text has been written for practicing safety professionals who may have had little training in the area of industrial health and hygiene but suddenly become responsible for these programs within their respective organizations. The *Basic Guide to Industrial Hygiene* provides the information upon which most industrial hygiene programs are based. Because its objective is to focus on only the basics of industrial hygiene, it is not the intent of this text to provide more than a novice's level of expertise in the field. Practitioners who require a more comprehensive knowledge will not be satisfied with the information presented herein. It neither practical nor feasible to expect a basic guide book to provide all there is to know on any given subject, especially one as complex as industrial hygiene. However, readers interested in acquiring a basic understanding of industrial hygiene discipline should find this material somewhat useful. The seasoned professional might find this material somewhat enjoyable or, at the very least, refreshing.

Shrinking corporate budgets continue to force organizations to implement increasingly innovative cost-saving measures. Streamlining the work force has surfaced as one method of reducing operating costs while (hopefully) improving efficiency. Occupational health should always be the shared responsibility of each member of the organization. Indeed, many forward-thinking companies have established successful occupational safety and health programs based upon the principal that safety is everyone's responsibility. In recent years, however, heightened emphasis on doing more with less has affected the way businesses choose to pursue their safety and health programs' goals and objectives. Now more than ever before, all members of an organization must have a fundamental understanding of the principals of occupational safety and health management.

This *Basic Guide to Industrial Hygiene* should, therefore, prove to be a quite useful tool for non-safety/health professionals as well.

In short, managers, engineers, occupational medical professionals (industrial nurses, doctors, toxicologists), safety professionals, academia, and professional consultants should obtain some level of benefit from the basic information provided in this text.

Acknowledgments

The author wishes to acknowledge those individuals and organizations who have assisted in the successful development of this publication.

First, thanks to my faithful proofreaders and reviewers, Susie Adkins, George S. Brunner, Steven S. Phillips, and Frank Beckage for their contributions and suggestions. As with the previous three titles of this Series, their help and assistance proved invaluable and is greatly appreciated.

Second, a special thanks to Bruce Kelly, CIH, and the other industrial hygienists who reviewed this manuscript prior to publication for helping to insure the accuracy and adequacy of the materials presented herein. I am certain that their suggestions and critiques help to make this a more usable publication.

Third, thanks to Bob Esposito, Ken McCombs, Barbara Mathieu, Kathy Bazane, and the rest of the staff at Van Nostrand Reinhold for their support and assistance in making this text a success and the Basic Guide Series a reality.

Fourth, thanks to Jeff Eckert and his staff at J.K. Eckert & Company, in Sarasota, Florida, for their competent and professional copy editing, composition, and layout work on this publication.

Finally, a very special thanks to James L. Barfield, CSP, MPH, for his contribution of materials that comprise Chapter 10 and most of Part III of this text, as well as for assuring the overall technical accuracy of the information presented herein.

I
Understanding Industrial Hygiene

INTRODUCTION TO PART I

The modern-day occupational safety and health professional must address a great number of issues and concerns on a daily basis to effectively ensure the preservation of a safe and healthy work environment. While this responsibility usually involves the recognition (identification), evaluation (assessment), and subsequent elimination or control of hazards that threaten both the safety and health of personnel, those who work in this field have generally specialized in either the *safety* aspect or the *health* aspect of the equation. In some cases the practitioner has had to serve in *both* capacities. This, as one can imagine, is an extremely cumbersome task. The occupational health profession, more commonly referred to as *industrial hygiene,* is examined in this *Basic Guide.*

Chapter 1 introduces the reader to the discipline of industrial hygiene, with a brief analysis of the history of its development and its place in the modern industrial setting. This will include a discussion of those professionals who specialize in the study and practice of occupational health, known as *industrial hygienists.* A review of the necessary organizational interrelationships that must occur to ensure proper implementation of an effective industrial hygiene program is also offered. This will help the reader understand the essential role of the occupational health professional. Also, since optimum human performance in the workplace is the key to the success of most business endeavors, this Chapter also examines occupational factors that can adversely affect the health of the worker.

Government regulations, a driving force in both the occupational safety and health arenas, mandate very specific requirements to ensure the provision of a healthy workplace. The reader must have a fundamental understanding of these regulatory requirements as a prerequisite to comprehending the importance of industrial hygiene in today's work environment. Therefore, a review of these regulations is the focus of Chapter 2. They are further discussed, as required, throughout the entire text.

Chapter 3 considers the basic requirements for establishing an effective industrial hygiene program. Implementation of such a program begins with proper *planning*. Successful organizations are those that develop and execute clear and effective operating plans to achieve their stated objectives. Planning for the implementation of an industrial hygiene program requires an equal commitment from management. Ensuring the worker's health is no less critical a task than any other that might be considered by management. *Management commitment* to the industrial hygiene effort is, therefore, essential to successful business operations. Finally, the importance of proper *record keeping and documentation* will also be briefly discussed.

Chapter 4 completes Part I with an extensive, albeit very basic, discussion on human health. To properly perform the industrial hygiene function, a fundamental understanding of human anatomy and the cause and effect relationships that exist is essential. Knowing how the human system will respond to various types of hazardous exposures is a major aspect of the industrial hygiene process.

On completion of Part I, the reader should have developed a basic understanding of and appreciation for industrial hygiene and its role in the occupational safety and health process. With this foundation, the information in Part II on the various types of health hazards that can exist in the workplace will further demonstrate the importance of industrial hygiene programs in the working environment.

1
Background and Overview

INTRODUCTION

Hazards to human health can exist in a variety of forms and degrees in any work environment. They can be insidious and not easily detectable, as is the case with exposure to asbestos. They can also cause instant irritation and/or damage to human health, as is typical on contact with corrosive chemicals. Whatever the case, it is important to understand that hazards to human health can be present in some form in every type of business endeavor. Even a workplace as seemingly benign as an office environment can contain health hazards associated with ventilation, illumination, temperature, and the like.

Over the years, the ability to recognize and identify such hazards in the work environment has become the basis of a specialized field of study known as *industrial hygiene*. The application of modern industrial hygiene is further concerned with the *control* of the occupational health hazards that may develop during the performance of normal (as well as non-routine) work. Industrial hygiene can be defined as both a science and an art. Those who practice it are concerned with the anticipation, recognition, evaluation, and the elimination or control of any environmental factors and/or stresses that may arise in or from the workplace. These hazards may cause sickness, impaired health, or significant discomfort among workers or among citizens of the community (NSC 1988). The *control* of health hazards is a major element of consideration. This aspect of the industrial hygiene profession is, perhaps, its most important contribution. The recognition and evaluation of these hazards are, of course, essential to the process. But developing and implementing effective control measures to either eliminate or reduce exposure to these hazards is the *primary objective* of the entire industrial hygiene effort.

HISTORY AND DEVELOPMENT

While concern for understanding health hazards in the work environment has been the focus of much attention in recent years, it is certainly not a new development. In fact, the relationship between the work environment and workers' health and well-being was recognized many hundreds of years ago. Evidence of this can be traced back as far as the early Roman and Egyptian Empires, where only slaves and prisoners would be used for tasks that were considered (or known) to be dangerous. Nearly four centuries before the birth of Christ, for example, Hippocrates is thought to have been the first to recognize and document the toxic effects of exposure to lead fumes. Subsequently, slaves were typically used during smelting operations. When they expired, others were brought in to replace them. Unfortunately, since slaves were generally considered expendable anyway, not much was ever done to address the problem further.

Later, during the first century A.D., the Romans made record of understanding the hazards of working with compounds such as sulfur and zinc. Pliny the Elder (Gaius Plinius Secundus, A.D. 23–79), a Roman of some renown during that period, described a protective mask, made from a bladder, that was used by workers in the dusty trades (Brief, 1975). Approximately 100 years later, Claudius Galenus (Galen) of Greece, a noted physician and writer on medicine and philosophy, accurately described the pathology of lead poisoning.

In 1473, Ulrich Ellenbog produced the first known publication to focus on occupational health. The pamphlet discussed the nature of occupational disease and injury among gold miners. His noted contributions also include writings on the toxic action of carbon monoxide, mercury, lead, and nitric acid. He provided training in occupational health and offered information on simple preventive measures for controlling exposure to these health hazards (Brief, 1975). During the next century, a German scholar named Georg Bauer Agricola (1490–1555), described the diseases that were common among miners and also prescribed simple measures that could be taken to prevent the occurrence of these diseases. Agricola, who was known as the *Father of Mineralogy* for his first-ever scientific classification of minerals, died a year before his writings on these health hazards were published.

In 1700 Bernadino Ramazzini (1633–1714) of Italy, the *Father of Industrial Medicine,* published the first comprehensive text on occupational medicine. His book, entitled *The Diseases of Workmen,* contained very accurate descriptions of many of the occupational diseases that were common during his time. For example, he was probably the first to correctly describe the pathological effect of silicosis. He also observed how workers who hammered copper suffered from hearing loss and eventual deafness. Unfortunately, although he provided a variety of suggested methods and ideas for preventing such problems, his work went largely ignored for the next several centuries.

By the end of the 1700s and into the early 1800s, the Industrial Revolution was in full swing in England. Charles Thackrah (1775–1833) devoted much of his professional life to the study and prevention of occupational hazards. Through his many published works, his influence extended across the Atlantic to a young America where, in 1837, the first article on occupational disease was published with heavy emphasis on the teachings of Thackrah.

In 1893, Alice Hamilton, M.D. (1869–1970), graduated from the University of Michigan Medical School. She is remembered most for her contributions to the field of occupational health. She founded the discipline of industrial medicine and, subsequently, saved thousands of worker's lives through the identification of the source of toxic substances in the workplace (Rhodes and Blanton, 1990). She worked tirelessly to convince factory owners that improving working conditions was a good business practice.

The tremendous changes that accompanied the Industrial Revolution intensified some existing concerns. It also ushered in new and different hazards to health and well-being, as well. By 1910, the Harvard University Medical School, together with the Massachusetts Institute of Technology, had developed specific courses to study the problems of industrial and occupational health. By 1918, the first known degree in Industrial Hygiene was granted.

The dramatic increase in the number of workers during this era resulted in a proportionate increase in hazard exposure potential, occupational illness, and injury. This alarming by-product of mechanization eventually led to the passing of the first legislation to focus on the protection of worker health and safety. The maturation of industrial hygiene in the United States is linked closely to the development of legislation aimed at protecting workers' health and safety and compensating them for occupational injuries and illnesses. Initially, workers who were injured on the job could only obtain compensation by suing their employer. Employees had to *prove* that their injuries occurred during the performance of their normally assigned duties. These early cases were decided based on the common law principles in existence at that time. For example, the *fellow servant rule* stipulated that employers could not be held liable for an employee's injuries if those injuries were, in fact, caused by the negligence of another employee. Similarly, the *contributory negligence rule* relieved an employer of liability if it could be proven that the injuries were the result of the employee's own negligence. Employees seeking compensation for injuries received at work would also have to negate the *assumption of risk rule* that basically stated that employers were not liable for such injuries because the employee accepted the job assignment with the full knowledge and understanding of the risks and hazards associated with that job. Furthermore, in those early years, an occupational illness was only compensable if it could be classified in some way as an accident. Of course, these examples are in marked contrast to most of today's State Worker's Compensation Laws. As today's laws matured, so did the need for methods, techniques, and procedures for protecting worker safety and health. In 1908, the United States passed a compensation act for certain civil service employees. Individual states (New York was the first) began passing their first worker's compensation laws in 1911. By 1920, 42 states had their own laws; when Mississippi's state law was finally enacted in 1948, workers in all states were covered by some form of worker's compensation legislation. Federal agencies, such as the U.S. Public Health Service, began the first official investigations of occupational diseases as early as 1910.

As the principle of compensation for the injured worker slowly matured in the United States, the importance of industrial hygiene in this country became the focus of increasing amounts of political and social attention. Today, nearly every State Worker's Compensation law allows for the consideration of occupational illness and/or disease as a compensable condition. U.S. industry has the *moral, ethical,* and, since the enactment of the Occupational Safety and Health Act in 1970, the *legal* responsibility to provide a healthy workplace for its employees. But, now that occupational illness and disease are covered by most worker's compensation laws, it also makes good business sense from an economic perspective. As with many business issues, ensuring the preservation of employee health has become a matter of protecting profit margins and decreasing expenditures (in the form of compensation and insurance premiums). This heightened awareness of the importance of a healthy work environment eventually led to the development of the study of human health in the workplace that we now know as industrial hygiene.

THE INDUSTRIAL HYGIENE PROFESSIONAL

Those who dedicate their professional careers to the practice of industrial hygiene are appropriately referred to as *industrial hygienists.* In general, these professionals are concerned primarily with solving occupational health problems that exist in the workplace. To accomplish this objective, the industrial hygienist must first be able to recognize that a particular problem exists. For example, certain stress factors that may be present in the work environment can endanger life and health or cause significant discomfort for the worker. Once such hazards to human health have been properly identified, the industrial hygienist must evaluate the various environmental factors and other influencing conditions that create or enhance those hazards. Proper evaluation of the magnitude of such hazards is performed through a combination of training, experience, and quantitative analyses of the chemical, physical, ergonomic, and/or biological stresses that may be present in the workplace. Finally, through the application of specific control measures, the industrial hygienist must mitigate the hazard to acceptable levels of exposure. For example, control procedures may include the substitution of harmful or toxic agents with less dangerous ones. Hence, the three general principles of recognition, evaluation, and control (or REC), as portrayed in Figure 1-1, provide the cornerstone for the effective management of industrial hazards. Each will therefore be discussed in greater detail in Chapter 3 under General Principles of Industrial Hygiene.

Because of the wide variety of potential threats to human health that can conceivably exist in the average industrial setting, today's industrial hygiene professionals must possess an adequate level of skill and training in a variety of diversified areas (i.e., management, engineering, physics, chemistry, mathematics, and biology) to successfully perform their function. Also, since occupational health hazards may cause legally compensable illnesses, the industrial hygienist must also be familiar with certain aspects of the legal process. These may include such principles as the preservation of evidence, discovery, sample integrity, and the like. In fact, the required level of knowledge in so many diverse specialty areas eventually created a demand for some type of mechanism to ensure professional competency in the practice of industrial hygiene. In an effort to standardize this

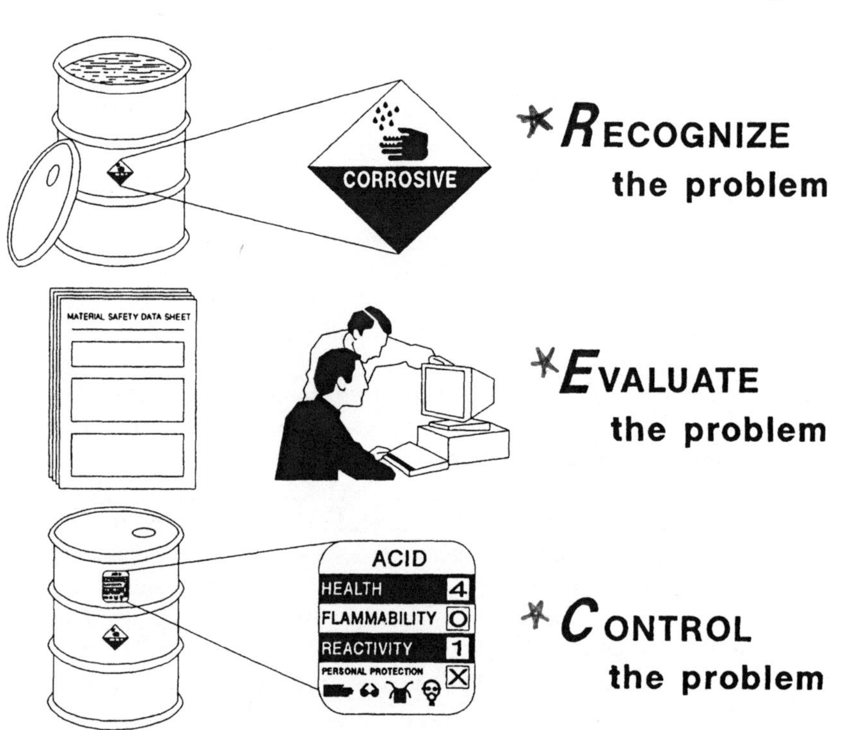

FIGURE 1-1 The three general principles of recognition, evaluation, and control (or "REC") provide the cornerstone for the effective management of industrial hazards.

elevated level of proficiency, the American Board of Industrial Hygiene (ABIH), a non-governmental professional and technical certification organization, has developed an examination process for industrial hygiene professionals. The designation *Certified Industrial Hygienist* (CIH) indicates that the bearer has demonstrated a specific level of scientific, technical, and practical expertise in the performance of the industrial hygiene function to warrant one of the highest available professional recognitions. This is not to suggest that practicing industrial hygienists who have not yet taken the examination are any less qualified. But, as with all such professional certifications, the CIH examination process does provide a means to measure the level of individual professional expertise demonstrated through a standardized and acceptable fashion. There are, of course, numerous other organizations offering certification examinations for practicing occupational health professionals. Each tests to a specified level of expertise. However, to date, few have equaled the level of proficiency that is measured by obtaining the CIH.

In today's highly complex work environments, the professional industrial hygienist has become a valuable asset to both the employer and the employee. For the employer, the industrial hygienist is the key to ensuring a healthy work environment. The hygienist must adhere to company policies and procedures while recognizing health hazards, evaluating them for severity, and recommending control measures. In some cases, industrial hygiene may have to go beyond company policies and procedures to ensure adequate worker health protection. The employer must rely on the professional judgment and expertise of the hygienist as these policies and procedures are formed, changed, modified, and implemented. Without the service of a competent industrial hygienist, the employer's ability to effectively provide a healthy workplace is significantly impaired. Of course, the level of service required will depend significantly on the nature of the company's business. Some will have a serious health hazard potential while others may have relatively little need for industrial hygiene services. Whatever the case may be, it is important for any company, as a management policy, to assess this need and act accordingly.

The industrial hygienist is the employees' watchdog for health hazards in the workplace. The hygienist is a valuable source of information, expertise, training, and consultation in those areas that can most affect employees' health and well-being while at work. Employees should, therefore, view the hygienist as an indispensable teammate who is uniquely qualified to assist them in performing their job duties in the safest and healthiest work environment possible.

Because their training requires impartial assessment and evaluation of occupational health hazards, the opinions and judgments expressed by the industrial hygienist ideally should be unbiased toward either management or labor. However, depending on the particular organizational structure and the existing relationship between management and labor, industrial hygienists may often find themselves caught between the special interests of each group. Unfortunately, in the real world of industrial operations, this is often the case. Therefore, the trained and experienced industrial hygienist must also possess the highest ethical standards to act with the integrity and honesty that the occupational health profession demands. This is perhaps one of the professional's greatest personal attributes.

ORGANIZATIONAL INTERRELATIONSHIPS

In a typical line and staff organization, personnel who are directly responsible and accountable for the daily operations of the business enterprise are considered *line management*. They typically have the authority to implement or change company policy and operating procedures. Personnel in this category usually range from first-line supervisors to executive management. Those who serve as advisors to line management can recommend policy changes and are considered staff management. In most large companies, service-oriented functions such as training, security, labor relations, safety, and industrial hygiene are generally referred to as *staff functions*. Depending on an organization's nature, smaller firms may have the safety and health function as a line element. However, in smaller, less-segmented organizations, sharing functions and responsibilities is usually the rule rather than the exception. In such cases, the individual performing safety and health services may also have many other unrelated responsibilities to perform, as well. Regardless of exactly how the organizational structure is established, the safety and health function (compliance assurance, program development, planning, monitoring, inspection, etc.) remains primarily with the staff while the responsibility (implementing, administering, managing) for safety and health in the workplace remains primarily with the line.

The advice and assistance staff organizational elements provide normally facilitates the line management function. In this regard, the industrial hygienist functions as a consulting expert within the staff management portion of the organization. Figure 1-2 shows an example of a line and staff organization and the typical location of the industrial hygiene function. Once again, depending on the particular company, the industrial hygienist may be placed in a location other than Human Resources. However, regardless of specific positioning, industrial hygiene remains a staff service to line management.

A successful occupational health program will undoubtedly require the interrelated services of many professional disciplines such as the industrial physician, the industrial toxicologist, the occupational health nurse, the safety professional, the health physicists, and the industrial hygienist. In addition, all must work effectively with both labor and management. Depending on the size and nature of any given company, each of these functions may be represented with on-staff support or contracted for external support services. Whatever the case, the indus-

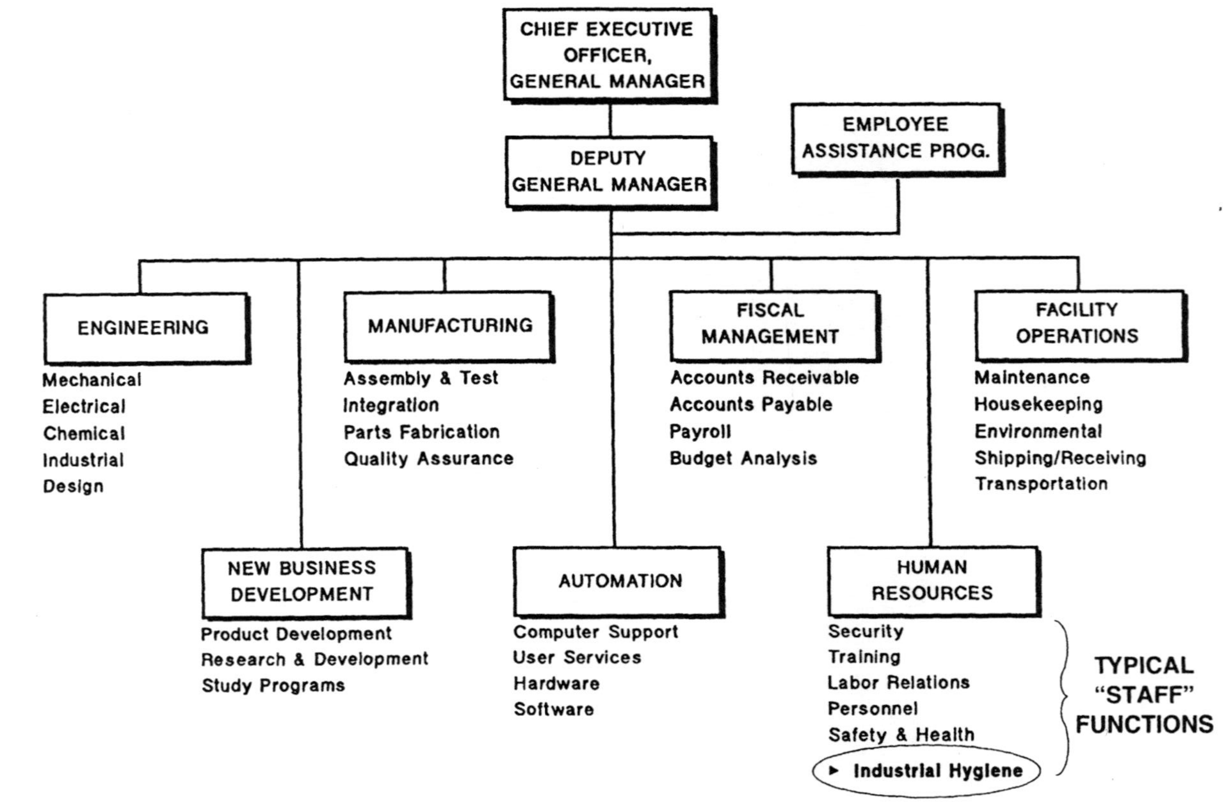

FIGURE 1-2 An example of a line and staff organization and the typical location of the industrial hygiene function

trial hygienist must know when to seek the support of appropriate professional services related to the preservation of worker health. While the industrial hygienist may be required to work with nearly all other organizational elements at one time or another, the following provides a brief discussion of the most important of these interrelationships (as shown in Figure 1-3).

The Safety Professional: The company safety professional will most likely have a great deal of interface with the industrial hygienist on a daily basis. In fact, with few exceptions, there has always been a close relationship between the specialties of occupational safety and occupational health (industrial hygiene). This relationship had grown even closer by the beginning of the 1990s, when an economic recession was in full swing in the United States. Competition for new business was always difficult, regardless of the industry. But during this particular period, the fight to retain existing business became a matter of life and death for many companies throughout America. Corporate budgets grew increasingly smaller and organizations "streamlined" their work force for survival. Forced diversification of responsibilities occurred across many companies' organizational structures. Suddenly, many safety professionals found that they were also

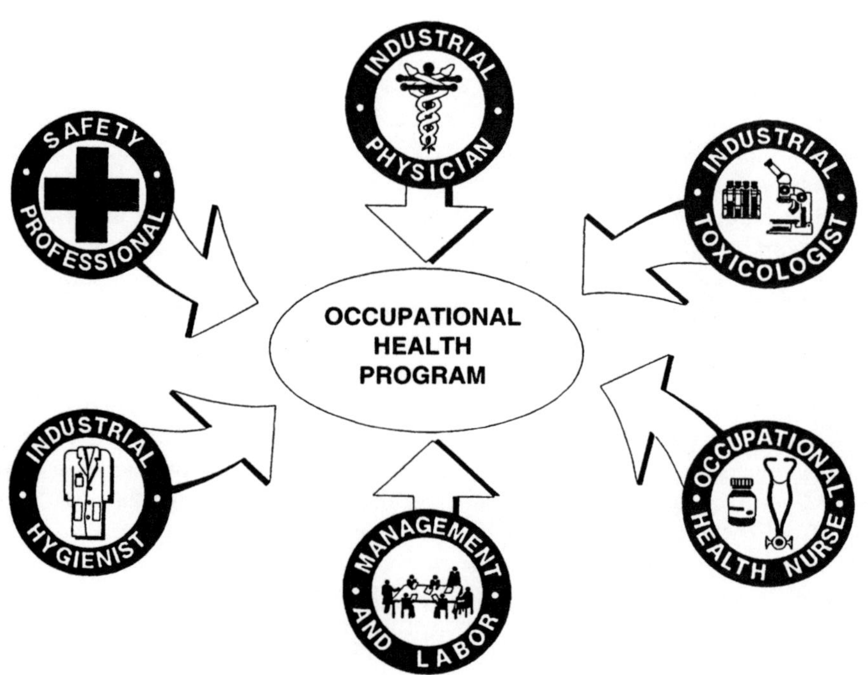

FIGURE 1-3 The important interrelationships in a successful occupational health program

responsible in some degree for industrial hygiene/occupational health, or vice versa. The merging of these specialties resulted in the creation of a much higher level of professional sophistication in the relationship between industrial hygiene and industrial safety. Even in those organizations that continue to maintain both as separate functions, the industrial hygienist and the safety professional are often professional allies, working together to maintain a safe and healthy working environment.

Safety professionals utilize many methods and techniques to achieve workplace safety. The approach is often very similar to that of the industrial hygienist. For example, they study job functions and operations, examining each task in search of existing or potential hazards to the operator (recognition). Once hazards are identified, safety professionals will assess each one to determine the safest approach to accomplishing the respective tasks (evaluation). Finally, they will make recommendations to eliminate or minimize exposure to these hazards (control). It is when these recognized hazards affect the health of the worker that the services of a competent industrial hygienist becomes essential. Working together, the safety professional and the industrial hygienist can often develop very specific methods to eliminate or control these hazards so that operational safety in the work environment can return to an acceptable level. This partnership has proven very effective throughout industry, especially in recent years as the sides of the safety and health equation have been driven closer and closer together. In fact, the two fields of study have become so complementary that the ABIH and the Board of Certified Safety Professionals (BCSP), an organization providing professional certification of safety professionals (the Certified Safety Professional, or CSP), came to an agreement in 1985 to offer a joint certification at the technician level. To recognize the professional competence of the technologist working in both the safety and health fields, the Certified Occupational Health and Safety Technologist (COHST) examination was developed. This action, however, was not meant to take the place of the CIH or CSP certification process. Rather, the two organizations recognized the close relationship between these fields of study and decided to cosponsor an appropriate level of professional certification. Further recognition of the growing number of individuals practicing both fields led to a recent agreement intended to facilitate the respective testing processes of each organization. Many individual practitioners who are responsible for both the safety and health aspects were attempting certification from both the ABIH and the BSCP. There proved to be much repetition in the first phase of each test. So, under a joint agreement the two organizations now allow a person who has successfully obtained one of the two certifications to by-pass the first phase of the second certification. The qualified applicant can now obtain professional certification in both the safety (CSP) and health (CIH) arenas without having to re-demonstrate professional competence on the basic level. This development further exemplifies the close relationship that typically and, in most cases, essentially exists between the safety professional and the industrial hygienist.

In general, being referred to as a safety professional has almost always implied some level of occupational health expertise as well. Until recently, however, being an industrial hygienist did not often carry the reverse implication. In most of today's industrial settings, the occupational safety and health professional has become a hybrid of both disciplines or, at the very least, someone who is capable of serving the interests of both occupational safety and occupational health. In organizations where these two functions remain separate (as well as in those that have combined safety and industrial hygiene), the primary purpose is still one of hazard recognition, evaluation, and control. Because this fundamental principle does not change, it remains the common thread between the two disciplines. It is the basis on which the practitioner provides the necessary safety and health services to ensure the protection of worker safety and the preservation of worker health.

The Industrial Physician: The occupational or industrial physician is primarily concerned with worker health. By evaluating the status of worker health through a variety of methods and techniques, the industrial physician can assess the level of adverse health affect caused by a particular work environment or condition. Relatively speaking, there are many more industrial hygienists on staff than there are industrial physicians. Except for the very largest of organizations, few companies will employ a full-time, on-staff industrial physician. It is more common to find that such services are provided on a contractual basis by an off-site, private medical doctor who specializes in industrial medicine or a related field. Whatever the case, it is imperative to the success of the occupational health program for the industrial hygienist to develop a close working relationship with the industrial physician. Since both are concerned with worker health, each should be familiar with the capabilities of the other. Then, when worker health concerns arise, the proper course of action can be quickly and effectively taken before any occupational illness occurs.

Since the objective of an industrial hygiene program is the preservation of worker health, industrial physicians should also be involved with the development of company policy and procedures before illness and disease occur. Their involvement should not be strictly limited to treatment. It should also include the assessment of unhealthy conditions and situations in the work environment, as well as the evaluation of health hazards identified by the industrial hygienist. In many cases, the industrial physician can also assist in the evaluation of certain control measures to verify adequacy. Here again, as in the case of the safety professional, we see the recurrence of the basic functions of recognition, evaluation, and control in the services provided by the industrial physician.

In general, a worker who is placed in an industrial environment will react to it in some way. This reaction will depend on a wide variety of variables, and may result in anything from little or no recognizable health effect to a total incapability to perform under existing environmental conditions. In the evaluation of these health effects, one must consider both the individual and the environmental variables that come into play in any given situation. Each person may react differently to the same stimulus. Physical make-up and condition, gender, age, stress, heredity, current health status, history of any previous exposure to toxic materials can all influence the level of health effect on the individual, as can many other variables. To understand how specific environmental stimuli impact human health, the industrial hygienist must understand how such materials can enter the body and the resulting effects in the individual. This is where the support of the industrial physician has its greatest impact on the industrial hygiene effort.

With the specialized capabilities of medical assessment and evaluation, the industrial physician can help identify specific health effects of toxins, chemicals, poisons, sensitizers, and any other materials that adversely affect human health. Using information on the presence of these materials in the work environment, coupled with the knowledge of an individual's health status, the physician can assist in the accurate identification of specific hazards that are causing (or are likely to cause) an adverse impact on the health of the worker. Basically, there are three major ways in which hazardous materials enter the human body: inhalation, ingestion, and absorption through the skin, as shown in Figure 1-4. A fourth, less-common route of entry is injection, which is not typically encountered during the performance of most job functions in general industry and is therefore not represented as a "common" or "primary" route of entry. Entry into the body through the three primary routes will generally result in some discernible effect on one or more of four specific human anatomic systems. These are the respiratory system, the digestive system, the circulatory system, and the central nervous system. The body's special sensory organs are also at risk of exposure to hazards in the workplace, even though no specific entry has to occur. These include the skin (damage to the skin itself can occur with or without absorption), the ears, the eyes, the nose, and the mouth. Chapter 4 contains further information on each of these organ systems. It is the evaluation of these specific organ systems that enables the industrial physician to identify the cause and effect relationship that exists between hazard exposure and human health. The industrial hygienist must be sufficiently knowledgeable to recognize when conditions exist that are conducive to the creation of such health hazards. It is the initial effort of the industrial hygienist that facilitates the medical review of workplace environmental conditions that, in turn, is the precursor to the proper evaluation and control of these conditions. Therefore, the importance of establishing an effective working relationship between the industrial hygienist and the industrial physician cannot be overemphasized.

The Occupational Health Nurse: Some companies may either employ or contract for the services of an industrial nurse, more commonly referred to as an

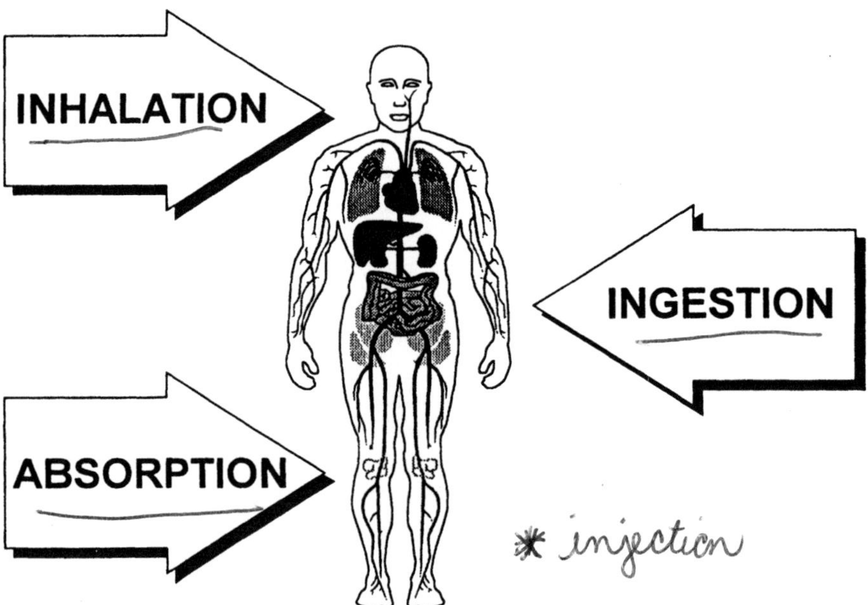

FIGURE 1-4 The three primary ways in which hazardous materials enter the human body

occupational health nurse (OHN). In addition to being a fully-trained registered nurse (RN), the OHN is also quite familiar with the specific health concerns that are common to most work environments. OHN medical training extends into many other areas of specialization such as the various regulatory requirements related to the health of the worker. For the industrial hygienist, OHNs can provide a practical medical perspective on issues such as hazard communication, hearing conservation, eye and respiratory protection, bloodborne pathogens, process safety, and other related regulatory concerns. They can also assist in the proper recognition, evaluation, and control of occupational health hazards such as heat stress, ventilation problems, exposure to radiation, occupational dermatitis, and so on. Because the practicing OHN must possess a sufficient level of expertise and knowledge in areas above and beyond that normally required of an RN, the American Board of Occupational Health Nurses (ABOHN) has developed a very stringent certification process for the OHN. Those who can first successfully satisfy the qualification criteria established by the ABOHN can then take a written certification examination administered by the Board. If the nurse passes the exam, the designation of *Certified Occupational Health Nurse* (COHN) will be granted. The COHN designation recognizes an individual who has met an established set of criteria and has successfully demonstrated a level of expertise consistent with that expected of a practicing OHN. As with the CSP, CIH, and most other professional certifications, the COHN is a method to standardize and recognize the degree of proficiency associated with the professional performance of that function. It does not in any way indicate that those who have not gone through the process are any the less qualified. Together with the industrial hygienist, the OHN can perform specific on-site evaluations of health hazard concerns. Hence, the OHN should be considered a valuable resource to the hygienist as well as the organization. The skills of OHNs should be utilized far beyond the day-to-day provision of first aid treatment of employees. Their specialized knowledge of the cause and effect relationship between health hazards and the exposed worker can be a key element of a successful occupational health program.

The Industrial Toxicologist: While most organizations do not typically utilize the services of an industrial toxicologist on a regular basis, there may be a need for such expertise at any given time. The industrial hygienist will probably be in the best position to determine if such services are necessary and should, therefore, be somewhat knowledgeable of the capabilities of the industrial toxicologist. Generally speaking, the industrial hygienist would look to the toxicologist whenever specific knowledge on the result of human exposure to a substance or material is not readily available or understood. The industrial toxicologist is specifically trained to evaluate the adverse health effects of substances that are present in the work environment. With a background in physiology, pharmacology, biochemistry, and other related health sciences, the industrial toxicologist can provide a technical and analytical determination of the specific interactions of these substances with living systems. In general, the objective of industrial toxicology is to provide a service of specialized evaluation. This evaluation provides very specific information concerning the protection of human health to both the industrial hygienist and industrial physician. The data will either support or refute theories on the relationship between specific substances and the development of adverse effects in living systems. According to Richard S. Brief (1975), the industrial toxicologist is concerned with three major aspects that he refers to as the interaction sequence. First, the toxicologist examines the capacity of a substance to induce injury, *or its toxicity.* Second, the probability that this substance will induce injury, or its *degree of hazard,* is considered. Finally, *safety aspects* are evaluated to determine if specialized handling or use of a substance will effectively eliminate its capacity to induce injury. The full realization of these aspects of interaction allows the industrial toxicologist to recommend appropriate guidelines that serve to protect individuals exposed to these substances.

With a focus on the objective of evaluation through this interaction sequence, the toxicologist will primarily study information obtained through documented studies on animals, as well as data available on known human exposures. For example, in some instances, extremely important information can be obtained from the medical records of exposed personnel. However, human exposures are not usually intentional, so the data generated is not based on any accepted or controlled experimental procedure. It is an unfortunate occurrence when any person suffers from exposure to any substance. But this type of data is generally all that exists in the human exposure category. Therefore, more often than not, the industrial toxicologist must depend on controlled animal experiments to obtain information on the reaction of living systems to specific exposures. In some cases, when no existing information can be located, the toxicologist may have to perform the required animal experiments to understand and properly evaluate the hazards of a given exposure. In any case, the experimental protocol must closely resemble the conditions likely to be encountered in the actual work environment. As a minimum, these should address the following:

1. The physical characteristics of the substance or material being evaluated
2. The biological characteristics (if applicable) of the substance or material being evaluated
3. The expected dose that will be encountered
4. The probable length of exposure likely to occur in the actual work setting
5. The most likely route(s) of exposure
6. The use of an animal species that most closely approximates the response of human exposure to the same substance

When analyzing animal experiments, the toxicologist must identify specific information that establishes the limits of simple irritation as well as the lethal dose of the substance in question. Once such parameters have been defined, the industrial toxicologist can then provide the industrial hygienist with specific recommendations on the maintenance of a safe and healthy work environment. They can often predict the nature of a specific toxicity that can be expected and help define the hazards associated with varying degrees of exposure. Depending on the extent of experimental data available, toxicologists can also provide an estimate of the level of exposure that can be safely tolerated without any irritating effects to those exposed, as well as the amount of exposure that is likely to cause serious health effects. They can recommend appropriate handling procedures to limit or reduce exposure, and they can even suggest antidotes to combat or reverse the effects of over-exposure to the substance in question.

In short, the services of the industrial toxicologist are quite specialized and, subsequently, may not always be necessary to ensure an effective occupational health program. However, it is the responsibility of the industrial hygienist to recognize when such services are required. Knowing when to seek the assistance of a qualified industrial toxicologist will allow the hygienist to better serve the occupational health interests of both management and labor.

Management and Labor: Without a close working relationship with both sides of the management/labor equation, the industrial hygiene effort would be reduced to nothing more than an exercise in futility. In this chapter, the importance of management commitment to the occupational health effort has been discussed. The staff industrial hygienist must be an outspoken proponent of management policy while maintaining a position on employee health that is both ethical and practical. The management/labor relationship often requires industrial hygienists to perform within three sets of operating criteria. First, they must always ensure that the occupational health program is conducted in accordance with the highest ethical and moral standards established by the industrial hygiene profession. Second, there is an inherent obligation to perform the duties and responsibilities of the staff position in consideration of the best interests of management. Third, as an agent for the employee, the hygienist must also ensure that occupational health measures taken by the company do not threaten the workers' health or well-being. As shown in Figure 1-5, the required working relationship between the hygienist, management, and labor can be portrayed as a triangle. Each consideration forms the three legs of the industrial hygiene program triangle. A breakdown in any one of the three sides would create a failure in the per-

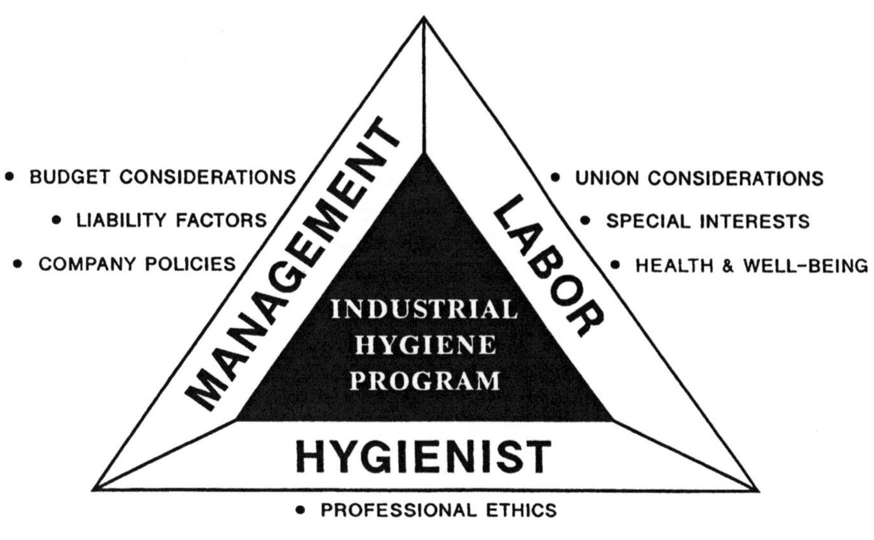

FIGURE 1-5 The industrial hygiene triangle shows the required working relationship between the hygienist, management, and labor.

formance of the overall industrial hygiene program. In short, the program would not succeed without a working relationship with management, with the employees, and without a firm foundation of professional ethics and standards.

HUMAN PERFORMANCE AND THE WORK ENVIRONMENT

Since optimum human performance plays such a key role in the success of any business endeavor, the factors that diminish that performance should be the focus of serious attention for every industrial hygienist. In fact, understanding the basic principles of industrial hygiene requires a fundamental appreciation for human performance and the many variables of the working environment that can impact this performance.

Every day, people are required to execute a variety of simple and complex tasks during the performance of their work duties and during their private lives. When human performance is somehow less than anticipated, expected, or desired in a given endeavor, the rate of progress is equally hindered. Ensuring the preservation of employee health depends, in part, on understanding the various factors that can impact their health and, subsequently, overall human performance.

These factors can be divided into two basic categories: physical and psychological. Table 1-1 lists some examples of physical and psychological factors that can adversely affect human performance. Since industrial hygienists have very little control over the mental stability of each employee, they cannot effect much change in such psychological factors. However, understanding these factors is nevertheless important to the evaluation of human performance at work. Becoming familiar with the potential existence of such stress factors will facilitate the evaluation of real and contrived occupational health concerns.

TABLE 1-1   Some examples of both physical and psychological factors that can adversely affect human performance

FACTORS AFFECTING HUMAN PERFORMANCE	
Physical	*Psychological*
Hunger	Stress
Thirst	Emotional distress
Fatigue	Preoccupation
Weakness	Lack of concentration
Strength	Learning disorders
Injury	Attention deficit
Height and weight	Ignorance
Physical illness	Fear/phobias
Vision inadequacy	Mental illness
Hearing inadequacy	Anxiety

Addressing the potential physical factors that can impact human performance is certainly a somewhat easier task for the industrial hygienist than attempting to resolve the many psychological factors that affect human performance. Physical factors are at least real and identifiable. The results of solutions and corrective actions are usually visible as either positive, neutral, or negative. Measurements of success or failure can be obtained and improvements can be tracked to determine the overall effectiveness of the industrial hygiene effort. For example, the industrial hygienist can identify and correct problems associated with illumination, noise, pressures, ventilation, vibration, temperature, and other environmental factors. Such action is not always possible when attempting to resolve psychological conflicts and problems that are also affecting employee health. Employees who are under stress resulting from personal conflicts such as financial or marital problems, difficulties with their children, the illness or death of a loved one, drug or alcohol use/abuse, emotional instability, mental illness, and so on, may seem to be quite normal based on outward appearances. But, as these factors continue to stress the employee, a decrease in their performance and a decline in their general health may become a visible manifestation of these adverse psychological influences. It is important that the industrial hygienist be capable of recognizing the difference between genuine industrial health concerns and psychosomatic indications that may not have anything to do with the actual work environment.

SUMMARY

Potential hazards to human health can be found in nearly every type of work environment. The ability to recognize, evaluate, and control these hazards has become the basis of a distinctive field of study known as industrial hygiene. Those occupational health professionals who dedicate their careers to the practice of industrial hygiene are appropriately referred to as industrial hygienists. In this chapter, the history and development of industrial hygiene as a separate and distinct field of specialization was briefly discussed. Although awareness of occupational health hazards is by no means a new development, great strides toward understanding and resolving health problems in the industrial arena have been made since the beginning of this century. The speed with which the industrial hygiene profession has grown in the United States is linked closely to the development and maturation of worker's compensation laws. As individual states began mandating compensation for occupational illness as well as injury, the necessity for a more definitive method to study and resolve such problems became essential to their prevention. Hence, industrial hygiene as a profession grew out of the necessity to prevent health-related problems in the workplace.

The industrial hygienist of today must be capable of responding to a variety of issues that often require training in areas such as management, law, biology, chemistry, physical science, ergonomics, and other related fields of study. In fact, the degree of expected proficiency in industrial hygiene practice led the American Board of Industrial Hygiene (ABIH) to develop a standardized method of evaluating the industrial hygiene practitioner. Once stringent qualifications have been met and the candidate successfully completes a written examination process, the designation of Certified Industrial Hygienist (CIH) is granted. The CIH designation is recognized as a premier certification of one's ability to properly perform in the industrial hygiene profession.

To accomplish the industrial hygiene objective of eliminating or reducing occupational exposure to health hazards, the practicing hygienist must work together with numerous other elements of the total organization, both internal and external. These include the safety professional, the industrial physician, the occupational health nurse, and the industrial toxicologist. The key to the success of any industrial hygiene effort is the establishment of positive working relationships with management and with labor. All must work together as a team with a common goal: ensuring the preservation of employee health and well-being.

In addition to understanding organizational interrelationships, the industrial hygienist must have a basic understanding of human performance in the work environment. The way people perform a task is influenced by many factors. These factors can be divided into two main categories: physical and psychological. Industrial hygiene focuses primarily on physical factors, since controlling these is much more probable than attempting to resolve the psychological factors that affect human performance. However, the hygienist should still understand that psychological factors exist. The industrial hygienist should be able to recognize the difference between legitimate physical health hazards and those that may be based on some psychological manifestation or influence that may or may not have anything at all to do with the work environment.

As a profession, industrial hygiene has evolved into a highly specialized field of science. The practitioner is concerned first with the preservation of employee health. This should be the basis for establishing an industrial hygiene program.

2

Regulatory Agencies and Their Requirements

INTRODUCTION

Regardless of the efforts of today's employers, the driving force of action to ensure a healthy workplace has been (and will continue to be) the *force of law*, as imposed by a myriad of regulatory standards. While such rules are created at the local, state, and federal levels, the intent of the enforcing agencies is the same: to ensure a safe and healthy workplace. For example, on the federal level there are industry-specific agencies such as the *Mine Safety and Health Administration* (MSHA), which is obviously interested specifically in the safety and health of the miner, and there are agencies concerned with general industrial safety and health, such as the *Occupational Safety and Health Administration* (OSHA). How regulations impact the industrial hygiene effort can be best explained through a discussion of the primary agencies concerned with employee safety and health in general industry. While it is true that many states (and even more local municipalities) have their own occupational safety and health requirements, it would not be feasible to discuss each in a limited volume such as this. However, by establishing a clear understanding of the federal scheme for assuring workplace safety and health, readers should have a better foundation on which to build a further, more detailed understanding of their own state and local requirements.

Figure 2-1 depicts the three federal entities that were created under the Occupational Safety and Health Act to ensure the protection of safety and health in the workplace. The first, OSHA, is concerned with the *promulgation and enforcement* of federal workplace safety standards. The *National Institute for Occupational Safety and Health* (NIOSH) conducts *research and tests* and makes *recommendations* on occupational health and, to a lesser extent, safety issues and concerns. Finally, the *Occupational Safety and Health Review Commission* (OSHRC) *adjudicates and interprets* regulations that may be under dispute or appeal. Each of these entities shall be briefly discussed below.

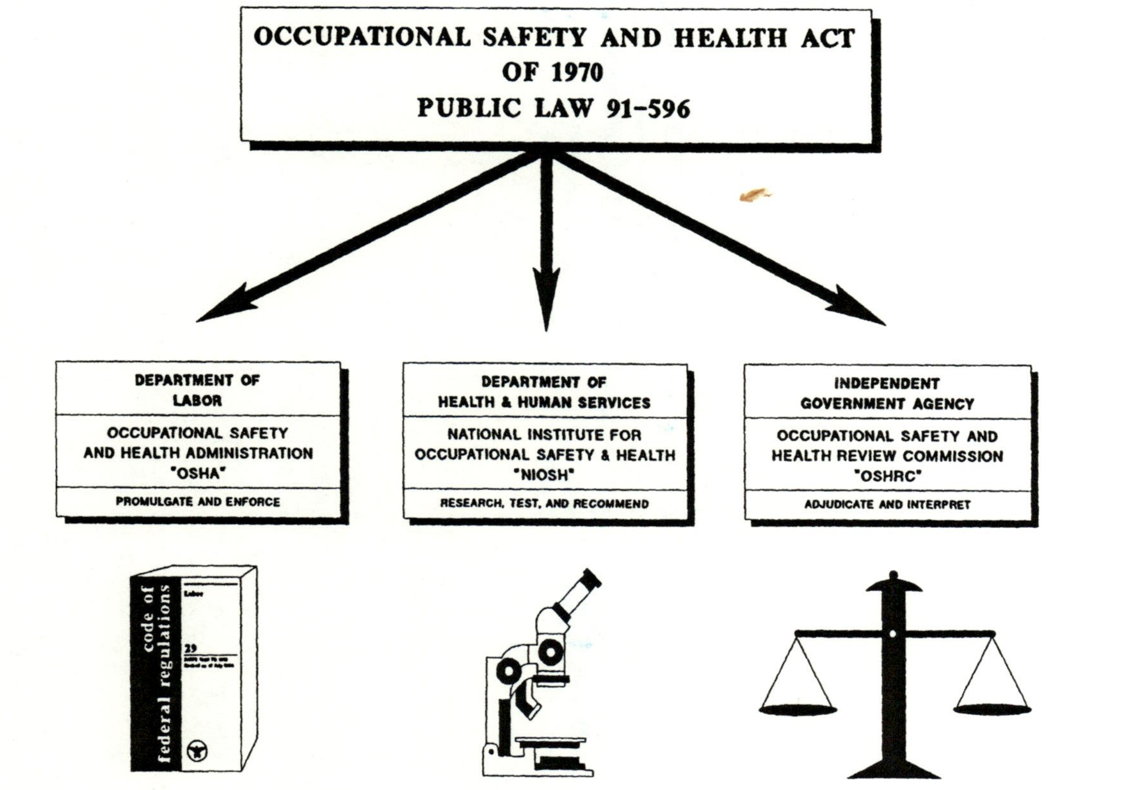

FIGURE 2-1 The three Federal entities that were created under the Occupational Safety and Health Act (OSHAct) to ensure the protection of safety and health in the workplace

THE OCCUPATIONAL SAFETY AND HEALTH ADMINISTRATION

Ensuring the safety and health of the general working population in the United States became a matter of law with the enactment of Public Law 91-596 on 29 December 1970. The Occupational Safety and Health Act (OSHAct), which became effective on 28 April 1971, was by far the most comprehensive federal statute to address the safety and health of the worker. While similar laws, such as the Federal Coal Mine Health and Safety Act (Public Law 91-173, 30 December 1969) and the Federal Metal and Nonmetallic Mine Safety Act (Public Law 89-577, 16 September 1966) came earlier, the OSHAct stretched across the wide gap of the general industrial work environment to cover all private employers, without exception (Note: In most cases, federal, state and local employees in states with their own approved safety plans are covered by separate occupational safety and health programs).

The stated purpose of the OSHAct of 1970 is to "assure so far as possible every working man and women in the nation safe and healthful working conditions and to preserve our human resources." To accomplish this objective, the Act authorizes the federal government to develop and set mandatory occupational safety and health standards that are applicable to any business that can affect interstate commerce. The Act placed the responsibility for the promulgation and enforcement of these occupational safety and health standards within the *U.S. Department of Labor* (DOL). The Act created the Occupational Safety and Health Administration (OSHA) within the DOL to oversee industry's implementation of the requirements set forth in the Act. Directed by the Assistant Secretary of Labor for Occupational Safety and Health, OSHA is the primary conduit for the interpretation, validation, and enforcement of those safety and health standards promulgated by the Congress under OSHAct. As granted to the Secretary of Labor by the U.S. Congress, OSHA has the authority to promulgate, modify, and revoke safety and health standards. They must conduct inspections (focused both on safety and industrial hygiene) and investigations, issue citations, and propose penalties when violations are discovered. OSHA requires employers to maintain specific records pertaining to workplace safety and health. They also have the authority to petition the courts to restrain the continuance of work when situations of imminent danger are evident.

THE NATIONAL INSTITUTE FOR OCCUPATIONAL SAFETY AND HEALTH

Because of its broad administrative and enforcement responsibilities, OSHA is clearly the primary federal agency that can affect the work of the practicing industrial hygienist. It is the safety and health regulations that are promulgated under OSHA that must be clearly understood and, if applicable, appropriately applied to the work environment. However, the particular decisions and findings of a second agency, also created under the OSHAct, can severely effect the direction of any industrial hygiene program.

The National Institute for Occupational Safety and Health (NIOSH) was created as the principal federal agency engaged in research and studies that attempt to eliminate occupational safety and health hazards in the American workplace. NIOSH was placed within the U.S. Department of Health, Education, and Welfare (now the *Department of Health and Human Services,* or DHHS) under its *Centers for Disease Control* (CDC) of the Public Health Service. It is important to understand that NIOSH is *not* a regulatory agency and, therefore, can provide only recommendations to improve workplace safety and health. Unless these recommendations are adopted or incorporated by OSHA (or some other regulatory agency), compliance is not mandatory. For example, NIOSH develops and periodically revises *Recommended Exposure Limits* (RELs) for hazardous substances or conditions in the workplace. They also recommend appropriate preventive measures designed to reduce or eliminate the adverse health and safety effects of these hazards. To formulate these recommendations, NIOSH evaluates all known and available medical, biological, engineering, chemical, trade, and other relevant information. These recommendations are published and transmitted to OSHA and MSHA for their use in the promulgation of regulatory standards. Once the regulatory agency acts on a NIOSH recommendation and incorporates it into a regulation, the recommendation then becomes law. To avoid any potential for a conflict of interest between NIOSH and the respective regulatory agencies, NIOSH was purposefully placed in the DHHS and not in the DOL. This successfully prevents any possibility of NIOSH acting under any direction (or pressure) from OSHA.

In addition to identifying occupational safety and health hazards and recommending regulatory changes, NIOSH is obligated to provide training for occupational health personnel. NIOSH maintains a training grant program to develop associate, baccalaureate, and graduate degree programs in our nation's colleges and universities. They also offer a wide variety of health-related short courses, seminars, and training sessions to help improve the knowledge and skills of our nation's occupational safety and health professionals. When contacted, NIOSH can also provide workplace evaluations and consultations, free of charge.

THE OCCUPATIONAL SAFETY AND HEALTH REVIEW COMMISSION

The OSHAct created yet a third agency, the Occupational Safety and Health Review Commission (OSHRC), to review and adjudicate challenges brought against the federal government concerning the conduct and performance of

OSHA. An independent agency, the OSHRC has final jurisdiction over the outcome of any legal challenges. This appeal system allows employers and employees to legally question certain OSHA actions. Since its inception, the actions of the OSHRC have been responsible for literally thousands of case law examples that have served to clarify, refine, exemplify, and even rewrite existing OSHA law. However, even though the OSHRC does, in fact, have the *authority* to resolve any concerns associated with the interpretation and/or applicability of a specific OSHA standard, it cannot act on such concerns without *justifiable cause*.

In the case of the OSHRC, justifiable cause translates into action on behalf of the people. Therefore, unless they are acting on an appeal (or if a particular OSHA regulation is otherwise challenged by an employee, employer, or other affected person) the OSHRC is powerless to act on any issue related to occupational safety and health law. This system effectively prevents any possibility of arbitrary review/revision of existing OSHA standards without just cause. Here again, to eliminate the possibility of a conflict of interest between the regulatory, research, and legislative aspects of the OSHAct, the OSHRC has purposely been established as an independent agency, separate in both function and location from OSHA and NIOSH.

OTHER REGULATORY AGENCIES

Of course OSHA, NIOSH, and the OSHRC are not the only federal entities concerned with the protection of worker safety and health. But they are considered the primary ones, since their actions affect the majority of the working population in general industry (including construction). Other regulatory agencies do exist; these are usually specific to individual industries and/or are concerned with very specific areas of worker health and safety. For example, the Mine Safety and Health Administration (MSHA), with its own Federal Mine Safety and Health Review Commission, was created under the Federal Mine Safety and Heath Act of 1977. The Act authorizes the Secretary of Labor to develop safety and health standards to protect miners, and it authorizes NIOSH to prepare criteria documents for the development of mine-related health standards.

The Federal Environmental Protection Agency (EPA) is concerned primarily with the protection and preservation of the nation's environmental resources. However, a number of its regulations have provisions that focus on public (including workers') health. For example, their Toxic Substances Control Act (TSCA) contains requirements for reporting incidences of employees' exposure to toxic chemicals. Title III of the Superfund Amendments and Reauthorization Act (SARA) of 1986 is specifically concerned with ensuring adequate public knowledge of hazardous materials and substances that are moved, stored, or used in their neighborhoods.

The U.S. Department of Transportation (DOT) also enforces regulations that contain requirements for the assurance of worker safety, especially those personnel involved in the transportation of hazardous materials.

Without a doubt, the many safety and health regulations promulgated and enforced by other agencies are of great importance to the practicing industrial hygienist. However, each of these industry-specific statutes cannot be properly discussed in this Guide. Therefore, because of their broad applicability, the remainder of this section will be dedicated to OSHA regulations that are primarily applicable to general industry.

REGULATORY REQUIREMENTS

The development of OSHA health regulations is a continuous process that generally begins with a recommendation from NIOSH concerning hazards to the health of a worker. Specific requests from industry and/or concerned representative employee groups (e.g., labor unions) can also prompt action from OSHA. Section 6(b)(5) of the OSHAct specifies the process for establishing safety and health standards as follows:

> The Secretary of Labor will promulgate standards based upon research, demonstrations, experiments, and such other information as may be appropriate. In addition to the attainment of the highest degree of health and safety protection for the employee, other considerations shall be the latest scientific data in the field, the feasibility of the standards, and experience gained in this and other health and safety laws. Whenever practicable, the standard promulgated shall be expressed in terms of objective criteria and the performance desired.

There is also a separate provision in the Act that will allow the Secretary of Labor to issue emergency standards, should conditions so warrant. When developing a standard, OSHA may look to a variety of sources for input, comment, and even assistance in formulating the new or modified regulation. One such source is known as a National Consensus Standard. These are usually a similar or identical set of working standards or rules that have already been developed by a nationally recognized non-regulatory authority on the subject. The American National Standards Institute (ANSI), the National Fire Protection Association (NFPA), and the American Society of Mechanical Engineers (ASME) are examples of organizations that have set consensus standards that were subsequently adopted in whole or in part by OSHA.

Specific OSHA regulations that pertain to the health of the worker can be found in the Code of Federal Regulations (CFR), Title 29 (labor), Part 1910 (general industry), Part 1915 (maritime industry), and Part 1926 (construction industry). To facilitate reference, Table 2-1 (parts a and b) provides a listing of the spe-

TABLE 2-1a   A listing of the specific health-related OSHA Standards for General Industry (*Source:* OSHA 29 CFR 1910)

SECTION NUMBER	TITLE OF STANDARD
Subpart C	**General Safety and Health Provisions**
1910.20	Access to employee exposure and medical records
Subpart G	**Occupational Health and Environmental Control**
1910.94	Ventilation
1910.95	Occupational noise exposure
1910.96	Ionizing radiation
1910.97	Non-ionizing radiation
Subpart I	**Personal Protective Equipment**
1910.132	General requirements
1910.133	Eye and face protection
1910.134	Respiratory protection
1910.135	Occupational head protection
1910.136	Occupational foot protection
1910.137	Electrical protective devices
Subpart J	**General Environmental Controls**
1910.141	Sanitation
1910.142	Temporary labor camps
1910.143	Non-water carriage disposal systems (reserved)
1910.146	Permit-required confined spaces
Subpart K	**Medical and First Aid**
1910.151	Medical services and first aid
Subpart Q	**Welding, Cutting, and Brazing**
1910.251	Definitions
1910.252	Welding, cutting, and brazing
Subpart R	**Special Industries**
1910.261	Pulp, paper, and paperboard mills
1910.262	Textiles
1910.267	Agriculture operations
1910.272	Grain handling facilities
Subpart T	**Commercial Diving Operations**
1910.421	Pre-dive procedures
1910.422	Procedures during dive
1910.423	Post-dive procedures
1910.424	SCUBA diving
1910.425	Surface-supplied air diving
1910.426	Mixed-gas diving
1910.427	Liveboating
1910.440	Recordkeeping requirements

TABLE 2-1b   (continued)

SECTION NUMBER	TITLE OF STANDARD
Subpart Z	**Toxic and Hazardous Substances**
1910.1000	Air contaminants
1910.1001	Asbestos, Tremolite, Anthophyllite, and Actinolite
1910.1002	Coal tar pitch volatiles; interpretation of terms
1910.1003	4-Nitrobiphenyl
1910.1004	alpha-naphthylamine
1910.1005	4, 4'-methylene bis (2-chloroaniline)
1910.1006	Methyl chloromethyl ether
1910.1007	3, 3'-dichlorobenzidine (and its states)
1910.1008	bis-chloromethyl ether
1910.1009	beta-naphthylamine
1910.1010	Benzidine
1910.1011	4-aminodiphenyl
1910.1012	Ethyleneimine
1910.1013	beta-proppiolactone
1910.1014	2-acetylaminofluorene
1910.1015	4-dimethylaminoazobenzene
1910.1016	N-nitrosodimethylamine
1910.1017	Vinyl chloride
1910.1018	Inorganic arsenic
1910.1025	Lead
1910.1027	Cadmium
1910.1028	Benzine
1910.1029	Coke oven emissions
1910.1030	Occupational exposure to bloodborne pathogens
1910.1043	Cotton dust
1910.1044	1, 2-dibromo-3-chloropropane
1910.1045	Acrylonitrile
1910.1047	Ethylene oxide
1910.1048	Formaldehyde
1910.1050	4, 4'-methylenedianiline
1910.1101	Asbestos
1910.1200	Hazard communication

cific health-related OSHA standards for general industry (Source: OSHA 29 CFR 1910).

While it is true that OSHA endeavors to identify and regulate all workplace hazards that threaten the health and/or safety of the employee, it is not entirely feasible that OSHA will act on every such hazard that could possibly or potentially exist. It is a fact that regulatory activities do not keep up with simple technological changes and advancements that, in some instances, can present a whole new and never before considered set of occupational hazards (e.g., robotics, use of new chemical compounds, increased computer capabilities, and so on). To ensure continued employee protection while new safe work standards are being researched, developed, and reviewed, the OSHAct also imposes a general duty on employers to provide a safe workplace. Specifically, the General Duty Clause, located in Section 5(a)(1) of the Act stipulates that:

> ...each employer shall furnish to each of his employees employment and a place of employment that are free from recognized hazards that are causing or are likely to cause death or serious physical harm to his employees, and (each employer) shall comply with the occupational safety and health standards promulgated under this Act.

Hence, in the absence of any specific safety or health standard, OSHA is still authorized to inspect, enforce, cite, fine, and penalize for unsafe or unhealthy conditions under the provisions of the General Duty Clause. It is important to note that OSHA cannot cite the General Duty Clause when a specific standard already exists that regulates the safety or health concern in question. Only when no existing standard can be specifically applied to entirely protect against the alleged hazard(s) will citation under Section 5(a)(1) be considered appropriate. This arrangement prevents double jeopardy situations, in which employers could be cited more than once for a single violation.

Because a violation of the General Duty Clause is a serious situation that can lead to a new or modified standard, the Act and subsequent rulings by the OSHRC have set guidelines under which employers can be found in violation of their general duties. According to the OSHRC, in order to establish a violation of Section 5(a)(1) of the Act, the Secretary of Labor (through OSHA) must demonstrate that:

1. The employer failed to render its workplace free from a recognized hazard
2. The occurrence of an accident/illness was reasonably foreseeable
3. The likely consequences in the extent of an incident was death or serious physical harm to its employees

A key to successful citation under the General Duty Clause is OSHA's ability to show that the hazard in question was, in fact, *recognized* as such. According to the OSHRC, a recognized hazard is a condition or practice in the workplace that is known to be hazardous either by industry (in general) or the employer (in particular). Furthermore, it is the hazardous condition itself (not the incident that may have resulted in injury or illness) that is relevant in determining the existence of a recognized hazard. The Secretary must also demonstrate that there are feasible means available to successfully abate the hazardous situation. If a feasible means of abating the recognized hazard is available (to industry or to the employer), the Act does not require that such means also be recognized by the employer in order to prove a violation. In other words, the means of abatement (unlike the hazard itself) does not have to be recognized by the employer or the employer's industry. The means of abatement must only exist, be available, and be feasible to implement. It is every employer's responsibility to recognize these hazards and determine whether appropriate means of abatement exist, are available, and are feasible to implement. Therefore, the Secretary need only prove that such a means, if utilized, would materially reduce the likelihood of injury or illness resulting from hazard exposure. Also, an employer is not excused from providing whatever protection is feasible simply because they felt full abatement could not be achieved by use of such protection.

✳EXPOSURE LIMITS✳

Most of today's OSHAct general industry health and safety standards were promulgated into law in 1971, just 30 days after the Act became law. These regulatory standards represent a compilation of many of the national consensus standards that existed at that time, plus other materials collected from existing federal standards. With the exception of some amendments, revisions, additions, and deletions, these 1971 regulations still comprise the majority of today's occupational safety and health standards.

Of particular interest to the industrial hygienist are the standards related to exposure to toxic materials and other harmful agents that were adopted by OSHA more than twenty years ago. Specifically, the 1968 American Conference of Governmental Industrial Hygienists (ACGIH) Threshold Limit Values (TLVs) were adopted directly into the regulations as performance standards and, therefore, have the effect of law. These performance standards, known under OSHA as Permissible Exposure Limits (PELs), prescribe specific levels of exposure that can normally be tolerated in an average work day before adverse impact on human health occurs. More precisely, PELs (and TLVs) are established based on an average work-time exposure of 8 hours per day, or 40 hours per week. This Time Weighted Average (TWA) has become the basis of acceptable exposure measurement under normal working conditions.

This PEL arrangement contains a noteworthy flaw. PELs do not take into consideration the synergistic effects that can result due to exposure to two or more chemicals. For example, a worker can be exposed to levels below the stated PELs for xylene, toluene, and methyl ethyl ketone (MEK) during a given work process. Even though individually these exposures may be below their respective PELs, together they can equate to a significant exposure.

There are also exposure criteria established for Short Term Exposure Levels (STELs) which are usually higher permissible levels for much shorter periods of time (e.g., 15 minutes). PELs establish specific exposure objectives that must be maintained. It is up to the employer to determine the best way to comply. Unfortunately, while the ACGIH updates their TLV ratings regularly (usually on an annual schedule), OSHA does not update its PEL ratings correspondingly. Therefore, with few exceptions, the PELs that are law today are the same PELs that were in effect more than twenty years ago. This has been one major point of contention between industry and the federal government in recent years, and it has been a significant precursor to the call for a complete revision and re-authorization of the OSHAct, commonly referred to as OSHA Reform. While the government and industry continue to debate the advantages and disadvantages of PEL update, the prudent industrial hygienist would be wise to consider all available information and documentation on exposure criteria when evaluating a workplace hazard. Always remember, it is not OSHA policy to cite an employer for being more stringent than the law dictates. While TLVs are not the law, they do provide a more accurate summation of health hazards than most existing PELs. Ethically and morally, when questions over health hazard exposure arise, the course of action should be more than obvious. However, economic feasibility and practicality must also be considered and, more often than not, a compromise between the TLV and the PEL must be made. This approach is also acceptable, as long as the exposure levels do not exceed the minimum requirements established by the regulations. The federal government understands that disparities do exist, and that they can cause significant questions about the most accurate protection of employee health. One tool that can help the industrial hygienist identify the exact point of exposure where corrective action must be implemented is appropriately referred to as the *action level*. OSHA defines the action level as half the PEL for a given chemical. While action levels are generally associated with contaminant-specific standards (such as asbestos, benzene, and formaldehyde), it is a suggested operating rule that the hygienist employ this concept for all known contaminants. Once exposure has been determined to have reached the action level, specific actions such as medical surveillance, full-time exposure monitoring and sample collection, and employee training can be initiated at the suspect work site. If a PEL is exceeded, more requirements must be met, including actual exposure measurements. On the other hand, if the employer can demonstrate that no exposure over the action level has occurred, they are exempt from the major provisions of a particular standard. Therefore, monitoring action levels established by the regulations becomes a very important task for the practicing industrial hygienist. Essential to this monitoring process is proper documentation and the maintenance of records that validate the employer's actions in the occupational health arena. Documentation and records become particularly important in the event of an OSHA industrial hygiene inspection.

OSHA conducts both safety and health inspections in accordance with its established field operations policies. Inspections are performed when complaints are received from a member of a particular company's work force. OSHA will also revisit a facility as a follow-up to any previous inspection where discrepancies were discovered. Finally, routine inspections are performed to determine the status of OSHA compliance for a given industry and/or a specified work force size. To standardize this process, OSHA has issued a *Field Operations Manual* (FOM) for use by its inspectors. The FOM establishes very specific protocol pertaining to inspection parameters and OSHA policy. This document is within the public domain and is therefore available to the private sector for review. It is an extremely useful resource because it affords the industrial safety and health professional a detailed understanding of the OSHA investigation process. Another OSHA document of particular interest to the industrial hygienist is the OSHA *Industrial Hygiene Technical Manual.* This manual, which is also in the public domain, establishes the technical industrial hygiene practices and procedures utilized by OSHA industrial hygiene compliance officers. It is highly recommended that both the OSHA FOM and the *Industrial Hygiene Technical Manual* be obtained for frequent reference by the practicing industrial hygienist. At the very least, the hygienist will be familiar with the operational philosophy and approach of the OSHA inspector. At most, the industrial hygienist will use this information to develop a more effective and practical industrial health program. Appendix A provides information on obtaining these documents from the government.

SUMMARY

The business of industrial safety and health is, by any definition, highly regulated and controlled at both the state and federal levels. Therefore, understanding regulatory agencies and their requirements is a major element of the industrial hygiene process. Agencies involved in the regulation and control of employee safety and health concerns exist on the local, state, and federal levels. This chapter discussed federal entities that have primary influence over the progression of occupational health requirements across most of general industry, including construction. The Occupational Safety and Health Administration (OSHA), the National Institute for Occupational Safety and Health (NIOSH), and the Occupational Safety and Health Review Commission (OSHRC), were all created under the Occupational Safety and Health Act (OSHAct) of 1970 and were purposely located in separate federal departments or established as independent agencies. OSHA promulgates and enforces health and safety standards. NIOSH conducts studies and makes recommendations to OSHA on occupational health-related concerns. The OSHRC adjudicates appeals and provides interpretation and clarification of OSHA law. The OSHA regulations published in 1970 were based primarily on existing federal standards and, for the most part, on existing National Consensus Standards developed by organizations such as the American National Standards Institute (ANSI). In the absence of a specific standard, the OSHAct still requires each employer to provide a safe and healthy work environment. This requirement, known as the General Duty Clause, has been the precursor to many new OSHA standards. It has also been an effective tool for OSHA to impose fines and penalties against employers who neglect their legal obligation to ensure employee safety and health even though they have not violated any specific law.

OSHA laws stipulate exposure criteria for specific chemical compounds and materials. Known as Permissible Exposure Limits, or PELs, these criteria assist the industrial hygiene professional in the evaluation of health hazards in the work environment. With few exceptions, today's PELs are based on the Threshold Limit Values (TLVs) developed by the American Conference of Governmental Industrial Hygienists (ACGIH) in 1970. TLVs and PELs are based on a Time Weighted Average (TWA), usually an eight-hour work day. Unfortunately, the regulatory update process is not as proficient as technological advancement. While TLVs are updated regularly, PELs are not. This has become a major focus of current OSHA reform legislation aimed at an update of the entire OSHAct. The prudent industrial hygienist should evaluate all available data to determine the true nature of health hazard exposure in the workplace. It is true that the PEL has the force of law under OSHA while the TLV does not, since the ACGIH is not a regulatory agency. However, TLVs are generally a more accurate assessment of true exposure criteria and, therefore, should be considered in any health hazard evaluation. OSHA does not normally find employers at fault when criteria more stringent than their own have been used to ensure employee health preservation. In fact, since OSHA regulations are generally considered the minimum requirements, the industrial hygienist will often find it necessary to implement more stringent controls to ensure the appropriate levels of protection.

3

Establishing an Industrial Hygiene Program

INTRODUCTION

To establish an industrial hygiene program, the professional must first determine the *level of effort* that will be required to successfully implement industrial hygiene within their particular organization. For instance, if no occupational health initiatives have yet been taken, the industrial hygiene program will have to be established almost from scratch. Other organizations have varying degrees of industrial and employee health programs already established. In such cases, a somewhat fragmented industrial hygiene program may already be in place. The hygienist must evaluate these existing initiatives to determine whether they are effective and complete, and make any necessary modifications to ensure a comprehensive industrial hygiene effort.

In either case, it is generally up to the hygienist to perform the *industrial hygiene assessment* of existing programs and working conditions, determine industrial hygiene needs based on that assessment, and develop the appropriate program initiatives for recommendation to management. Since management will take action on the basis of these recommendations, it is essential to the program's success that the industrial hygienist perform a complete and unbiased assessment of the organization's true industrial hygiene requirements. To accomplish this, the hygienist must establish a comprehensive plan that demonstrates the exact approach to implementing the proposed program. The plan should clearly establish the *goals and objectives* of the industrial hygiene program as they relate to the business objectives of the entire organization. The importance of obtaining *management support* for the industrial health effort cannot be overemphasized, especially at the early stages of program development. As this program development matures, the hygienist must clearly define the specific *roles and responsibilities* that will be critical to the overall success of the program.

Once implemented, the hygienist must verify program effectiveness through frequent assessments or audits. To facilitate the audit process, a system of *documentation and record keeping* must be established as an integral aspect of the industrial hygiene effort. Proper record keeping becomes even more crucial to program success during government agency audits or when there is any possibility of litigation arising out of allegations of poor employee health programs. Indeed, from a *legal perspective,* the hygienist must be constantly aware of the ever-present threat of litigation brought by employees and/or their representatives seeking relief for alleged health hazard exposures. Unfortunately, such concerns are commonplace in the United States which, by most accounts, is the most litigious society in the world.

On completion of Chapter 3, the reader should have developed a fundamental appreciation for the *basic elements* of an industrial hygiene program, how to implement these elements, and how to ensure program success.

THE PLANNING PROCESS

As is the case in any business endeavor, planning is essential to the success of the industrial hygiene program. Without proper planning, there will be little chance of successfully achieving the objective of controlling health hazards in the work environment. In fact, unless all potential program distracters are properly considered during the planning process, the industrial hygiene effort may be nothing more than a costly exercise in futility.

Figure 3-1 depicts the two basic phases of a typical *planning process flow* for establishing an effective industrial hygiene program. On analysis, there are, of course, many additional process considerations that can be added to this sample flow if a more detailed program is anticipated. However, as an example for discussion, the planning process provided in Figure 3-1 should suffice. Each element of the planning process, from establishing program objectives through the implementation phase, will be briefly discussed to facilitate further understanding and appreciation of this most important aspect of establishing an effective industrial hygiene program.

The Planning Phase

1. *Establish Program Objectives.* There is an ancient Chinese proverb that roughly translates as, *"If you do not know where you are going, any road will take you there."* This wisdom is certainly appropriate when discussing the importance of goals and objectives. Without a doubt, establishing the objective(s) of any effort is the critical first step toward accomplishing any given task. This is also a primary reason why many initiatives fail in busi-

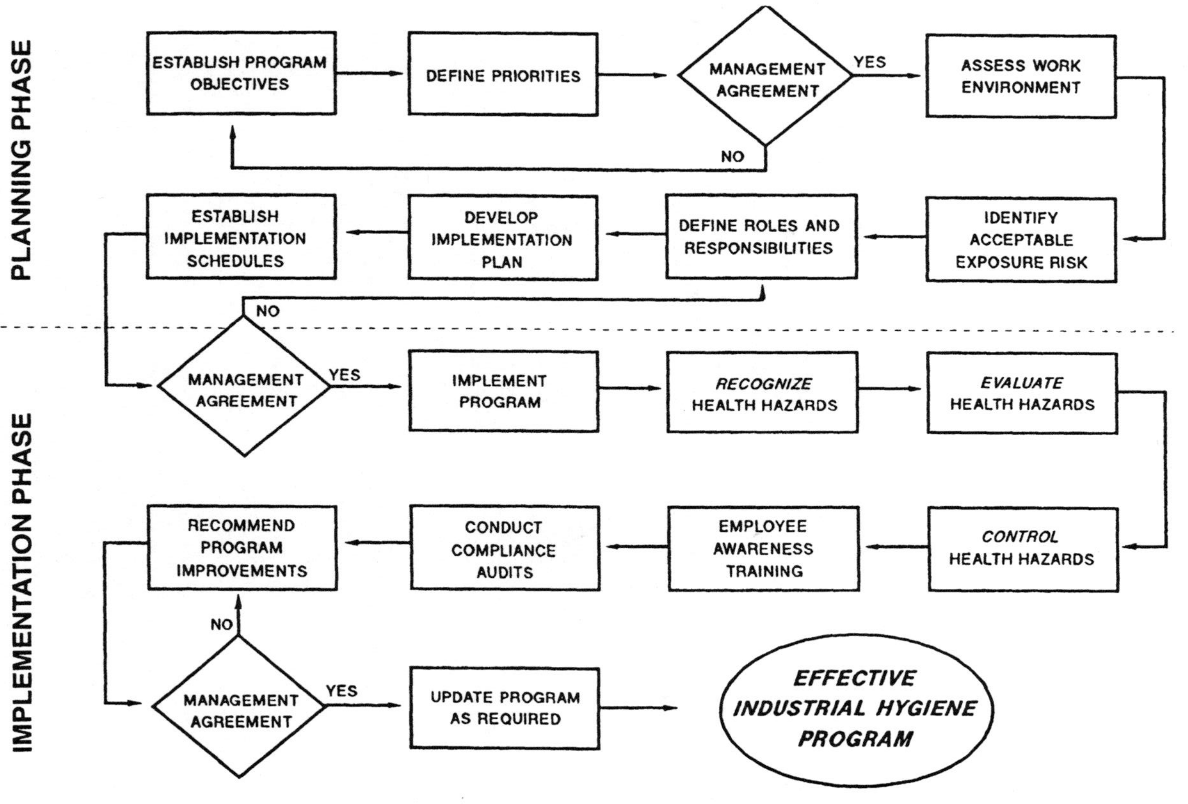

FIGURE 3-1 The two basic phases of a typical planning process flow for establishing an effective industrial hygiene program

ness today. Unless the intended direction and purpose are clear and understood, achieving success is not likely. In establishing an industrial hygiene program, the identification of program objectives should be done at the beginning of the planning process to ensure that each subsequent phase of the process is properly focused. Throughout Chapter 1, the objective of the industrial hygiene effort was discussed as being the elimination or, more commonly, *gaining control* over hazards present in the work environment. During the planning process, this objective (or any variation) must be formally established and documented as a clear management policy. This will provide positive direction for the program and ensure a focused effort by all involved in its implementation.

2. *Define Priorities.* Priorities must also be understood at the early stages of program development. While objectives are primarily driven by the task itself and are usually easy to define, priorities are more oriented toward the particular desires of those responsible for the program. They are, therefore, more difficult to identify. For any given program, priorities can be economical, political, social, personal, and/or cultural in nature. When priorities appear to disagree with program objectives, the success of that program is endangered. For instance, when an objective of preventing human exposure to occupational health hazards is mixed with a management priority that is primarily concerned with tight economic controls and restrictive budgetary guidelines, establishing an effective industrial hygiene program becomes somewhat more difficult. When such controls are excessively restrictive, the adequacy of the industrial hygiene program may be further jeopardized. This example is not intended to suggest that an unlimited budget is essential to establishing a working industrial hygiene program. In fact, many excellent programs exist throughout corporate America under extremely tight budgetary controls. The only message here is for the reader to understand that priorities do not often parallel objectives. The further apart these two elements are from each other, the greater the effort must be to ensure program success. Understanding this basic fact of business will greatly facilitate any attempt to implement an industrial hygiene program. If the industrial hygiene program must operate under strict financial controls, so be it—as long as this requirement is understood from the beginning. Likewise, if management priorities are more focused on the prevention of social or political ramifications, so be that, as well. The task of the industrial hygienist is not to change management priorities but to understand them. Once understood, boundaries and limitations can be identified. Hopefully, the hygienist can work successfully within these boundaries to accomplish the industrial hygiene effort's objectives.

3. *Management Agreement.* Throughout the planning process, coordination with upper management is essential. As the process flow indicates, obtain-

ing management agreement once specific phases of the process have been completed will help to ensure management buy-in of the program as a whole. Management has the right to know the status of any initiative or undertaking being implemented in their organization. Obtaining their agreement at critical junctions during planning prevents the waste of valuable time and resources. There is no sense in planning, designing, or developing any process if management will not approve the final product. Coordination *during* the implementation stage of the process reduces the likelihood of any surprises upon completion.

4. *Assess Work Environment.* Once management has agreed to the objectives and priorities, the planning process allows the hygienist to conduct a general assessment of health hazards that may be present in the work environment. It is during this assessment that the hygienist will begin to define the level and the nature of the program that will be required to control these hazards. When the tasks and functions performed in the work environment are properly evaluated, the hygienist can identify specific areas of concern where health hazards are likely to exist. This assessment includes a review of the operating procedures and instructions associated with the performance of all work tasks and functions. The hygienist should examine historical records, exposure monitoring results, past OSHA citations or notices of alleged violations, OSHA 200 logs, and industrial hygiene data for companies that manufacture similar products (if applicable). This assessment allows the hygienist to define the areas that will require attention. It also helps eliminate unnecessary focus on non-issues and concerns. For example, if no work areas produce annoying levels of noise, the industrial hygiene program will not generally have to address hearing conservation requirements. Similarly, if there is no task, function, or process that utilizes volatile chemical commodities or other nonvolatile substances that can pose respiratory health hazards (i.e., asbestos), no effort need be wasted evaluating respiratory hazards. This initial assessment provides direction and helps define specific parameters for the implementation of the industrial hygiene program.

5. *Identify Acceptable Exposure Risk.* As the assessment defines the required parameters for the industrial hygiene program, the hygienist should begin to identify the hazard exposure levels that are applicable to the work environment. Since no actual exposure monitoring may have yet been taken at this point in the planning process, the hygienist is primarily concerned with the identification of applicable exposure criteria. For instance, if exposure to high noise levels is felt to be a potential health hazard, the hygienist must review the latest regulatory standards pertaining to hearing conservation programs and identify the action level as well as the actions that will be taken. For any possible health hazard revealed during the assessment,

the hygienist must identify the appropriate exposure parameters that will be enforced to ensure the preservation of employee health. These parameters establish specific levels of exposure at which the risk of health damage is considered to be acceptable, based on available information (as established by NIOSH, ACGIH, OSHA, etc.). Identifying acceptable exposure risk further quantifies the specific objective of the industrial hygiene effort: controlling health hazard exposure. This, however, cannot be accomplished until a determination of what exactly constitutes a health hazard is established.

6. *Define Roles and Responsibilities.* No program will succeed unless the specific roles and responsibilities associated with program implementation are clearly defined from the onset. After the work environment has been properly assessed and the level of health hazard and exposure risk determined, the industrial hygiene program plan must also explain the role to be played by each member of the organization. From top executive management to every employee, the role of each teammate must be made clear. For example, upper management must issue policies and approve procedures that are consistent with the overall objective of the industrial hygiene effort. It must provide the appropriate delegation of authority to lower-level management so that it, too, can properly execute decisions related to the protection of employee health. Supervisors and managers who are responsible for the work areas identified during the initial industrial hygiene assessment may require additional training so that they fully understand the potential health hazard risks present in their assigned areas. Their responsibility to obtain this training is as essential as it is for the industrial hygienist to provide it. The plan should go on to address any additional roles and responsibilities that may be incumbent on specific personnel in areas such as safety, purchasing, medical (i.e., physician, nurse, toxicologist), engineering, fiscal, and any other organizational element. Finally, the plan should also establish any general and specific employee responsibilities as they relate to the industrial hygiene effort. For instance, employees are generally the first to notice abnormal conditions in the workplace. They must notify their supervisor(s) immediately when such things happen so that the proper precautionary actions can be taken. As a first line of defense, they must follow established rules and always practice good habits of personal hygiene and housekeeping.

7. *Develop Implementation Plan.* Perhaps one of the most important aspects of the overall planning process is to develop a plan for the implementation of the entire industrial hygiene program. The best approach to implementation of any project is to provide answers to the basic questions. Specifically, *who* will do what and when will it be done? Where in the facility will it be done? Why must it be done? How will it be accomplished? Although

this approach may sound somewhat oversimplified, it really does help facilitate the development of the implementation plan. Using this method, the hygienist may find that many of these questions have already been answered during other aspects of the planning process. For example, who will do what should have already been established while defining roles and responsibilities. If this was properly done, pulling this information into the implementation plan should be relatively easy. Although the when portion of the equation has yet to be developed and should be based on degree of hazard risk (i.e., priority), doing so here will facilitate the establishment of implementation schedules. Where in the facility these things must be done should have been identified during the initial assessment of the work environment. Using this previously established information, the development of the implementation plan should not be too difficult. The answer to why these things must be done should be more than obvious at this point—to control exposures to health hazards in the workplace. In fact, this should be a recurring theme through the entire implementation plan. The final question to be answered, how will these things be accomplished, is where the services of a competent industrial hygienist become paramount. It will be left up to the hygienist to determine, based on all available information and their individual expertise, the most appropriate methods, techniques, and strategies to eliminate or control the hazards to human health. The implementation plan establishes the overall approach to the industrial hygiene program. It provides direction, identifies policy, and forms the basis for management's commitment to the health of their work force.

8. *Establish Implementation Schedule.* Once the plan has been finalized, a schedule for implementation should be established. The implementation schedule should clearly demonstrate priority by establishing the time frame in which specific activities are to occur. The schedule should be flexible enough to allow for work slippage or other contingencies that may effect adherence to an otherwise rigid schedule. When developing the schedule, the hygienist should also consider all possible factors that can influence achievement of the dates and milestones defined in the schedule. Factors such as cost, resource management, equipment calibration requirements can affect the accomplishment of specific activities per established schedules.

9. *Management Agreement.* Once management has agreed to the roles and responsibilities defined in the plan, and to the implementation plan itself and its associated schedule, the planning process moves into actual program implementation. However, before going forward, it is imperative that management develop and define program accountability. This can be as important to the program's success as the basic planning activities discussed above.

The Implementation Phase

10. *Implement Program.* Implementing a well planned industrial hygiene program is not extremely difficult, but it can present many challenges, even to an experienced industrial hygienist. Challenges like personality conflicts, resistance to change, lack of understanding, and other personnel-driven barriers are not easy to predict during the planning phase. Unexpected work process changes, equipment modifications, market and/or customer-driven task modifications, and the like can also impact the implementation of a planned program, its schedule, and even its costs. While challenges like these can effect the implementation process, they will rarely cause the total failure of an industrial hygiene program. The combination of a competent industrial hygiene staff and planned program flexibility is generally sufficient to meet these challenges and any others that may arise during or after program implementation.

    Implementing the program requires a coordinated effort from all levels of the organization. Each member of the work force must be cognizant of the program, its requirements, and their part in its overall success. The best way to accomplish this is to issue written policies, procedures, directives, and other documentation (including the implementation plan and schedule) that explains the intent of the industrial hygiene program. If simply stated, communicating your intent will foster understanding. With understanding comes awareness, cooperation, and, ultimately, program success.

    The next three steps deal with the concepts of hazard recognition, evaluation, and control that will be discussed in greater detail later in this Chapter under the section entitled General Principles of Industrial Hygiene.

11. *Recognize Health Hazards.* Recognition of health hazards is the first aspect of the implemented program taken by the industrial hygienist. The industrial hygienist should now revisit those work areas where potential health concerns were first identified during the initial assessment of the work environment. Hazardous conditions should be recognized and documented at this point so that the degree of exposure can be determined.

12. *Evaluate Health Hazards.* After the threats to employee health have been recognized as such, the hygienist must evaluate each to determine their exact nature. This includes a summation of exposure levels, the risks associated with such exposures, and the methods that are available to reduce, eliminate, or control the health hazard risks.

13. *Control Health Hazards.* If the objective of the industrial hygiene program is to be achieved, the hygienist must develop specific control measures that will effectively address each recognized health hazard in the workplace. At the very least, these measures should reduce, to the lowest possible level, all health hazard exposures. Control measures may include the use of

equipment (i.e., personal protection and/or physical barriers), administrative controls (i.e., procedures and policies), or process changes (i.e., alternative working methods and product substitutions). Implementing controls means implementing the corrective actions necessary to ensure a safe and healthy work environment. Hence, control is the essence of the overall industrial hygiene effort. However, regardless of the specific control measures that have been identified, management must make the final decision regarding implementation. Because cost is always a consideration, it is highly recommended that, whenever possible, management be presented with more than one option to control workplace hazards.

14. *Employee Awareness and Training.* For any control measures to be truly accepted and effective, employees that are involved (exposed) must be made aware of the health hazards and the control methods available to protect themselves from such exposures. In fact, the Federal OSHA Hazard Communication Standard requires such training. Specifically, OSHA 29 CFR 1910.1200(h)(1-2) requires employers to provide their employees with information and training on hazardous chemicals located in their work areas, including the measures employees can take to protect themselves from these hazards. By incorporating the appropriate training into the industrial hygiene effort, the program will be more effective and will meet these OSHA requirements.

Since the line employee is usually the end user of most chemical commodities or other substances that can pose a health threat, proper employee awareness and training is the key to the success of the industrial hygiene program. However, not all health hazards that may be present in the work environment can be linked directly to some specific work process, task, operation, or procedure. In many instances, insidious hazards associated with the work environment itself can be just as threatening to employee health as any related to a specific work process. For instance, the air flowing through the workspace can contain physical hazards such as asbestos or fiberglass, or biological contaminants such as spores or mold, each of which can affect the health of those exposed. Similarly, insufficient light levels have been the cause of numerous complaints of headache, fatigue, and eye strain, all of which can lead to injury or illness if the hazardous condition is not properly recognized and corrected. Employee training must, therefore, include information on the potential existence of these hazards, any special observation methods and techniques that can be used to initially recognize when they are present, and the proper reporting procedures to be used to notify supervision that a suspected problem exists. Adequate employee training should be viewed as the first line of defense against health hazards in the workplace. Its importance to total program success cannot be overemphasized.

15. *Conduct Program Audits.* Once the program has been implemented, the hygienist should conduct periodic audits or reviews to ensure that the objectives of the program are being met. These audits should be objective and encompass all aspects of the program, from policies to training. The audit should evaluate each aspect of the industrial hygiene initiative to determine overall effectiveness and adequacy. Also, if any work processes have recently changed, the audit will reveal any new concerns as well as eliminate those that, as a result of the change, are no longer posing a health threat to employees. This audit can be performed internally by the industrial hygienist with the assistance of other members of the organization such as safety personnel, affected managers, the industrial nurse, or any other employee. An alternative is to utilize the services of an external agency or consulting professional to perform the audit. While this is not always necessary, it will help ensure a completely unbiased assessment. Larger organizations, for example, often make it a practice to utilize other personnel (corporate experts, personnel from other locations, etc.) to ensure objectivity. Whatever the method used, periodic audits are an essential element of any initiative and the industrial hygiene program is no exception.

16. *Recommend Program Improvements.* Discrepancies or shortcomings discovered during the program audit are areas for improvement. Also, any modifications to the program that are necessary due to changes in processes, additions or deletions to work steps, or possible regulatory changes, must be addressed at this time. The industrial hygienist should determine the appropriate actions necessary to resolve any of these concerns. Before implementing any program changes, however, the hygienist should demonstrate to management how these recommendations will improve the industrial hygiene program. Management approval is especially important when additional costs are associated with planned improvements.

17. *Management Agreement.* Before moving on to actually implementing any recommended updates to the program, management has the right to review and approve these changes. The industrial hygiene program is, after all, a management initiative. Since management's commitment to the program is essential to program success, it should be in agreement with any suggested changes before any updates are initiated.

18. *Update Program as Required.* The final step in the implementation phase of the planning process is to take the actions necessary to update the program. These actions can include anything from simple procedural changes to the purchase of new protective equipment and the provision of the training required to safely use such equipment. What ever the extent of the update, it is essential that optimum employee health preservation be ensured through a vigilant, responsible, comprehensive, and evolving industrial hygiene program.

The planning process requires full consideration of every aspect of the entire industrial hygiene effort. Beginning with the pre-implementation phase (planning requirements) and moving through the planned implementation activities, each step must be carefully thought out to ensure program adequacy and completeness.

GENERAL PRINCIPLES OF INDUSTRIAL HYGIENE

Chapter 1 introduced the concept of REC (Recognition, Evaluation, and Control). The process of recognition, evaluation, and control are described as the general principles of industrial hygiene, since these three steps summarize the true purpose of the industrial hygiene effort.

Recognition

As stated in Chapter 1, it is obvious that no hazard to human health can be controlled unless it has first been recognized as such. Because of their specialized training and experience, industrial hygienists are most responsible for the recognition of occupational health hazards in the work environment.

To accomplish this goal, a hygienist can employ a variety of relatively simple methods and techniques that are generally available and rather easy to perform. For example, reviewing historical records of past performance may help identify concerns that were previously reported or investigated that may or may not have been fully resolved. Also, regular walk-through inspections or surveys of the work environment will help the industrial hygienist become more familiar with the various work processes and operations that occur on a daily basis. Understanding the work that is performed will also facilitate recognition of any associated health hazards. Also, when changes occur or work procedures are modified, the hygienist will more readily recognize any potential health threats associated with these changes. To the astute hygienist, the inspection can reveal at least the following basic data:

1. Work process familiarization
2. Potential or existing chemical hazards
3. Potential or existing physical hazards
4. Number of employees potentially or actually exposed
5. Assessment of existing control measures
6. Basic requirements for *personal protective equipment* (PPE)
7. General employee understanding/awareness of health hazards
8. Work area layout and emergency procedure requirements
9. Employee health complaints related to work processes

The amount of basic information obtainable during a simple walk-through observation inspection of the workplace cannot be overemphasized. Recognition, the first step toward the elimination of health hazards in the workplace, cannot occur unless the hygienist is a keen observer.

Conducting a regular review of the company's chemical inventory is another simple technique that can be used to recognize potential health hazards. These inventories should be readily available and current since they are generally required to meet the compliance provisions of other programs, such as the OSHA Hazard Communication Program and the EPA SARA Title III Community Right to Know requirements. Aside from the physical inventory of commodities on hand, additional chemical data and health hazard exposure criteria can be obtained from individual chemical material safety data sheets (MSDS), purchase orders/specifications, work process descriptions, and other available information. Since chemical manufacturers are required to list health hazard data on the MSDS, hygienists already have a great deal of specific information at their disposal. Even when the information provided by the manufacturer is too vague to be useful, using the emergency contact telephone number on the MSDS will usually resolve any remaining health hazard concerns. Also, the industrial hygienist should review the health and safety data of any new chemicals entering a facility before placing that chemical into service.

Another method of recognition involves a regular review of industrial health/medical records of those employees who are assigned to work areas where exposures are likely or suspected. Frequent reports or complaints on record regarding specific health problems (dizziness, nausea, gradual hearing loss, headache, cumulative trauma, and so on) may help the hygienist recognize the existence of insidious health hazards not otherwise easily identifiable in the workplace. In such instances, an occupational health nurse on staff or available for consult will certainly help expedite the health hazard recognition process.

Finally, a well educated and trained work force can prove invaluable in terms of health hazard recognition. Generally, no one knows a particular job function better than the employee who performs it on a daily basis. Proper training and understanding of the nature of health hazards may be all that is required for the employee to recognize when something is not quite right in their work area. Just knowing when to stop working and call for assistance will not only alert the hygienist of a potential problem, it may also help avoid a needless exposure situation. Also, when employees know that they are an essential element of their company's occupational safety and health assurance program, the productivity of the workplace and of the workers themselves almost always improves.

The chapters in Part II focus on some of the major hazards that can be found in the work environment. The various methods of the recognition process discussed above should be applied to each work situation to determine if any such hazards exist. Once recognized, the hygienist can move on to evaluate the degree of harm the hazard(s) may pose to human health.

Evaluation

After the hazard has been recognized as such, the industrial hygienist must proceed with a professional evaluation assessing the risks associated with personnel exposure to that hazard. On completion of this assessment, the hygienist provides an informed opinion on the degree of risk that the hazard imposes on the workplace.

During this evaluation process, the hygienist must consider at least the following questions to properly assess the risk of hazard exposure:

1. **What is the nature of the hazard?** It could be associated with the transfer of (or contact with) energy stressors such as vibration, illumination, temperature, noise, and so on. Or, the hazard exposure could be caused by contact with agents that produce certain degrees of biological stress such as chemicals, air contaminants (including irritants such as dusts, mists, fumes, smoke, etc.), asphyxiants, and the like, with subsequent toxic effects on human health. It may even be a combination of both types of stressors, depending on the circumstances. Reviewing exposure records, including employee medical records, can often yield specific information concerning the nature of past exposures. Since OSHA requires employers to maintain copies of such records for a period of not less than 30 years (OSHA 29 CFR 1910.20), the industrial hygienist may discover significant information concerning exposure history.

2. **What is the degree of exposure?** This should include the potential as well as the actual magnitude of the hazard exposure. The hygienist must evaluate the levels at which the exposures are occurring or can occur. Additional concerns that should be addressed include assessing whether the exposure levels exceed any established guidelines, such as OSHA PELs, and to what extent.

3. **What is the typical duration of the exposure?** The amount of time a person is exposed to a health hazard is an extremely important factor in evaluating the overall risk associated with that hazard. The duration, coupled with magnitude, will determine the dose an individual has received. Here again, knowledge of any established exposure guidelines such as the PEL or, in the case of duration, the Short Term Exposure Limit (STEL), will assist the hygienist in determining the specific risks associated with the exposure event.

4. **What is the route (or routes) of exposure?** For an exposure to a hazardous material to effect a person, the material (or its by-products) must enter the human body in some way. There are basically three primary avenues of entry. *Inhalation,* the most direct route, is the quickest entry into a person's body. Transfer into the blood stream and distribution throughout the body

is relatively rapid, and the subsequent effects on health are almost always acute (immediate). However, depending on the substance and the person exposed, chronic exposures can also occur without any visible (or otherwise obvious) indication that anything is wrong until some time in the future. Whatever the case, the hygienist must assess the presence of such exposure potential during the evaluation process.

Absorption through intact skin is considered the second quickest route of entry. Although many substances are not prone to this route, the hygienist must be aware of those in the workplace that are.

The third route, ingestion, must also be evaluated. However, since ingestion of hazardous materials usually occurs as a result of careless work or personal hygiene habits that vary from individual to individual, the hygienist may find it somewhat difficult to isolate this route of entry as the cause of exposure. Nevertheless, the question of exposure route must be fully explored.

In some occupations, a fourth, less-common route of exposure may have to be considered as well. Exposure by *injection* may be of particular concern in the medical profession, for example, where a careless needlestick could subject the individual to a variety of health hazards, ranging from poisons to bloodborne pathogens.

5. What is the individual sensitivity to the exposure? In general, people react differently upon exposure to the same substances. For example, one person may have a severe reaction when exposed to a skin irritant (such as fiberglass, a nuisance dust) while others exposed to the same substance have no visible reaction whatsoever. This is especially true with materials known as *sensitizers*. An employee may have worked with the same substance for a very long time, perhaps over a period of years, with no adverse reaction. However, at some point in time the body may finally become sensitive to this frequent exposure, resulting in some adverse health effect or reaction. Once a person becomes sensitized to a particular substance, the condition generally remains for life. Each time a subsequent exposure occurs, the health effect will undoubtedly become more severe. While it does not bother some people to work with, for example, fiberglass, others who are sensitive can develop an irritating rash following even the slightest exposure. This allergic reaction, known as *sensitization,* develops as a result of frequent exposure to specific materials. The hygienist must evaluate carefully to determine the true risks that are present in the workplace. Such a hazard is insidious and therefore difficult to isolate until an actual indication of exposure occurs. It is difficult because nothing may have changed in the work process or the work environment when signs of the exposure become apparent. Determining the cause of the problem and the risk of exposure requires detailed knowledge of the work processes as well as an

understanding of each substance used in the workplace. The material safety data sheet (MSDS) for each substance normally provides information on health hazards and the effects of exposure. If the substance is a known sensitizer, it should be listed as such on the MSDS. Also, employees with predisposed physical conditions such as anemia, high blood pressure, diabetes, asthma, bronchitis, or other respiratory aliments (heavy smokers), psoriasis or other dermatitis, and so on, may also be effected by certain exposures when other employees are not. Therefore, a detailed review of an employee's medical records as well as a consultation with the occupational physician or nurse can be a key element of the evaluation process.

Once each of these basic questions have been explored, the hygienist can determine the next step in the evaluation process. The answers to these five questions should begin to narrow the focus of the industrial hygiene program. In addition, the hygienist will know what specific areas require further investigation and what concerns no longer need be considered. If necessary, physical sampling of the air, work surfaces, and any other element of the work environment can proceed, with specific attention placed on those areas identified during the initial evaluation process.

Each of the general techniques and principles of evaluation discussed here can be applied to the hazards discussed in Part II. When adequate evaluation of the recognized hazards is complete, the hygienist can develop specific recommendations to control such exposures and ensure the preservation of employee health.

Control

Depending on the conclusions reached during the evaluation process, the hygienist must identify the next appropriate course of action. This may range from no action to complete modification of work practices and procedures to ensure the proper control of any exposure hazards. For example, the evaluation may have demonstrated that a given hazard possesses no real threat to human health under the conditions observed in the workplace. As a result, no exposure potential exists and very little must be done to ensure the threat of hazard is controlled. On the other hand, the hygienist may have discovered that a serious exposure hazards is occurring in the workplace and that immediate actions are required to eliminate or control the extent of employee exposure. Whatever the case may be, the hygienist must determine the most appropriate level of control required to properly protect the worker from the health hazard(s). It is often beneficial to the control process if the hygienist seeks input (formally or informally) from plant maintenance personnel, engineering, equipment operators, and the medical staff (if applicable). Many excellent suggestions can be obtained using this team approach to hazard control. Once the control measures have been formulated, a recommendation to management should be made with full disclosure of the nature, level, and degree of hazard and how the implementation of the suggested controls will reduce or eliminate the exposure potential. Essentially, management must be provided with all the determining factors and pertinent information that led to the selection of specific control measures. Remember that hazard exposure control is also a management decision. The hygienist must, therefore, ensure that management has the information it will need to properly assess the problem and accept the solutions as recommended.

Under extreme circumstances of health hazard exposure, a control measure might be the total elimination of the process involved in the exposure. While this method certainly solves the exposure problem, it is seldom a viable option in the real world. It is more probable that the hygienist will suggest specific actions designed to control the nature, level, and degree of exposure. Depending on the hazard, these control actions may include:

- Substituting less hazardous commodities for those that pose a significant threat of health exposure risk
- Substituting less hazardous machinery or equipment for those that produce the health hazard
- Redesign of equipment to reduce the threat of hazard
- Implementation of administrative and procedural controls to ensure minimum personnel contact with hazardous conditions (e.g., changing a work process so that exposure time is kept to a minimum, staggering work shifts, establishing controlled access areas, using warning signs, etc.)
- Requiring the use of *personal protective equipment* (PPE), such as respirators, gloves, goggles, and so on, to reduce the likelihood of exposure
- Installation of equipment designed to reduce the potential for hazard exposure (e.g., ventilation hood, oxygen level indicators, alarms, warning lights, etc.)

Once the specific controls have been determined and approved by management, the hygienist must develop these measures into written requirements and proceed with implementation. Without the successful implementation of effective controls, the recognition and evaluation of hazardous conditions becomes meaningless. Ultimately, these controls will become the key elements of the organization's written industrial hygiene program. This is why the control of health hazard exposure is often referred to as the industrial hygienist's primary objective.

The three general principles of the industrial hygiene process (recognition, evaluation, and control) are summarized in Figure 3-2. As the various types of common industrial health hazards are presented in Part II, the reader should refer back to the general principles discussed here since, regardless of the hazard itself,

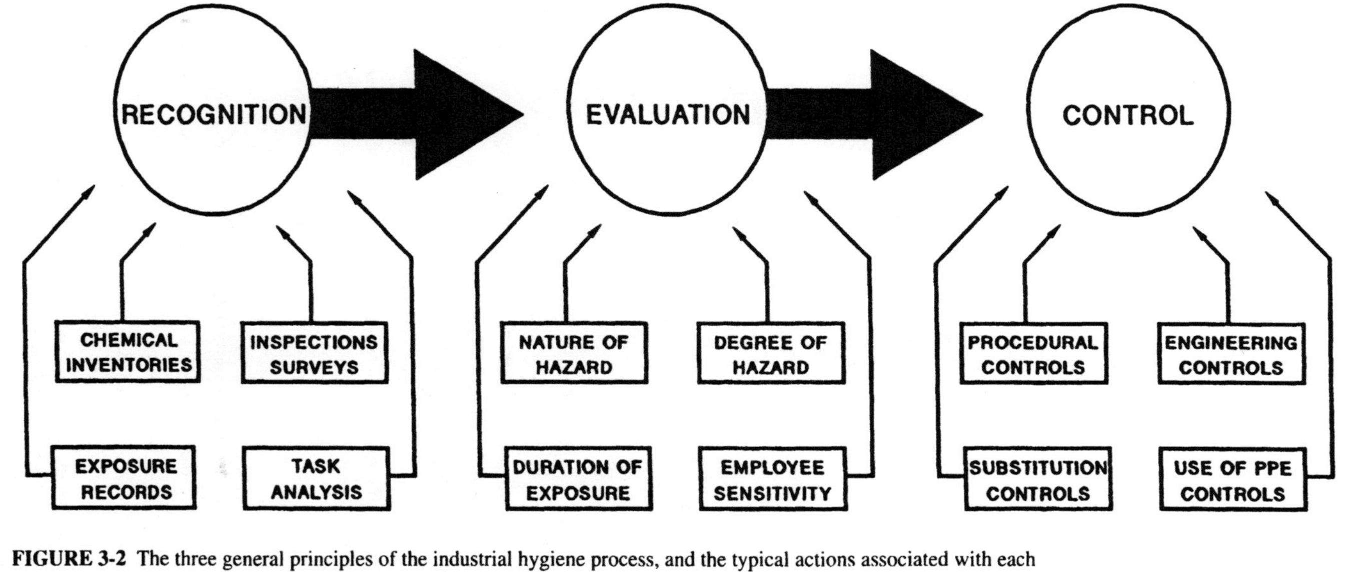

FIGURE 3-2 The three general principles of the industrial hygiene process, and the typical actions associated with each

an industrial hygiene program will always require the completion of this recognition, evaluation, and control process.

HEALTH HAZARDS VS. PHYSICAL HAZARDS

In terms of human exposure, most would agree that a *hazard* is something capable of creating a risk of injury or illness. Its existence, therefore, is dangerous and any situation, task, or function that contains such conditions are said to be *hazardous* in nature. In the average workplace (if there really is such a thing), the practicing safety and health professional can most likely identify (or recognize) a variety of hazards or potentially hazardous conditions or work practices. In general terms of recognition, evaluation, and control, hazards in the workplace can be divided into two groups or categories: health hazards and physical hazards.

To facilitate this discussion, it shall be established that exposure to hazards that can result in *illness* will be considered *health hazards,* while exposure to those that can result in *injury* will be termed *physical hazards.* It is acknowledged that arguable exceptions to this premise can exist, depending on the specifics of any given exposure situation. For example, exposure to large doses of radiation will result in damage (injury) to specific organs and/or the total body itself making radiation a physical hazard under this definition. However, small doses over a prolonged period could result in illness that may or may or may not lead to death, which would mean radiation is a health hazard. The point is, depending on the circumstances of exposure, a hazard may be considered either a health or physical hazard (or perhaps both in some cases). Examples of physical hazards under this definition would include missing machine guards that expose pinch points, or oil spilled on a shop floor creating a slipping hazard for workers. Health hazards can be equally obvious, such as the presence of noxious air contaminants, or they may be unnoticeable, as in the case of low-level radiation exposure. In the practice of industrial hygiene, knowledge of both physical and health hazards is important. But, with some exceptions, the industrial hygienist's profession specializes in the practice of occupational health, and is concerned primarily with the health hazards that may be present in the work environment.

Health Hazards

As established above, a hazard that causes or has the potential to cause illness of any nature and to any degree can be considered a health hazard. Health hazard exposure can cause obvious indications of illness, such as nausea, head or stomach ache, dizziness, drowsiness, dermatitis, and so on. Health hazards can also cause not-so-obvious indications that an exposure has occurred. Exposure to some known health hazards may not result in any noticeable symptoms for quite some time, perhaps weeks, months, or even years. A coal miner, for example, does not contract black lung disease the first day at work in a coal mine, with the first inhalation of coal dust. It may take many years of chronic, long-term exposure before there is any indication that a health hazard exposure has occurred. This is also true of other hazardous agents such as asbestos, cotton, and silica. Similarly, a person who works with organic solvents every day may not display any immediate signs of adverse reaction. At some point, however, that person may develop a sensitivity to that solvent and develop a skin rash or some other more serious symptom of contact dermatitis (localized inflammation of the skin tissue).

To become ill subsequent to such exposures generally requires either a subtle/insidious or sudden/obvious adverse reaction involving one or more anatomic systems. In other words, the effects of exposure to any given health hazard is usually evidenced by the presence of illness or damage in one or more components of the human anatomy. The resulting illness or damage, therefore, is often referred to as *systemic*. Specifically, exposure to health hazards may have a direct or indirect impact on any of the following systems, depending on the route (or routes) of exposure:

1. The respiratory system (through inhalation)
2. The circulatory system (through inhalation, ingestion and/or absorption
3. The digestive system (through ingestion)
4. The central nervous system (through inhalation, absorption)
5. The human sensory organs (including the skin, eyes, and ears), which are also at particular risk of exposure to both health and physical hazards in the workplace

In addition to understanding the nature of health hazards in the workplace, the hygienist must understand how systemic illness can occur following health hazard exposure. These two factors are essential to the overall recognition process. Part II of this *Basic Guide* focuses on some of the common health hazards that can be found in the industrial work environment. Information on recognizing the results of exposure on the human anatomic systems is provided in subsequent chapters of this Part.

ESSENTIAL ELEMENTS OF AN INDUSTRIAL HYGIENE PROGRAM

It has been clearly established that recognition, evaluation, and control are the basic ingredients of the industrial hygiene process. Any attempt to establish an industrial hygiene program must be based on these principles. While industrial hygiene programs differ from one company to the next, the essential elements depicted in Figure 3-3 should remain characteristic of even the most basic pro-

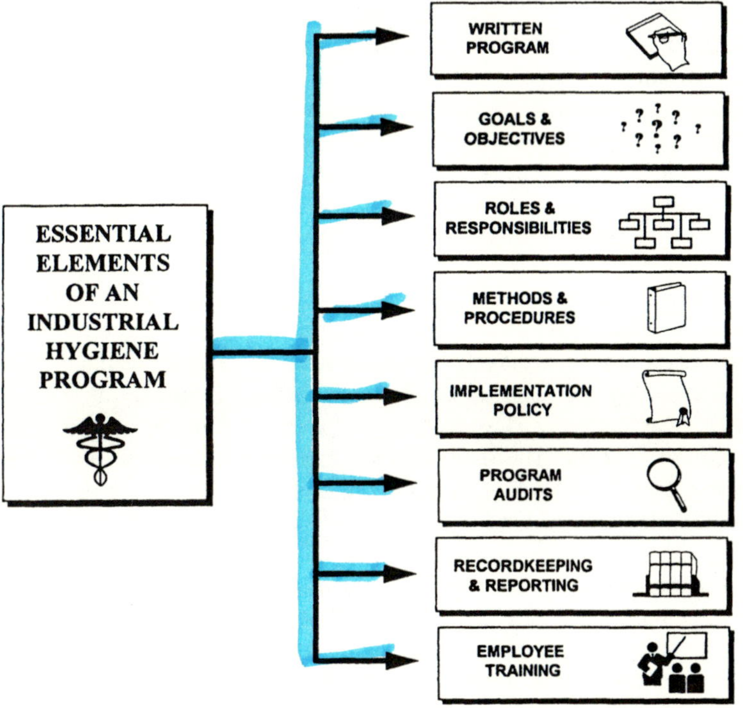

FIGURE 3-3 The essential elements of an effective industrial hygiene program

grams. Although the type of industry, management approach, size of the company, nature of work performed, organizational structure, and specified job responsibilities are all variables to be considered when designing a program, the essential elements of a basic industrial hygiene program should include at least the following:

1. *Development of a written program:* It is common practice throughout industry to document company policies in written form. Industrial hygiene is no exception. As with any company initiative intended for implementation and enforcement across the entire organization, development of a written program is essential. Written programs give management the opportunity to review and approve each aspect of the program prior to implementation. The written program will establish intent, goals, objectives, roles and responsibilities, program parameters, expected results, contingencies, as well as specific information about the overall implementation of the program. It is highly recommended that the written program be reviewed by

appropriate legal representatives prior to implementation. This should ensure the proper consideration of any liabilities that may be associated with the program. Once written, the program should be a usable, living document. In other words, it should be referred to frequently, reviewed regularly, and updated as required to ensure its ongoing adequacy.

2. *Established goals and objectives:* From the onset, the program must clearly define the purpose and intent of the organization's industrial hygiene effort. Purpose and intent are often referred to as goals and objectives. It is here that recognition, evaluation, and control of occupational health hazards should be established as the foundation (i.e., purpose and intent) of the overall program. Also, depending on the nature of the organization's structure, the scope of the program should also be provided. If the program will apply to the entire organization, this should be stated. Likewise, if an industrial hygiene program is only necessary for a small operating division within the organization, the defined scope should clearly establish the limited applicability of the program. Whatever the case may be, it is important that program goals, objectives, and, if required, scope be well defined to avoid potential misunderstandings after implementation.

3. *Defined organizational roles and responsibilities:* As with any other management initiative, who in the organization will be responsible for what aspects of the program must be clearly defined. From top management to the lowest-level employee, responsibilities should be spelled out in writing. Each functioning element of the organization that will be required to perform some action or fulfill some requirement of the implemented program must know the exact nature of their responsibilities to ensure the success of the program. The chief executive officer, the chief financial officer, the safety professional, purchasing agents, line managers, line employees, and, of course, the industrial hygienist are all examples of responsible organization personnel who may have specific roles to play to ensure successful program implementation.

4. *Approved methods and procedures:* Approved methods and procedures will define the operating parameters of the industrial hygiene program. Descriptive information pertaining to the specific methods and procedures that will be used by the industrial hygienist during the recognition, evaluation and control process should be provided. For example, in large organizations, corporate procedures may already exist that require individual components to perform the industrial hygiene function following specific guidelines. Depending on the situation at a given facility, federal or state regulations or even national consensus standards may prescribe certain methods of health hazard evaluation that might be incorporated into the program as a standard (and legally defensible) approach.

5. *Defined program implementation policy:* Exactly how the program is to be implemented must also be clearly defined. An implementation policy may be nothing more than a simple statement of intent that takes into consideration all other program elements such as scope, objectives and goals, roles and responsibilities, operating parameters, and any other program element specific to that organization. While this may appear to be nothing more than a summation of the other elements, it is still important to define the organization's implementation policy.

6. *Ensured method to evaluate program effectiveness:* As with most other initiatives, the effectiveness of the effort must be measured periodically to determine level of program success as well as areas needing improvement. Regularly scheduled program audits (e.g., annual) should therefore be performed as a method to measure the industrial hygiene program's overall effectiveness relative to its established objectives. Audits can be conducted internally by company personnel such as the industrial hygienist, the safety professional, or other designated representatives of management. Advantages of using company personnel include their familiarity with the company, its products and services, and the industrial hygiene program. Disadvantages include the possibility of unintentional (or intentional) bias during the audit. For example, some may dismiss specific program concerns as unfounded because they happen to know that actions are planned or will be taken to correct a problem and they perceive no need to report a non-compliant situation during the audit. While this may be a rare circumstance, indiscriminate decisions not to report all program concerns could seriously jeopardize the success of the audit. Also, some internal auditors may not wish to report problem areas to avoid documenting known violations of occupational health standards. While this may seem a valid concern, it has been demonstrated time and again throughout industry that documenting violations is not generally considered self-incriminating by regulatory agencies. Fines and penalties may be expected, however, when the employer fails to act to control a recognized hazard. Another option is to contract an external expert such as a consultant to audit the program for effectiveness. The resulting audit will usually be detailed, comprehensive, and unbiased. As a result, it will provide an accurate assessment of the organization's industrial hygiene program at the time of the audit. Disadvantages include cost (use of external consultants can be an expensive venture) and time (it may take longer if the consultant is not familiar with the organization, its policies, its products, or its services).

7. *Record keeping and reporting requirements:* Another essential element to an effective industrial hygiene program is the maintenance of accurate records. Many federal and state occupational safety and health agencies have specific requirements regarding record keeping. These requirements

include the documentation of occupational injuries and illness, records that document personnel exposures (to noise, chemicals, vibration, temperature extremes, and so on), medical records, material safety data sheets (MSDS), and numerous other materials. Also, depending on the degree of an exposure and the number of personnel effected, agencies also mandate specific reporting and notification requirements. This critical element will be discussed in greater detail in the next section of this chapter.

8. *Employee training provisions:* Employee training is critical to the success of the industrial hygiene program. Here again, both federal and state agencies require the employer to provide a minimum level of employee training. Under most "right to know" laws, employees must be made aware of the health and physical hazards with which they must work during the normal course of their assigned duties and, in some instances, during extreme emergencies. In addition to meeting these minimum requirements, employees should also be trained on the specific elements of their company's industrial hygiene program. Employees who feel that they are an important part of their organization's efforts to provide a safe and healthy work environment will often work in a more conscientious, safe, and informed manner.

While the specifics of any given industrial hygiene program will vary from one organization to the next, adequate consideration of the elements briefly discussed above are generally essential to the success of any program.

RECORD KEEPING AND DOCUMENTATION

As mentioned above, the importance of proper record keeping and the maintenance of proper documentation cannot be overemphasized. In fact, according to Federal OSHA, citations for failure to comply with the required record keeping standards continues to rank among the leading violations in general industry today.

Tracking and recording workplace injuries and illnesses is a key provision of the OSHA record keeping requirement. While this task may fall to a variety of personnel within any given organization (e.g., a safety professional, an industrial hygienist, a human resources representative, an occupational nurse, etc.), the industrial hygienist should still play an important role in assuring the accuracy of these essential records. When claims of industrial illness are reported, the hygienist must not only investigate the matter—it is also necessary to ensure its accurate documentation. OSHA requires employers to document this data on the Log and Summary of Occupational Injuries and Illnesses, also known as the OSHA 200 Log. With regard to injuries, the industrial hygienist should know that OSHA requires employers to record injuries that resulted in medical treatment, loss of consciousness, restriction of work or motion, or transfer to another job function.

When illness occurs, the employer must maintain records of employees exposed to potentially toxic materials or other harmful agents.

The OSHA 200 Log is used to record each case of injury or illness, classify each case by degree and type of lost time, and briefly describe the injury or illness. Additional background information on each case must be recorded on the OSHA Form 101 entitled Supplemental Record of Occupational Injuries and Illnesses. Form 101 provides detail on how the accident or illness occurred and the determined cause(s), the nature of the injury or illness including body parts involved, and other specific information about the employee, such as the number of hours worked prior to the incident, how long employed, and so on. It is important to note that OSHA will permit substitute use of other forms for both the 200 Log and the Form 101, as long as the substitute forms meet the mandatory recording requirements specific by OSHA. Whatever forms are used, OSHA currently requires the employer to maintain the data for a minimum of five years.

In addition to the retention of the OSHA 200 Log and the Form 101, OSHA requires employers to maintain other industrial health records and related information, some for specified periods of time. For example, records pertaining to the maintenance and calibration of equipment and instrumentation used to support the industrial hygiene effort (such as air sample pumps, temperature probes, illumination level meters, etc.) as well as the results of any workplace sampling activities must be kept readily accessible for inspection and review.

Under OSHA Standard 29 CFR 1910.20, Access to Employee Exposure and Medical Records, the law requires employers to retain employee exposure and medical records for a period of at least the length a person's employment plus 30 years. There are also regulations that pertain to very specific exposures (such as benzene, ethylene oxide, and asbestos) that may stipulate their own record keeping requirements. The industrial hygienist should review Subpart Z to Part 1910 of Title 29 of the Code of Federal Regulations, that lists specific standards pertaining to Toxic and Hazardous Substances, for additional information on record keeping requirements.

Incident Rate

Documenting a company's performance is standard operating procedure when it comes to sales, profit and loss, adherence to established and forecasted budgets, and the like. Documentation of safety and health performance is equally important. The data, recorded as an incident rate, can be used to measure the success or failure of an organization's occupational safety and health programs. Since the methods used to calculate incident rates are generally the same throughout industry, any company in any line of business can measure their rate against that of others in the same or similar business. OSHA also examines incident rates to target specific industries whose safety and health programs may require closer examination and inspection. Each year, the Bureau of Labor Statistics (BLS) compiles and publishes workplace incident rates for nearly every type of business by industry category and employer size.

The incident rate is calculated using the following formula:

$$\text{Incident rate} = \frac{\text{number of injuries and illnesses} \times 200{,}000 \text{ hr}}{\text{employee hours worked}}$$

The 200,000 hours is a constant value and represents 100 employees working 40 hours per week for 50 weeks. Use of this constant in the formula will result in a small and more comprehensible incident rate value. Since it remains the same while other values in the formula change from one employer to the next, it enables more meaningful comparison with the rates of other companies, large or small. Also, it should be noted that "employee hours worked" represents the actual hours worked during that time period, not the hours paid. For instance, vacation pay, sick leave, and other such benefits where hours not actually worked are still paid, are not included in the calculation. As an example, if an employer (Company 1) experienced a combined total of 15 occupational injuries and illnesses during a given period of time and worked a total of 586,294 hours during that same period, their incident rate is calculated as follows:

$$\text{Company 1 incident rate} = \frac{15 \times 200{,}000}{586{,}294} = 5.12$$

If a larger company (Company 2) in the same industry worked 1.5 million hours during the same period of time and experienced a total of 22 injuries and illnesses, there incident rate would compare as follows:

$$\text{Company 2 incident rate} = \frac{22 \times 200{,}000}{1{,}500{,}000} = 2.93$$

In this example, Company 2 has a much more impressive incident rate than Company 1, even though Company 1 had a lower combined total of injuries and illnesses during the same period of time. Use of the standard formula for calculating incident rates enables a more equalized comparison between companies and between industries.

Lost Time Frequency Rate

The rate at which employees incur lost time occupational injuries or illnesses can also be measured using a simple, industry standard formula. It is noted here that "time lost" from work does not include time that would not have been paid anyway, such as weekends, normal days off, or holidays. Also, since most companies will pay the employee for the day that the injury or illness occurred, the industry standard is to begin calculating the time from the very next shift that the employee would have worked.

Similar in structure to the incident rate, the lost time frequency rate is a way to measure the rate at which lost time injuries occurred during a period of time relative to the number of hours worked during that same period. As with the incident rate, the frequency rate is used to track progress, enable comparisons, and facilitate the measurement of program success. The formula uses two of the three values that are used to calculate the incident rate (i.e., employee hours worked and the 200,000 hour constant). The total combined incidents however, is replaced by the total number of lost time incidents, as follows:

$$\text{Frequency rate} = \frac{\text{number of lost-time incidents} \times 200,000 \text{ hr}}{\text{employee hours worked}}$$

Therefore, if a company experienced 10 lost time injuries or illnesses during a period where a total of 385,000 hours were worked, their lost time frequency rate would be calculated as follows:

$$\text{Lost-time frequency rate} = \frac{10 \times 200,000}{385,000} = 5.19$$

The lost time frequency rate is a common measurement used to gauge an employer's safety and health program success (or failure) against that of other employers in the same or similar industry. It is also used by many insurance carriers to rate the relative risk of the employer in terms of worker's compensation premiums.

Lost Time Severity Rate

Together with the frequency rate discussed above, the lost time severity rate provides a clear summary of overall safety and health performance for any given period of time. The hygienist should therefore be equally familiar with these rates so that an accurate assessment can be made of the company's occupational safety and health program success.

The same formula is used once again to calculate the severity rate, with one exception. Instead of the total number of injuries/illnesses incurred, the total number of days lost as a result of those injuries/illnesses will be used in the numerator of the equation. The resulting value is an expression of the severity that has been incurred as a consequence of the company's lost time experience. Hence:

$$\text{Lost-time severity rate} = \frac{\text{number of lost days} \times 200{,}000}{\text{employee hours worked}}$$

If 98 days were lost as a result of the 10 lost time injuries in the previous example, the severity rate would be:

$$\text{Lost-time severity rate} = \frac{98 \times 200{,}000}{385{,}000} = 50.9$$

Each of the various rates briefly discussed here are used as measures of performance. It is important that hygienists be aware of these performance measures for their company. Typically, any proposed improvements to existing procedures require some justification before management approves the change. The tracking, recording, and reporting of incident rates are essential aspects of industrial hygiene record keeping. They also provide an excellent way to demonstrate the overall need for programmatic improvements to a sometimes skeptical and always cost-conscious management.

SUMMARY

The focus of this chapter has been on the establishment of an industrial hygiene program and the level of effort required to successfully implement such a program. To determine the actions that will be required for a particular company, the hygienist must first assess the adequacy of program elements that may already be in place. In some companies, the hygienist may find that absolutely no industrial health initiatives have been taken. In this case, the establishment of an effective industrial hygiene program will require a great deal more effort than if the company already had a somewhat fragmented program in place.

Whatever the case may be, the hygienist must develop a comprehensive plan, based on assessed need, to show the exact approach that will be taken to implement the industrial hygiene program. In the plan, the goals, objectives, and scope of the program, as well as the roles and responsibilities associated with its implementation, must be clearly addressed. It is here where the importance of obtaining management support for the program must be emphasized.

Throughout the program development phase, it is important to focus on the three general principles of the industrial hygiene process: recognition, evaluation, and control. Understanding the difference between health hazards and physical hazards is also important to the success of the industrial hygiene effort.

Once an industrial hygiene program has been approved and implemented, the hygienist must take specific actions to ensure its adequacy (i.e., that the program is meeting established objectives and that the objectives themselves are appropriate for the particular needs of their company). To accomplish this, program audits or assessments should be conducted on a regular, scheduled basis. As companies change, modifications in the scope, goals, objectives, and other elements of the overall program may change as well. Specific records and program documentation should also be evaluated during the audit process. In fact, since violations in federal occupational safety and health record keeping requirements continue to be a frequently cited condition during federal OSHA investigations, the importance of proper record keeping cannot be overemphasized. Moreover, the ever present threat of litigation against employers for alleged safety and health violations makes proper documentation an even more essential element of the industrial hygiene program.

4

Understanding Human Health

INTRODUCTION

It has been established that the practice of industrial hygiene is concerned primarily with ensuring the preservation of employee *health*. While physical injuries such as lacerations, abrasions, sprains, strains, and the like are certainly serious concerns for anyone involved health and safety, exposures to occupational stressors that can lead to *ill health* are the focus of the industrial hygiene effort. Of course there are also health hazard exposures that can result in injury, such as the physical damage inflicted on the human ear when exposed to high noise levels. While the resulting effect of such exposures may be considered more an injury than an illness, the hygienist is usually involved in the evaluation of the problem from a health perspective.

In this regard, the hygienist must possess an elementary understanding of the various components (or *systems*) of human anatomy. Without an appreciation for the *normal* function of these systems, assuring that employees remain free from exposure to *recognized* hazards to the health of those systems will be greatly hindered. Therefore, this chapter provides discussion on the basics of human anatomy in terms of occupational health and the industrial hygiene process. The more we can learn about the subject of human performance on the anatomical level, the better will be our ability to properly perform the industrial hygiene function. In fact, the recognition that is so basic to the industrial hygiene process demands a certain level of knowledge of the human system. Identification of hazards in the work environment fulfills only half of the recognition process. Being able to recognize the health effects of exposure is often critical to the proper assessment of health hazard risk.

FUNDAMENTAL PRINCIPLES OF CAUSE AND EFFECT

In literal terms, anything that acts either as a voluntary or involuntary agent to bring about some *effect* is referred to as a *cause.* This basic relationship is the principle upon which most everything that happens during everyday life is based. When a certain desired end-result or effect occurs, it is most likely due to the application of some causal factor or combination of factors. Similarly, when some undesired or unwanted effect is observed, such as occupational illness or disease, it is likely the result of exposure to specific hazards. Hence, understanding how exposures to hazards (the cause) can harm human health (the effect) greatly facilitates health hazard recognition.

In terms of human exposure to occupational health hazards, it is difficult at best to understand effect since, in most cases, all things are not equal. Reactions differ from one person to the next following exposure to a given hazard. While the effects of skin exposure to an organic solvent may be severe rash (or dermatitis) in one case, another person may show little or no adverse effect under the same conditions. In fact, human exposures to inhalation hazards, absorption hazards, or ingestion hazards may result in anything from no or negligible effects on health to a severely adverse reaction, up to and including death.

In an effort to simplify the health hazard recognition and evaluation process, specific standards of measurement have been developed to assess the highly complex relationship between cause and effect. These standards, which are based on accepted scientific research, focus on *dose* and *concentration.*

Dose And Concentration

When attempting to determine the specific levels of hazard posed by a given toxic exposure, the hygienist should become familiar with the concept of dose and concentration. As stated in the previous paragraphs, it is not only the cause and effect relationship that must be considered when attempting to determine the level of health hazard in any given exposure situation. Because individual reaction to an exposure will vary from person to person, the effects of exposure cannot always be accurately anticipated. Other variables of condition also affect the level of toxicity and, subsequently, the reaction of the individual. For example, a gram of salt will kill a rat; five grams of caffeine will kill a human; a single tablet of penicillin can kill a person who has certain allergies.

In simple terms, the *dose* measures the amount of exposure to those hazards that can be *absorbed* or *ingested* into the body. *Concentration* is a measurement of exposures that enter the body through *inhalation.* Chapter 6 provides more detailed discussion on the specific use of dose and concentration measurements in the industrial hygiene process.

Real vs. Perceived Hazards

Any discussion concerning the recognition and evaluation of hazards to human health that can exist in the workplace would be remiss without consideration of the *psychological aspects* of hazard exposure. Being able to recognize the difference between hazards that are *real* and those that are only *perceived* by the worker to be real can be extremely important to the success of the entire industrial hygiene program. In this regard, the understanding of a specific cause and effect relationship can become quite distorted and even more difficult to evaluate.

When word that a particular health hazard may be present begins to spread through the work force, many employees may experience anxiety or fear. This is especially true when there is evidence that an actual exposure did, in fact, occur somewhere in the facility. Depending on the nature of the hazard and their individual understanding and/or perception of it (which may not always be accurate), some employees may develop symptoms of exposure that can affect their health in much the same way as an actual exposure. This *psychosomatic* reaction has often been referred to as the *"I'm sick, you're sick"* syndrome. There is insufficient space in this basic volume to explore the peculiarities of the human psyche that are responsible for such reactions. Suffice it to say that ill health *can* result from the mere perception that health hazards in the workplace exist. The industrial hygienist must take such phenomena under careful consideration during the recognition and evaluation process. While physical illness (the effect) resulting from psychological perceptions (the cause) are worthy of resolution, the hygienist must be able to *recognize* the difference to properly assess the true nature of the problem.

Exposures and Human Health

In this basic study of industrial hygiene, the emphasis has been on the fundamental recognition, evaluation, and control (REC) process. Proper execution of an industrial hygiene program requires an understanding of basic human health as well. Simply speaking, understanding only the cause of a particular health hazard exposure does not complete the industrial hygiene process. The effect must also be understood to properly implement a successful industrial hygiene program.

Since inhalation provides the most direct route of exposure to health hazards, the respiratory system is quite vulnerable to the effects of many hazards. An understanding of the respiratory system and its components is essential to appreciate how that system can be harmed. However, in terms of cause and effect on the total human system, an equal understanding and appreciation of the interrelationships between the respiratory system and other such anatomical systems is also important. The circulatory system, for example, plays a key role in the distribution of inhaled toxins to the rest of the body. Likewise, there is also a relationship between the central nervous system (CNS) and the respiratory system. When inhaled (or ingested through the digestive system) certain hazardous substances can compromise the efficiency of the CNS to a point that becomes critical to the health of the exposed worker. Even certain muscles of the musculatory system (i.e., the diaphragm, intercostals, and abdominal muscles) are important to the effective operation of the respiratory system.

To facilitate the recognition and evaluation of hazards to the overall human system through inhaled exposures and their subsequent effects, a somewhat detailed review of the respiratory system and its relationship with other anatomical systems is provided in the next section. Ingestion hazards, although less common in the workplace than inhalation hazards, will also be discussed in this chapter, with focus on understanding the human digestive system.

The hygienist must also be somewhat knowledgeable of the proper function of the special sensory organs (skin, eyes, ears) and how occupationally induced injury or illness to these can be recognized through a cause and effect evaluation. Therefore, the anatomy of the human sensory organs will be explained in terms of health hazard risk.

Under the right conditions, many toxic or otherwise hazardous substances found in the average workplace can pose a serious threat to the health of any or all of the human systems/organs and, subsequently, to the overall health of the worker.

INHALATION HAZARDS AND THE RESPIRATION PROCESS

As previously established in this text, inhalation (or the act of drawing air into the lungs) is the most common route of exposure to harmful agents or pollutants in the air. Since inhalation is also the quickest route of entry, the respiratory system is particularly vulnerable to assault by a variety of common health hazards found in the workplace. The remarkable efficiency of the respiratory system will bring inhaled toxins, contaminants, poisons, and the like into the body to either directly affect the respiratory system itself or, through its association with the circulatory system, effectively distribute these substances to the rest of the body.

Simply speaking, people have to breathe. Because of this basic fact of human existence, virtually anything that happens to be in the air people breathe can easily enter their bodies during the respiration process. In any work environment where harmful agents are present in the air, personnel in the general work area are susceptible to exposure.

When the respiratory system itself suffers the direct effects of exposure to harmful agents in the workplace, the medical profession refers to these agents as *occupational lung hazards.* The resulting effects of exposure—occupational lung diseases—are illnesses that people can develop due to exposure to contaminated air in their working environments. Coal dust, ammonia gas, solvent vapors, cereal dusts, welding fumes, and degreasing sprays are a few of the thousands of pollutants found in the workplace today. While occupational lung disease can affect all types of people in many different lines of work, it is statistically most common to find this type of health hazard exposure in blue collar employees working in industrialized areas (American Lung Association, 1983). According to the American Lung Association (ALA), it is also more common to find these exposures in small- and medium-size businesses than in larger companies. This may be due to larger budgets, better staffing capabilities, or easier access to technology.

Some types of occupational lung diseases are easily recognizable and are also very well publicized. Asbestosis, for example, is a lung disease caused by the inhalation of asbestos particles. It generally afflicts workers who, at one point in their lives, worked with or around asbestos or asbestos containing materials (ACMs). It is relatively easy to label asbestosis as occupational because the worker's lungs will contain asbestos in levels common among those who have worked with the substance. Other occupational lung diseases are also easy to recognize as such, either because a workplace substance accumulates in the worker's lungs, such as silica in silicosis, or because the disease pattern is found only in certain occupations, such as the metal fume fever (see Chapter 6 on hazardous metals) that can afflict welders. Diseases that are common to certain work groups are sometimes referred to as *occupation-specific diseases*. There are other equally debilitating lung diseases that may not be attributable to any specific type of work or occupation but are still a threat to employees who may be exposed to lung hazards. Conditions such as bronchitis, asthma, emphysema, or even cancer can be caused or aggravated by many different occupations and by such non-occupational exposures as tobacco smoke or atmospheric air pollution. For instance, according to the ALA, many employees who work in the mining industry or who are employed making fertilizers are diagnosed with a higher number of cases of chronic bronchitis than the general working population. However, while this may seem occupation-specific to some observers, it is noted that chronic bronchitis is also a common disease in cigarette smokers. Therefore, chronic bronchitis can be an occupational disease but not an occupation-specific disease. The hygienist must be able to recognize the difference so that respiratory hazards that may be causing the problem can be properly evaluated and controlled.

The threat or presence of occupational lung hazards in the workplace is a serious consideration for any industrial hygiene professional. In addition to the threat of occupational lung disease, the hygienist should be aware of the potential for numerous other detrimental health effects that can develop as a result of exposure to respiratory hazards. It must be remembered that anything inhaled can potentially be distributed to very specific locations in the body or evenly dispersed throughout the body. In either case, the effect can range from chronic (long-term) illness to acute (immediate onset) illness.

Recognition of respiratory hazards requires some basic knowledge of both occupational and non-occupational lung diseases. While it is true that the occupational health nurse or physician are the experts on such diseases, not all organizations have this talent in residence. Therefore, understanding that a particular lung condition may not be occupational in origin is as important to the hygienist in hazard recognition as knowing those diseases that are, in fact, the result of occupational exposures. The hygienist must be capable of recognizing when a respiratory health risk exists in the workplace, and must also be able to recognize the difference between occupational and non-occupational respiratory conditions. While this may seem more a function of evaluation than recognition, it is obvious that a great savings in time, money, and resources can be realized if the astute hygienist can determine these differences at the beginning of any industrial hygiene assessment. Whenever possible, working together with a toxicologist, an occupational health nurse, and/or an occupational physician, the hygienist should be able to identify when true occupational exposures have resulted in any particular employee illness.

Anatomical Overview

Through the respiratory system (Figure 4-1), and its close relationship with the circulatory system (Figure 4-2), anything that is inhaled finds a quick and direct route to the rest of the body. Therefore, anything that can affect the respiratory system, ranging from a lack of oxygen to inhalation of a contaminated gas, can affect the entire human system. Understanding how this occurs is essential to the recognition, evaluation, and control of respiratory hazards.

The respiratory system is composed of a series of separate and distinct airways. It is responsible for supplying oxygen to the blood and expelling waste gases, of which carbon dioxide (CO_2) is the primary constituent, from the body. The initial airways in the upper structures of the respiratory system are combined with the sensory organs of smell and taste (in the nasal cavity and the mouth) and the digestive system (from the oral cavity to the pharynx). At the pharynx, the specialized respiratory organs diverge into a separate airway. The larynx, or voice box, is located at the head of the trachea, or windpipe. The trachea extends down to the smaller subdivisions of the airway known as bronchi that branch off at the tracheal bifurcation to enter the left or right lung. The lungs contain even narrower passageways, or bronchioles, that carry air to the functional units of the lungs, the alveoli. There, in thousands of tiny alveolar chambers, oxygen is transferred through the membrane of the alveolar walls to the blood cells in the capillaries within. Likewise, waste gases diffuse out of the blood cells into the air in the alveoli, to be expelled upon exhalation.

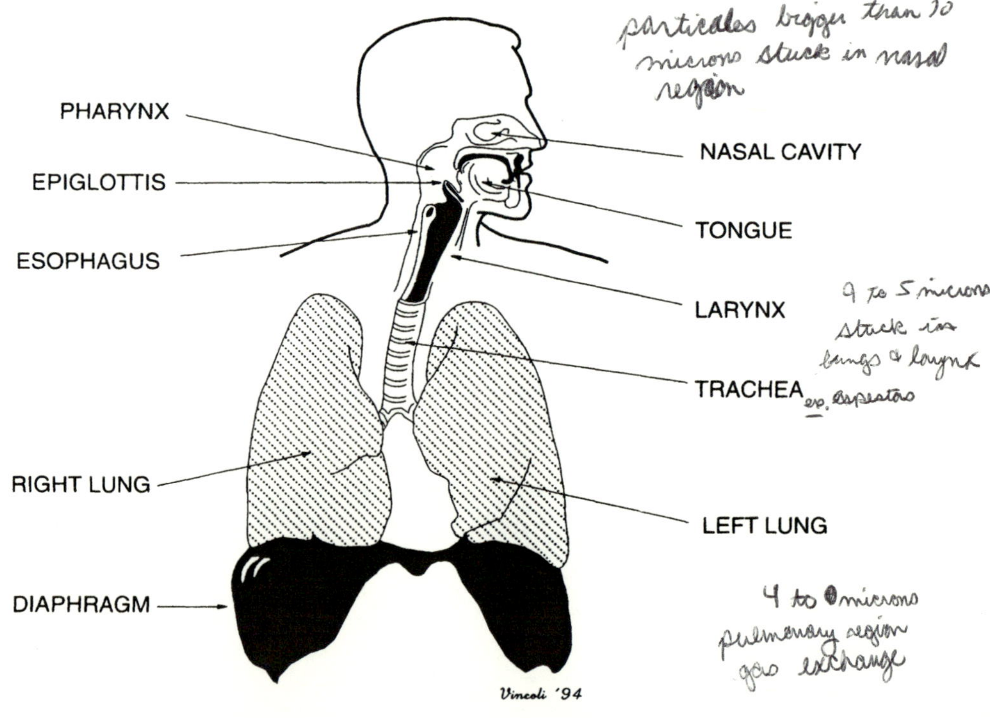

FIGURE 4-1 The major components of the human respiratory system

The following brief discussion is intended to familiarize the reader with only the basic features and functions of the respiratory process, beginning with the nose and proceeding down through the remaining elements of the respiratory system.

The Nose

The nose is located at the beginning of the uppermost airway of the respiratory system. Along with the mouth, it provides the primary passage of inhaled air into the body. Normally, the nose is evenly divided into two nasal cavities by a structure known as the nasal septum. Deviation of the septum to one side or the other can occur in some people as a result of a minor accident to the nose or due to some other source of trauma or even a congenital defect. It should be noted that some airborne chemical contaminants (many chromium compounds, for example) are particularly damaging to the nasal cavity, including the septum. Employees with existing abnormal conditions such as a deviated septum may be somewhat quicker to react to such contaminants than the person with a normal nasal cavity.

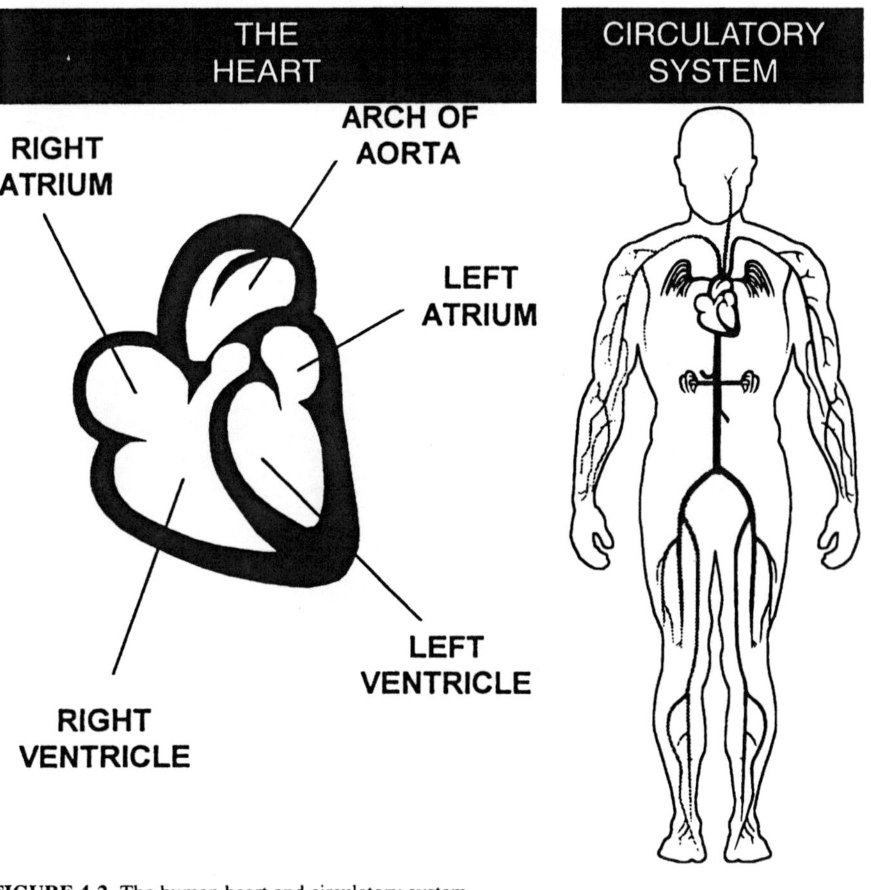

FIGURE 4-2 The human heart and circulatory system

The two small openings at the base of the exterior nose are known as the external nares. Small hairs that act as filters for the inspired air grow just inside the external nares. These are considered the first line of defense against any particulate matter that may be present in the air (see Chapter 5 on air contaminants). If the particles are large enough, they will be trapped in the hair and prevented from further passage into the respiratory system. However, microscopic particulate matter can escape the hairs in the external nares. Once the air is filtered through the hairs, it passes along a layer of mucus secreted by cells located in the nasal tissue. This mucus layer serves to filter out particulate matter too small to be trapped by the hairs. The mucus layer rides atop a series of microscopic hair-like projections known as cilia that beat back and forth in a coordinated fashion. The primary purpose of the cilia is to transport any trapped particulate matter back toward the external nares for discharge (sneezing, coughing, etc.). If the particles are deep enough, they will be mixed with mucus and transported, via the cilia, to the pharynx, where they are swallowed. In some cases, however, there may still be particulate matter so tiny that the particles escape even this sticky layer of mucus and make their way further into the respiratory system. For example, asbestos fibers, which can be so small that they cannot be seen with the naked eye, can pass through the nares, slip past the mucus layer with relative ease, and end up lodged deep inside the lungs. Infection or inflammation of the nasal cavity, known as rhinitis, is part of the condition referred to as the common cold or hay fever. However, some people that are sensitive (or even allergic) to certain airborne particulate matter at work may exhibit signs of rhinitis. If such symptoms are present only when the employee is at work or only during the performance of certain dusty job functions, the hygienist should recognize the possibility that the condition may not be simple hay fever. It may, in fact, be the first sign of an occupational health hazard exposure.

The Pharynx

After leaving the nasal cavities, the inspired air enters the pharynx. Seven tubes enter this small area inside the throat: two from the nasal cavity, two Eustachian tubes (from the ears), the mouth cavity, the opening to the larynx and esophagus, and the opening of the windpipe (trachea). The pharynx itself is a tubular passageway. Its walls are constructed of skeletal muscle lined with a ciliated mucus membrane for lubrication and for the transport of any trapped particulate matter back into the pharynx for swallowing.

The two passageways at the bottom of the pharynx—the esophagus and the trachea—provide a crossroads for anything entering the body through ingestion or inhalation. Food and liquids are routed to the esophagus and the stomach, while air is passed through the trachea to the lungs. A thin cartilage-like plate behind the tongue known as the epiglottis acts like a lid when swallowing to prevent food or drink from entering the trachea through the larynx.

The Larynx

The voice box, or larynx, is the passageway that allows air to pass between the pharynx and into the trachea. In addition to housing the vocal cords, the triangular shaped larynx and its attached epiglottis act as a valve at the entrance to the windpipe. It controls airflow and allows only air to enter the lungs. Since exhalation of air through the larynx is controlled by voluntary movement of muscles, the larynx becomes the primary mechanism for the creation of voice. It is also a mucus-lined structure whose cilia fibers attempt to move any trapped particulate matter up toward the pharynx, whereupon it is swallowed. It is important to understand that damage to the larynx can be caused by injury, overuse, viruses, or inhalation of irritating air contaminants. Impairment of the larynx can prevent it from functioning properly. For example, when an employee shows indications of laryngitis (inflammation of the larynx), the hygienist should assess the work environment for any signs of unusually dusty or dust-generating conditions. When considering the health of an employee, it is not an overreaction to investigate the possibility that the laryngitis may be due to some previously unrecognized hazardous condition in the workplace. The illness may be personal, as opposed to occupational in origin. But the industrial hygiene professional will never know without an assessment of the work environment.

The Trachea

The trachea, or windpipe, is a fibrous hollow tube approximately 12 centimeters (4.5–5.5 inches) long in the average adult (Figure 4-3). It extends from the bottom of the neck, directly in front of the esophagus, to the chest cavity. A series of 16 to 20 "C"-shaped cartilage-like rings are embedded in its front and side walls to support the trachea and prevent it from collapsing. Without such support, the soft, pliable trachea would certainly collapse on itself, causing death by asphyxiation.

The internal walls of the trachea are also lined with a mucus membrane containing thousands of cilia that fan upward toward the throat to move dust particles trapped by the mucus. At the termination of the trachea, it subdivides into the left and right main-stem bronchi. This fork in the road is located in an area directly behind the midpoint of the breastbone. A respiratory infection that extends to the trachea are referred to as *tracheitis*. When the walls of the trachea are inflamed, breathing is very harsh and difficult and is accompanied by a deep and resounding cough. Inhalation of corrosive chemical vapors can be extremely damaging to the trachea.

The Bronchi

At the tracheal subdivision of the left and right main-stem bronchi, the airways continue to divide, forming a tree-like structure of air tubes that serve to ventilate all regions of the lungs. It is important to note that the right main bronchus remains relatively in alignment with the trachea, whereas the left bronchus deviates upward and at a more acute angle. This is why fluids inhaled (or aspirated) into the lungs will generally enter the right lung first. When the continued branching results in the formation of an airway with a diameter slightly less than 1 millimeter, the tube is referred to as a *bronchiole*. The small bronchiole continues to divide into even smaller and smaller airways. The final subdivision, the *alveolus*, is where respiration finally occurs.

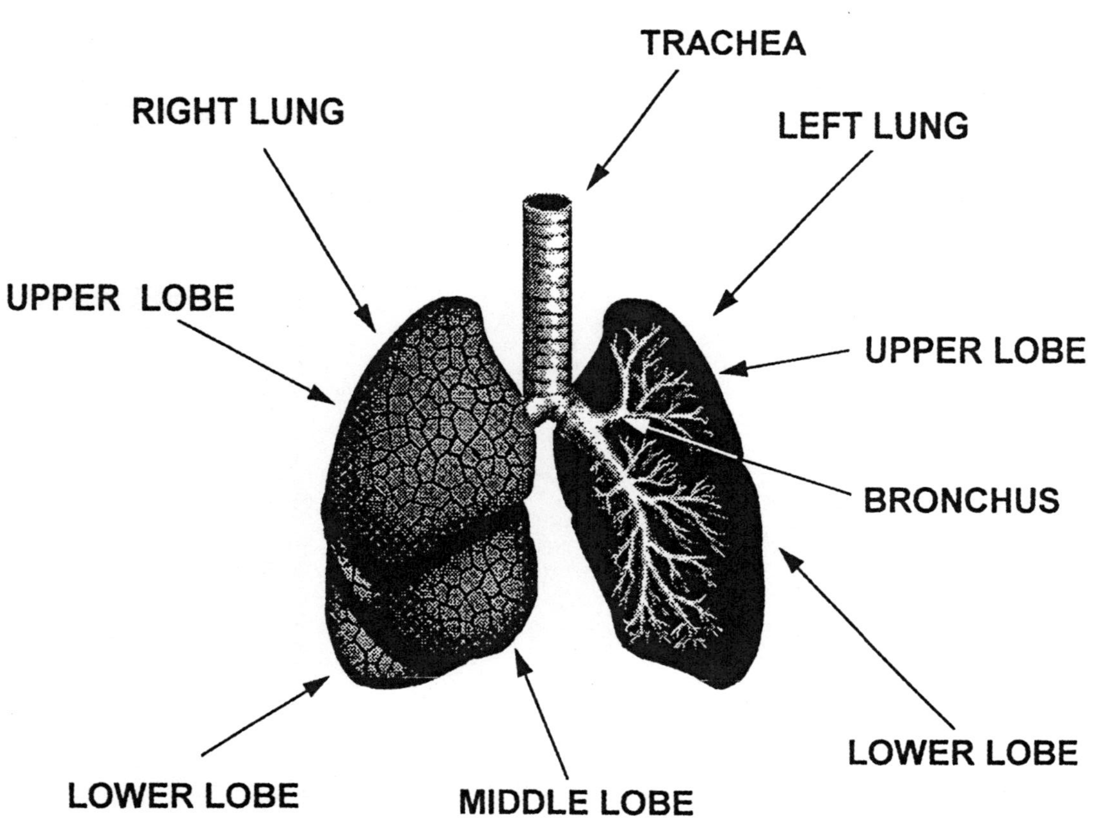

FIGURE 4-3 A general view of the various characteristics of the lungs

The Lungs

There are two lungs. Figure 4-3 provides a general view to familiarize the reader with the various characteristics of the lungs. The left lung consists of two main lobes—the left upper lobe and the left lower lobe. The right lung is divided into three major lobes referred to as the right upper, middle, and lower lobes. The right lung is shorter, due to its position over the dome of the liver. However, it is wider than the left lung. From a three-dimensional perspective, the lungs are considered cone-shaped.

The inner surface of the lung tissue provides containment of the alveoli air sacs. The approximately 300 million alveoli cover a surface area ranging from 60 to 75 square meters, roughly the equivalent of 600 to 700 square feet. This extremely large surface area, compacted into such a small space, enables gas exchange to occur on a scale necessary to maintain normal metabolic processes in the human body. When an abrasive dust such as asbestos fibers reach the alveoli, the dust can be lodged in the sac and remain there for many years. Over time, scar tissue that develops around the dust slowly begins to disable the gas exchange process. If enough inhaled fibers reach the alveoli sacs, asbestosis, a disabling or even fatal lung disease, can result.

In addition to the defense provided by the cilia and the mucus layers located throughout the respiratory system, tiny, highly specialized mobile cells called *alveolar macrophages* will engulf and digest bacteria and viruses that have entered the lungs. In the process, the macrophages themselves may also be destroyed. The resulting debris is swept away by the cilia to be discharged from the body. However, when macrophages attempt to engulf and digest such lung hazards as toxic dusts, they are often destroyed and the dust remains in the lungs.

The Respiration Process

Having briefly discussed the basic anatomy and physiology of the primary components of the respiratory system, the reader should have an appreciation for the complexities of the human respiration process.

Respiration is the act of inhaling and exhaling air. As air is inspired through the nose (or mouth), it is warmed, filtered, humidified, and delivered to the lungs through a series of passageways. The purpose of respiration is to sustain life by providing a system to deliver oxygen to the bloodstream while removing carbon dioxide (a waste by-product) at the same time. The blood carries the oxygen, along with other nutrients, to each cell in the body. Once there, it picks up waste products and returns them to the lungs for discharge. The exchange of oxygen and carbon dioxide occurs between the blood and atmospheric air inhaled into the alveoli.

The exchange of oxygen for carbon dioxide in the alveolus is known as external respiration. The exchange of oxygen for carbon dioxide between the blood and tissue cells is called *internal respiration*. The blood contains a chemical that is part protein and part iron pigment, known as hemoglobin. It is the hemoglobin of the blood that binds with oxygen when the blood flows through regions of the body that are oxygen rich, such as the alveoli (external respiration), and releases it to tissues that are using or consuming oxygen (internal respiration). Likewise, the carbon dioxide produced when body cells burn their fuel is dissolved into the blood stream (diffusion) as it flows through the tissues where carbon dioxide content is high and releases it in the lungs where carbon dioxide content is relatively low.

Oxygen-poor blood is pumped to the lungs from the circulatory system by the heart through the pulmonary artery. This artery contains branches that go to each lung. These arteries subdivide and eventually lead to networks of even smaller vessels, called capillaries, that pass through the alveolar surface. The capillary network in the alveolar tissues allows for the transmission of gases between the air in the alveoli and the blood cells within the capillaries. The capillaries are so small that they only allow blood cells to pass through them one at a time. The actual gas exchange occurring at this level is, therefore, quite remarkable.

Gases diffuse rapidly from areas of higher concentrations to areas of lower concentrations. Since the concentration of oxygen is higher in the alveolar air than it is in the blood coming to the lungs from the heart, the oxygen will diffuse quite readily into the blood from the alveolar air through the capillaries in a process known as osmosis. Osmosis is the act of a substance (in this case oxygen and carbon dioxide) passing through a semi-permeable membrane (the alveoli) from a region of high concentration to one of lower concentration. The thin walls of the alveoli are only two cells thick. The blood cells passing through the alveolar capillaries are depleted of oxygen and rich in carbon dioxide and other waste gases. As a result, the carbon dioxide osmoses through the membrane to the air within the alveoli (that have relatively low concentrations of carbon dioxide). Similarly, the oxygen-rich air in the alveoli osmoses through the membrane to oxygen-poor blood cells. In this process, the blood gets rid of excess carbon dioxide and is replenished with oxygen. The blood is then routed to the heart through a series of arteries where it is pumped via the circulatory system to the rest of the body.

Any gas other than oxygen, or any minute particulate matter that happens to be in the breathed air can seriously undermine the respiration process. For this reason, much emphasis in this chapter has been placed on understanding both the respiratory and circulatory systems.

To accomplish breathing, certain muscular controls are required. While numerous muscles participate in the respiratory process, the most important muscle of respiration is the diaphragm. The thin diaphragm forms a domed structure. When the diaphragm muscle contracts, it lowers to a more flattened arrangement. This flattening causes a negative pressure that creates a vacuum effect in the lungs.

The vacuum is filled by expanding lung tissue with inhaled air. When the diaphragm relaxes back to its original domed shape, the air is exhaled and the lungs contract. Though the intercostal muscles between the ribs and the abdominal muscles are also used in respiration during sleep, respiration is primarily due to the contraction of the diaphragm.

In some lung diseases, including occupational lung diseases, air can remain trapped in the lungs due to a decreased breathing capacity. If air is trapped in the lungs, the volume is abnormally increased and the diaphragm remains in a constantly lowered position. This condition seriously affects a person's ability to inhale since there is little room available for the motion of the already lowered diaphragm. Understanding this process is important to the hygienist because injury to the abdominal muscles can occur as a result strain or overuse. Although such injuries may not appear directly related to the respiratory process, it is the act of forced expiration under strain that actually causes some injuries.

The process of respiration is controlled by areas within the body's nervous system known as respiratory centers. At the base of the brain, the area known as the medulla contains one such center. Nerve impulses that originate in the motor areas of the brain's cerebral cortex (Figure 4-4) are transmitted to the respiratory center of the medulla. This process enables people to voluntarily alter the rate and depth of breathing. This is important when preparing to lift heavy articles, for example. Also, some workers may be required to wear personal protective equipment (PPE), such as respirators or a self-contained breathing apparatus (SCBA), during the performance of certain job functions. If their work also requires particularly strenuous activity while wearing such equipment, the employee's rate of breathing will quicken. This, in turn, alters the duration and effective use of the respirator and/or the SCBA.

It is important to note that voluntary control of breathing is not unlimited. Humans can hold their breath voluntarily for only a short period of time before the respiratory center in the medulla ignores the command from the cerebral cortex and forces the body to begin breathing again. The body's basic need is to obtain oxygen and discharge carbon dioxide. By holding one's breath, carbon dioxide is built up quickly in the blood and oxygen is depleted to extremely low levels. This action stimulates the respiratory control center in the brain and forces the individual to expel the carbon dioxide and immediately draw in a fresh supply of oxygen. Likewise, and of particular concern to the hygienist, a resting person who continuously takes in extremely deep breaths over a relatively short period (approximately two to four minutes) will begin to feel the affects of a condition known as hyperapnea (over-breathing). Under normal conditions, a resting person will inhale and exhale 10 to 15 times per minute (each breath lasting between 4 and 6 seconds). Therefore, in 1 minute a person will take in between 4.3 and 5.7 liters of air. Each day, a person will breathe approximately 12,500 liters (3,300 gallons) of air. Exertion causes the rate of breathing

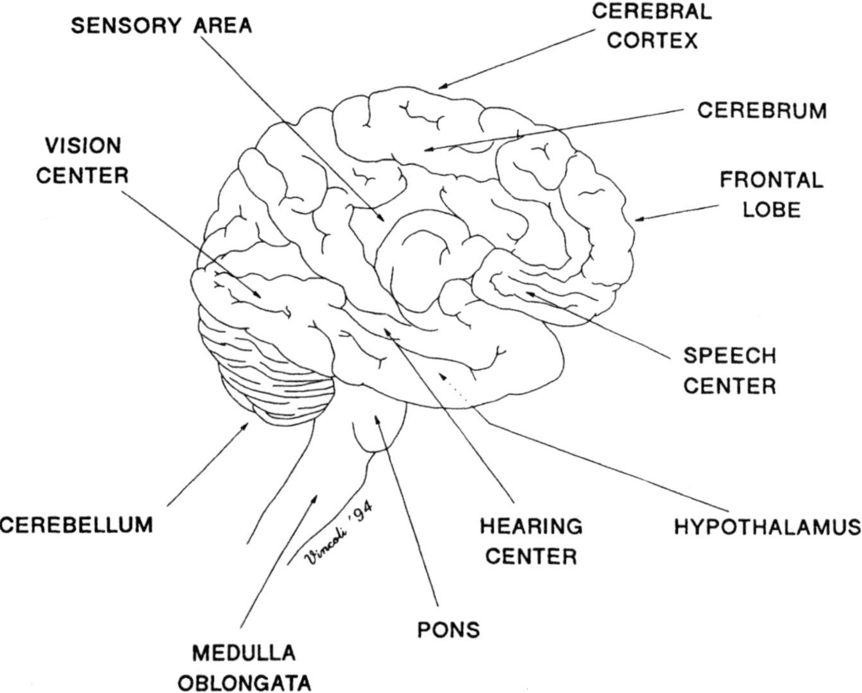

FIGURE 4-4 Simplified view of the human brain and its major functional areas

to increase to as much as 1 breath per second and a total intake of 120 liters (31 gallons) of air per minute. However, over-breathing or panting under normal, resting conditions causes a rapid buildup of blood oxygen content with an equally rapid discharge of carbon dioxide in a process known as hyperventilation. After a short period of time, the cells of the body will not be responsive to the excessive flow of oxygen, since the blood will contain much more than the cells require. This is why the quick remedy for hyperventilation is to breath into a paper sack. The intake of the CO_2-saturated air helps to equalize the oxygen-rich blood and stabilize the respiration process. The results of hyperventilation can range from a lightheaded feeling and numbness or tingling of the lips, fingers, and toes to dizziness, nausea, and unconsciousness. These also happen to be symptoms of exposure to some chemical solvent vapors. The hygienist must be able to recognize the difference to properly assess the nature of any hazard in the workplace. Hyperventilation can be caused by stress, anxiety, or some other mental or physical anomaly that may or may not be related to job tasks or the work environment.

Effects Of Inhalation Exposures

The effects of exposure to health hazards through inhalation can be direct or indirect, chronic (long-term) or acute (immediate). Direct effects result in some level of damage or some degree of illness to the lungs or any of the other members of the respiratory system. Indirect effects result in illness or damage associated with some other part or parts of the body.

For instance, in pure, clean air the lung-protecting mechanisms work at a minimum level. Mucus secretion is generally just sufficient to keep the lining of the respiratory system moist. Whenever hazards such as dust, fumes, smoke, vapors, or mists are inhaled, the lungs react directly by increasing the amount of mucus secreted in an effort to clear the lungs of the offending substances. An infection in the lungs or any other member of the respiratory system will also result in an increase of mucus secretion, as well as an increase in the quantity and activity of the macrophages. When an irritating pollutant has been removed or the infection has been eliminated, mucus secretion and macrophage activity will return to normal levels (which vary from one person to another). Some people will produce more mucus than others, and some people will have a more marked response to irritants than others. Therefore, the overall efficiency of the lung clearing mechanisms (cilia, mucus layer, macrophages) also varies between individuals. Two people can work side by side under identical conditions for many years and have different degrees of lung disease. The ability to handle exposure to hazards and to resist disease varies with age, physical condition, smoking habits, nutritional status, and other less understood factors such as stress, inherited traits (predisposed susceptibility), and emotional stability.

On the other hand, the inhalation of a gas may have an entirely different effect, especially if that gas happens to be odorless, tasteless, and colorless. Whether such gases are poisonous, such as carbon monoxide, or asphyxiating, such as nitrogen, there is typically no noticeable biological reaction of protest from the lungs (an indirect effect). When low levels of the gas are being inhaled over a period of time, people may not even be aware that they are inhaling a poisonous or asphyxiating gas until the gas exchange process in the alveoli transfers the gas to the blood which, in turn, distributes it to the cells and tissues throughout the body. In the one case, the tissues will be poisoned and, in the other, the tissues will be deprived of oxygen. In either case, depending on the level and degree of exposure, the effects can range from illness to death.

Because of the close relationships that exist between the various anatomical systems, inhalation hazards can cause harmful effects to any or all of these other systems. For example, certain solvent vapors, once inhaled into the respiratory system, can interfere with the proper function of the central nervous system. CNS depressants can cause serious impairment of nerve response to a point where vital body functions (both voluntary, such as motor control and respiration, and involuntary such as heart beat sequencing) do not perform properly.

Inhalation hazards clearly present the most threatening of exposure hazard risk. The critical importance of the REC process cannot be overemphasized when such hazards exist in the workplace.

Evaluating The Degree Of Harm

There are some specific tests that can be performed on employees to assist in the evaluation of any impairment to the respiratory system. These tests, often referred to as ventilatory measures, can show any decrease in the capacity of the respiratory function. This is another example of the close interrelationship that must exist between the hygienist and the occupational medical professional. Whenever possible, the hygienist should consult the occupational health nurse or physician while assessing the work environment to ensure an accurate determination of any respiratory hazards that may be present.

Under such medical supervision, tests can be performed to measure the largest volume of air on complete expiration after the deepest inhalation without forced effort. This is known as the vital capacity (VC). When the test is done under forced conditions, it is referred to as forced vital capacity (FVC). Once these values are established, additional variations of these tests that introduce time as an element of measure can assist the evaluator in determining the extent of any decreased capacity in the respiratory function. When assessing the nature of a particular respiratory hazard, such information can be valuable to the hygienist. Knowing how the respiratory system has been affected can help isolate the nature and degree of hazard to the worker.

INGESTION HAZARDS AND THE DIGESTIVE PROCESS

The various organs of the digestive system work together to perform an extremely vital function—that of preparing food for absorption and use as energy by the millions of body cells. The principal hazard posed to human health in this regard is the ingestion of materials that can be harmful to the organs of the digestive system or the human system as a whole. For example, in a work environment where extremely large quantities of dusts are generated, employees could actually eat (ingest) airborne dust particles every time they open their mouths to talk. More insidious (and more common), however, is the hazard of ingestion that occurs when workers neglect to properly wash their hands before preparing food, eating, or drinking. Contaminants (metal particles, dusts, chemical residues, and the like) can be easily ingested when an employee's personal hygiene habits are less than adequate.

Once a harmful substance is taken into the body through ingestion, a process of absorption and transfer occurs that can effectively distribute the substance to the rest of the body tissues. Understanding the basic anatomical features and functions of the digestive system will facilitate the recognition of hazards to that system, as well as the effects such hazards can have on overall human health.

Anatomical Overview

The digestive system has essentially one purpose: to break down food, both solid and liquid, into a form that can be used by the body. This process includes the separation of any unusable constituents in the foods and the preparation of these materials for discharge from the body as wastes, in the form of feces and urine. The food left behind is then processed and used as energy to fuel every cell in the body. The digestive process actually begins in the mouth where salivary glands saturate food materials with saliva, a slightly acidic fluid. After leaving the mouth, foods entering the body pass through the throat (pharynx) and, with the assistance of the epiglottis at the base of the throat, is routed away from the windpipe (trachea) and into the esophagus. From here the food moves directly into the stomach. Inside the stomach chamber, a combination of acids and enzymes quickly liquefy all solid materials (3–5 hours) and continues with partial digestion before the food is passed into the small intestine. Absorption of the food materials occurs in the small intestine. After passing into the large intestine, where excess water is absorbed, waste products are deposited in the rectum for discharge as feces through the anus.

Introduction of a material that can cause harm or damage to any of the organs of the digestive system can seriously impede the digestion process (i.e., the conversion of food to energy) and cause ill-effects on a larger scale. The opportunity for ingestion of such materials can exist at home as well as at work. It is a function of the industrial hygiene effort to recognize when ingestion hazards exist in the workplace and to attempt to control such exposures. As with inhalation hazards, proper recognition and evaluation of ingestion health hazards requires some understanding of the digestive system, its associated organs, and their proper (intended) function. With particular emphasis on the function of the liver, the following brief discussion is provided on the digestive system.

The Digestive Tract

Beginning with the mouth and extending to the anus (Figure 4-5) the digestive tract, or alimentary canal, is a long and continuous tube in which food passes through the digestive process. Although the straight-line distance from the mouth to the bottom of the trunk is only two or three feet, the distance along the canal is

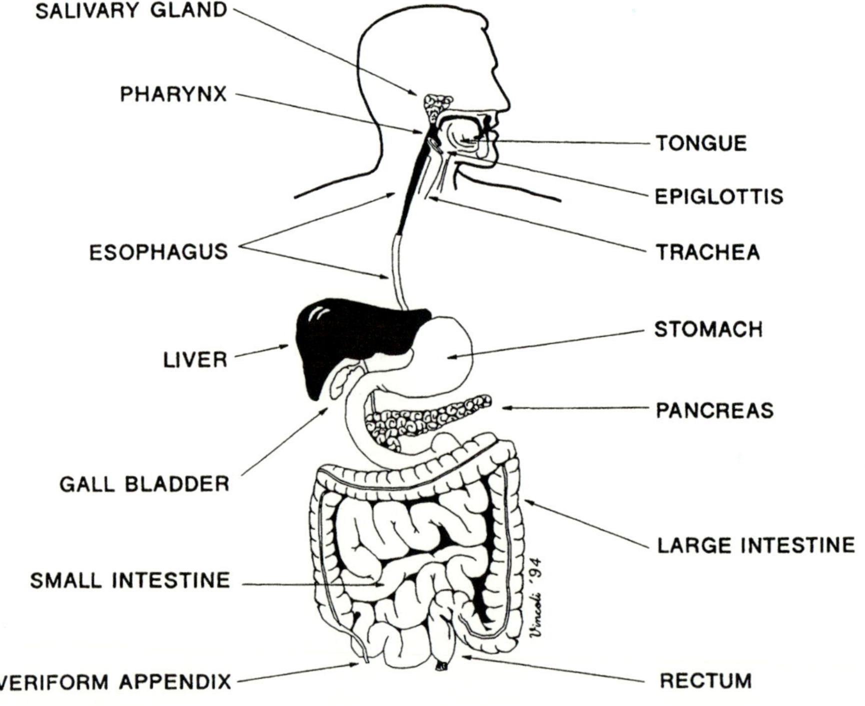

FIGURE 4-5 The human digestive system, showing the basic components of the digestive tract, or *alimentary canal*

approximately ten times greater. The canal comprises a winding, twisting, looping passageway of up to 30 feet or more that can accommodate a great quantity of foods in various stages of digestion. Any materials entering the digestive tract other than digestible food will make the same progression along with the food through the canal.

The Stomach

About ten inches down the esophagus, swallowed food that is fairly diced and smashed at this point enters the stomach. Once food enters the stomach, it will be subjected to the first serious elements of the digestive process. When empty, the stomach appears flat with numerous wrinkles and shrivels. When full, it is a plump, J-shaped bag approximately a foot long and six inches wide, holding about two quarts of food and drink. When food is released to the small intestine from the stomach, it is no longer recognizable. At this stage, the food is referred to as chyme.

The Small Intestine

The small intestine is so named due to its diameter (one to two inches) relative to the diameter of the large intestine (three to four inches). However, its actual length (more than twenty feet) renders its name as something of a contradiction. The preparation of food particles to pass into the body through absorption is completed in the small intestine.

The Large Intestine

Also know as the large bowel, the large intestine is a broad, corrugated tube that accepts the by-products of digestion from the small intestine and passes them along to be excreted, processing the material along the way. Its principal function is to absorb water from the remaining food materials.

The Liver

Four pounds of highly efficient chemical processing tissue, the liver is the body's largest solid internal organ. More than any other organ, the liver enables the body to obtain the full benefit from eaten food. Without the liver, digestion would be impossible and the conversion of food into living cells and energy practically nonexistent. The liver's principal functions can be divided into two groups: those that break down food molecules, and those that build or reconstitute these nutrients into a form that the body can use or store efficiently. In the former, the liver produces bile, a bitter, viscid, yellow or greenish alkaline liquid that aids the digestion process. In the later, the liver reconstitutes many of the proteins and carbohydrates destroyed by the bile back into the forms of nutrients that can be used by the body. This action alone is quite remarkable. However, of greater importance to the industrial hygienist is the liver's ability to detoxify virtually everything that enters the body for digestion. As the liver breaks down molecules with its bile secretions, the fats, proteins, and carbohydrates are separated for reconstitution. Any leftover toxins are then discharged along with other unusable by-products of ingestion. The liver also acts as a filter for the blood as well. The liver removes toxic substances from the blood and renders their poisons harmless. It literally picks up spent red blood cells from the circulatory system, and, in simple terms, dismantles them, removes harmful constituents on the molecular level, and continually manufactures new blood elements.

The close interrelationships between various anatomical systems become evident once again: harmful agents that enter the body through inhalation (respiratory system) or absorption (through skin) are transferred into the blood for distribution (circulatory system), where they will be processed by the liver, a major organ of the digestive system. There are many other functions of the liver, such as the production of cholesterol, that are beyond the scope of this publication. From the industrial hygiene perspective, it is important to understand the functions explained here since, because of its role in the digestion process, the liver is quite susceptible to harm and damage.

The Gall Bladder

The primary function of the gall bladder is to store excess bile. However, the bile found in this three-inch organ is much more concentrated and thicker than that which is fresh from the liver. This thickening process can cause problems in the form of painful gallstones that are solidified bile. While gallstones are hardly considered an occupational illness, proper diagnosis by the occupational nurse or physician is necessary to ensure accurate assessment of true health hazards in the workplace.

The Pancreas

This manufacturer of powerful digestive enzymes, only six inches in length, resembles a branch heavily laden with ripe berries. Numerous industrial chemical agents and toxic compounds can interfere with the proper operation of the pancreas. The gland secretes a clear, watery, alkaline fluid (the pancreatic juice), that passes to the intestine. The juice is one of the principal chemical agents involved in digestion, for it contains enzymes that break down starch, sugar, and fats into glycerin and fatty acids, and proteins into peptones and amino acids. This important role is often overshadowed by the fact that the pancreas also produces the hormone insulin. Pancreatic insulin is essential to the regulation of carbohydrate metabolism. When an imbalance of blood sugars in the blood and urine occurs, a person may suffer the symptoms of diabetes. These include increased appetite plus weight loss, accompanied by excessive thirst. Such symptoms are alleviated by injection of insulin on a scheduled basis. When the hygienist suspects an exposure based on the presence of any similar symptoms, he or she should not rule out the possibility of a non-occupational cause. A conference with an occupational nurse or physician becomes particularly critical during the evaluation of such symptoms. While every employee's health is of concern to the hygienist, the primary objective is to recognize, evaluate, and control occupational hazards. In the process of doing so, however, the hygienist may actually help an individual recognize a previously undiagnosed personal illness in need of attention.

The Digestive Process

Having provided basic information on the anatomy and physiology of the primary members of the digestive system, the reader should have developed a basic understanding of the digestive process. There are also other organs and processes not discussed here that participate in this process. However, from a fundamental perspective, the digestive process relies primarily on the components discussed above.

Digestion begins in earnest once food reaches the stomach and is converted through a series of chemical reactions into chyme. The various chemicals found in the stomach are produced and secreted into the stomach cavity by some 40 million gland cells that line the interior of the stomach walls. The constant movement of the stomach helps to distribute these chemicals thoroughly into the food. A combination of enzymes (pepsin, rennin and lipase) along with hydrochloric acid and watery mucus make up the chemical constituents of the stomach. From an industrial hygiene perspective, it is not entirely essential to understand how these various chemicals are produced or used. However, it is somewhat useful to know that they exist and that certain harmful agents that may be ingested can interfere with their proper function.

Up to this point, the food has traveled a distance of only two feet or so and has been in the system for approximately five hours. However, once inside the small intestine, chyme will begin from 6 to 20 hours of processing inside the small intestine.

By the time it has passed through the first 12 inches or so, the chyme is nearly completely ready for absorption. By the time it leaves the small intestine, chyme has given up virtually all its nutrients. In other words, the process of absorption (or, more accurately, assimilation) has taken place. The nutrients (along with any harmful agents that may have made it this far) have left the digestive tract for distribution to the rest of the body through tiny hair-like projections called villi. The villi absorb proteins and carbohydrates into their capillaries while the lymphatic nodules absorb fats. The villi pass the proteins and carbohydrates to the liver for metabolic processing, and the lymphatic nodules pass the fats through the lymphatic system into the bloodstream. What passes onto the large intestine is principally water and waste products. Solid materials such as metal dusts and fibers, that have not been digested will be discharged with the rest of the wastes. However, some toxic chemicals (including metals) can be absorbed into the blood or passed onto the liver along with the food nutrients. When this occurs, systemic poisoning can be the end result of the ingestion hazard.

On leaving the small intestine, unabsorbed food materials are stored in the large intestine until the body can partially reabsorb water from it. Anything left is then passed on to the anus for discharge. Some exposures to chemicals and otherwise toxic materials can affect the proper function of the large intestine. Specifically, the over-absorption of water from the waste materials may lead to hard, relatively dry feces that can become impacted, making discharge painful in a condition known as constipation. On the other hand, if not enough liquid is absorbed, as is often caused by a viral infection, malnutrition, or some type of exposure to certain chemical agents, the control of fecal discharge becomes difficult and, in most cases, painful. This condition, called diarrhea, can result in serious (even fatal) consequences due to excessive water loss. When employees seek treatment from a company nurse or physician for such symptoms, and the exact cause is not easily diagnosed, the hygienist should be alerted to the possibility of an exposure. Consequently, any hygienist who becomes aware of an employee suffering from such conditions must carefully evaluate the known effects of any substances that the employee may have been exposed to in the work environment.

Together with the functions of the liver, gall bladder, and pancreas (as described earlier in this section) the body's digestive process is complete. For the hygienist, it is important to understand the proper function of this process and how it relates to other important anatomical processes (such as respiration and circulation). This understanding coupled with knowledge of those hazards in the workplace that can adversely affect these systems will greatly facilitate the performance of the overall industrial hygiene effort.

Effects Of Ingestion Exposures

As with inhalation hazards, the effects of ingestion hazards can be either direct or indirect, chronic, or acute. Direct effects may cause illness or injury to the organs of the digestive system itself. For example, excessive and prolonged (chronic) exposure to alcohol (a solvent) will eventually diminish the liver's ability to detoxify the solvent and render it harmless. If exposure continues, the liver will become damaged to a point where it cannot perform many of its other critical functions. Serious illness or even death can result.

Indirect effects of ingestion exposures occur when harmful agents make it into the blood through the lymphatic system and become distributed to the rest of the body tissues to some degree before the liver has a chance to properly detoxify these agents. In either case, ingestion hazards can pose a serious health threat of which the industrial hygienist must be aware.

Evaluating the Degree of Harm

Specific tests to evaluate the function of the liver and other organs of the digestive system can be conducted under medical supervision. For the hygienist, knowledge of specific testing protocol is not entirely necessary. However, knowing when such tests might be in order can be essential to early diagnosis and treatment as well as to the subsequent recognition of any risks posed by ingestion hazards in the workplace. Testing usually begins with blood analysis for the presence of elevated enzymes that might indicate a problem in the liver, bladder, pancreas, and so on. Also, exposure to certain toxins (such as lead, for example) through ingestion or inhalation often results in elevated blood levels of the toxin. Such information can greatly facilitate the hazard recognition process. As an example, some electricians often make habit of chewing on the color-coded plastic insulation they have stripped from the ends of copper wires. Lead is used in many coloring processes. Over the course of many years, blood lead levels have been elevated and indications of chronic lead poisoning diagnosed in electricians. The industrial hygiene process becomes much easier when the practitioner is aware of the potential hazards in the work environment and the signs (or symptoms) of health hazard exposure. But knowing how to recognize when one such hazard has been the cause of some specific health effect is a key to the successful implementation of any industrial hygiene effort.

ABSORPTION HAZARDS AND THE SKIN

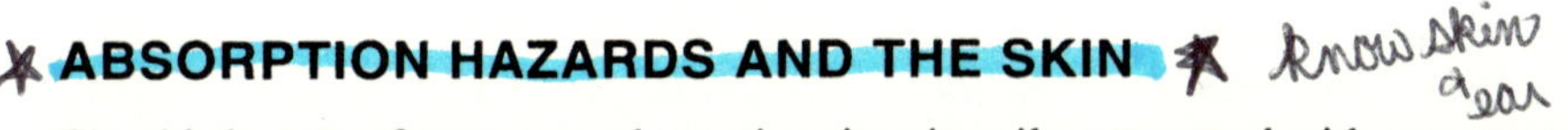

The third route of exposure, absorption, is primarily concerned with exposures to the skin and, in some cases, the eyes. Because internal body temperature is regulated by blood flow (the circulatory system) close to the surface of the skin, foreign agents that can be absorbed through the skin can easily enter the bloodstream and be distributed throughout the body. Likewise, in rare cases, exposure through absorption can occur when such agents contact the eye and enter the body. In one such case, for example, a nurse contracted the human immunodeficiency virus (HIV) that causes acquired immune deficiency syndrome (AIDS) after blood from a patient accidentally sprayed into her eyes.

While the previous section established that the liver is the body's largest internal organ, the skin is the largest organ overall (approximately 20 square feet in surface area). It comprises up to 15 percent of total body weight. Because of its size alone, the worker's skin is especially susceptible to a wide variety of exposure potential in the average working environment. Understanding how the skin works, what its primary functions are, and how skin can be damaged or affected by exposure to various agents is critical to the recognition, evaluation, and control of such exposures.

Anatomical Overview

Despite the visible differences in the physical appearance of the skin from one part of the body to the next, the body's entire outer wrapper, more technically known as the *integument,* is similarly constructed. In fact, even though the hair and fingernails appear to be quite different from the skin itself, they are all made of the same materials.

The skin serves three primary functions. Its varying outward appearance on different parts of the body reflects to some extent which of these functions a certain area of skin primarily serves. The first function is to protect the body from invasion by germs as well as from damage due to physical contact with other surfaces or structures. It also serves to protect its owner against exposure to various forms of ultraviolet (UV) light energy, including solar UV (sunlight). Temperature control, the second function, keeps the body's internal organs at their normal operating temperature of 98.6° F through a process known as perspiration. Associated with this function is the skin's role as an organ of secretion (the elimination and subsequent evaporation of water and other substances). Third, the skin provides the sense of perception so that the owner can experience sensations such as touch, pain, heat, cold, and so on.

As shown in Figure 4-6, the skin has three distinct layers. The outermost (i.e., that which is visible) is called the epidermis (sometimes referred to as the cuticle); the middle layer is the dermis (or the true skin); and the innermost layer is known as the subcutaneous (underskin) tissue. Each serves very special functions. All are susceptible to damage and/or disease due to occupational exposures to harmful agents.

The Subcutaneous Layer

This innermost layer of skin is really a rather vague border between muscle and bone tissues on one side and the dermis on the other. It is primarily a fatty padding that gives bounce and a look of firmness to the skin above it. With age, replacement of the fatty cells of the subcutaneous layer slows dramatically and the layer begins to thin. As these cells die out with none to replace them, skin appears to wrinkle and sag. In fact, these fat cells are the distinguishing feature of the subcutaneous layer.

This layer also contains the lower portions of some sweat glands as well as millions of nerve endings and blood vessels.

The Dermis

The dermis, also known as the corium or true skin, is serviced by millions of tiny blood vessels and nerve fibers that reach it through the subcutaneous tissue. In addition, many special structures and tissues that enable the skin to perform various functions are found in the dermis layer. Sebaceous (meaning "skin oil") glands and sweat glands, as well as minuscule muscles and the roots of hairs encased in narrow pits called follicles are all located in the skin's dermis layer.

Under a microscope, the topmost layer of the dermis, interconnecting with the epidermis above it, resembles a rugged, ridge-crossed landscape carved with valleys, caves, and tunnels. The basic forms of this microscopic terrain are cone shaped hills called papillae. The average adult's dermis contains from 100 to 200 million papillae. The dermal papillae serve as the bedrock for the surface layer of the skin, the epidermis. The papillae are also associated with a dense concentration of nerve endings. Fingertips, for example, have more papillae than most other parts of the body. These not only provide the uniqueness of fingerprints but, because of the excessive nerve endings, the fingertips are also considerably more sensitive to touch and feel than most other parts of the body.

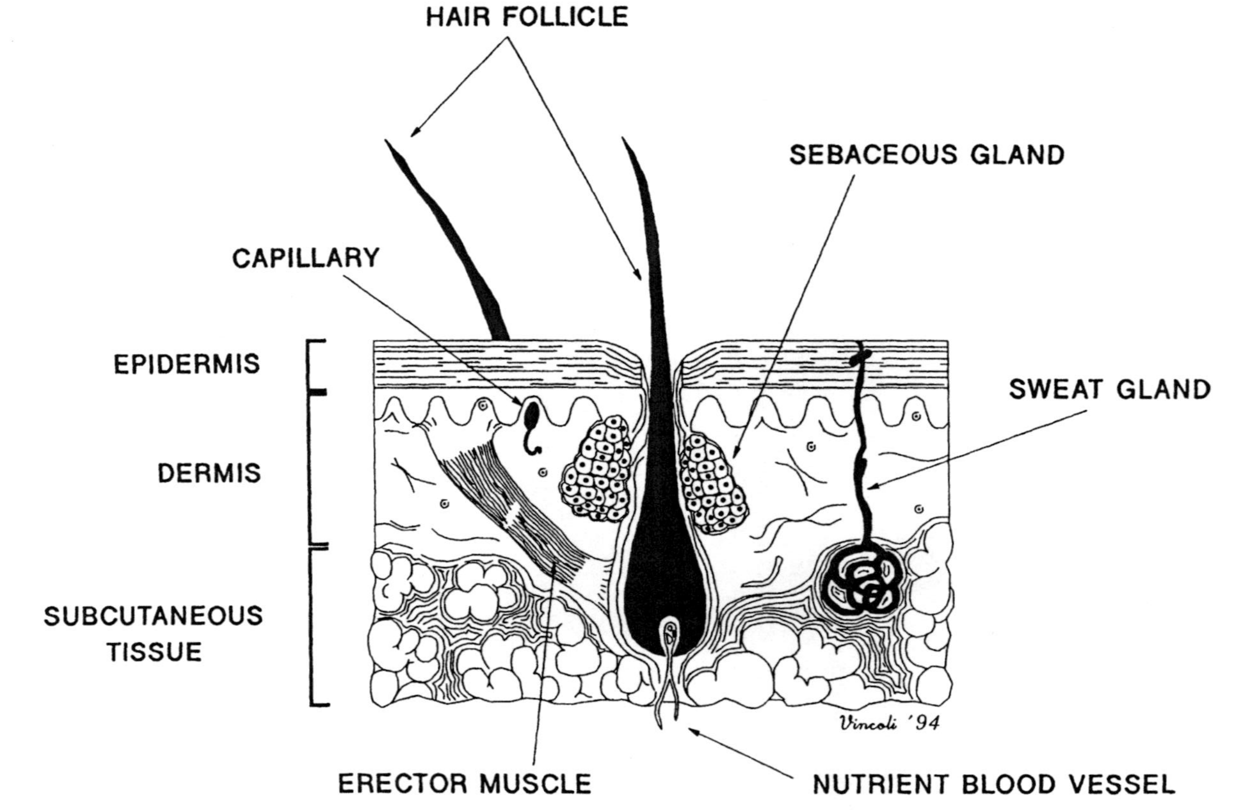

FIGURE 4-6 A cross-section of human skin, showing the basic components of the epidermis, dermis, and subcutaneous tissue

The Epidermis

At the bottom layer of the epidermis, its papillae fit snugly into the top layer of the dermis. This area of the epidermis is occupied by new young cells. These cells gradually move upward toward the skin's surface. As they approach the external surface of the skin, these cells die and become tough, horny, lifeless tissue. This outermost layer of the epidermis is called the stratum corneum (or "horny tissue") because of its microscopic appearance. It is continually being shed, largely unnoticeably, and replaced as younger cells move toward the surface.

Hair and Nails

Certainly the most noticeable of the specialized forms of skin are the hair and nails. They are actually a dead tissue called keratin, similar to the dead skin cells that are constantly being shed by the body, but much more firmly packed together. However, hair and nails both originate in cells that are very much alive; hence the pain of a torn fingernail or plucked hair follicle. In fact, the growth of the nail and hair actually occurs in this living region, with new cells literally pushing the dead, hard hair and/or nail stalks upward.

The bottom end or root of a hair is lodged, as noted above, in a follicle, a hollow shaft resembling a rounded bottle with a long narrow neck slanting toward the skin's surface. Each follicle is supported by a hummock of papilla and is serviced by tiny oil glands that lubricate the shaft through which the hair pushes toward the surface. The follicles of the long hairs of the scalp, groin, and armpits may be found deep in the subcutaneous layer while others may be no deeper than the upper layers of the dermis. Attached to a follicle are microscopic muscle fibers that, if stimulated by cold temperatures or some emotional factors, can contract around the follicle. The result is goose flesh or goose pimples that may also give the sensation that the hair is "standing on end." It is not a proven fact, but many believe this reaction is a holdover from ancient man when outward appearance was relied on as an intimidation factor during conflicts. When the human body was covered with hairs much longer and more dense than most of today's population, such a reaction would have made the individual appear much larger and perhaps more threatening than they actually were.

A nail's living, growing part is found beneath the whitish half moon, or lunula, at its base. The lunula is sometimes obscured because a layer of epidermis (cuticle) may have grown over it. Except at its very top, the nail is firmly attached to the ridged upper layer of the dermis, a region richly laced with tiny blood vessels.

Glands

The skin basically contains two types of glands, both of which have been previously mentioned. The skin's oil-producing (or sebaceous) glands are almost always associated with hair follicles, into which they seep their oils (or sebum). The oily substance works its way up toward the surface, lubricating both the hair and outer layers of epidermis. Both need continual lubrication to stay soft and flexible. Also, and perhaps of greatest importance to the industrial hygienist, skin oils serve as a protective coating against painful drying, chapping, and cracking of the skin. Most industrial grade solvents such as methyl ethyl ketone or even alcohol will effectively remove this protective oily coating leaving the skin dry and open to infection.

Most everyone is aware that the rate of perspiration is related to the atmospheric temperature of the working environment. However, an equal number of people do not realize that there are countless tiny blood vessels in the skin (some 15 feet of them coursing beneath every square inch of skin) that also react to outside temperatures. Sweat glands and blood vessels have the important role of keeping internal body organs near their normal temperature. To accomplish this, body heat must be conserved when atmospheric temperature is colder and released when it is warmer. The portion of the brain responsible for regulating this heat conservation/release process is known as the hypothalamus (Figure 4-4). Blood circulating through the tiny capillaries near the surface of the skin (Figure 4-6) is warmed (gains heat) or is cooled (loses heat) according to the outside temperature. Understanding this function of the skin is fundamental to the recognition of the hazards to human health posed by exposure to temperature extremes (heat stress/cold stress), as discussed in Chapter 7.

The skin's myriad blood vessels constrict when the outside air temperature is colder. This means that less blood can pass close to the surface of the skin and be cooled by the outside air and, therefore, the overall body temperature will remain warmer in colder climates. Conversely, when the blood needs to lose heat, the skin's blood vessels dilate (expand or swell) to allow more blood to flow close to the skin's surface. This phenomena accounts for the heat flush or reddening of the skin that some fair-skinned people exhibit when they are overheated.

The sweat glands aid in temperature regulation by secreting moisture that cools the skin as it evaporates and subsequently cools the blood flowing beneath the surface as well. Moisture that does not evaporate but remains as liquid on the skin or runs off in rivulets is not efficient to the cooling process. Humid air tends to prevent evaporation while moving air (wind) aids it. Sweat that evaporates as soon as it reaches the skin's surface usually goes unnoticed. Fresh sweat actually has no odor; but, if it remains without evaporating, bacteria begin to give the odor known medically as bromidrosis. There are two million sweat glands in the skin. Each consists of a coiled, corkscrew-like tube that tunnels its way up to the surface of the skin from the dermis or from the deeper subcutaneous layer.

Skin Absorption and Its Effects

The primary agents of harm through absorption are by far the hundreds of chemical substances that are used in modern industry today. Other exposures can also pose a significant health risk, such as that associated with the unprotected exposure of the skin surface to radiation. The skin's surface is generally coated with a lipid film consisting of mostly oils (sebum), decomposing products of keratin (from the hair and nails), and sweat. Although its primary purpose is to provide lubrication for the skin surface to keep it soft, flexible, and pliable, the lipid film serves no real barrier function in terms of exposure to harmful agents. The first line of defense against such attack is the outer horny layer of the epidermis. Obviously, whenever damage to the skin surface occurs (abrasion, laceration, etc.), the rate of absorption would increase dramatically. While some agents can also pass through the skin via the open ducts leading to the sweat glands or by migrating down a hair follicle shaft directly into the subcutaneous layer, absorption through the epidermis offers by far the largest opportunity for exposure.

As with exposures to other human systems and organs previously discussed, the results of exposure through absorption can be direct or indirect. Direct results of exposure usually cause some degree of damage to the skin surface itself, while indirect results will affect other human systems or organs (such as the nervous system, the liver, or the kidneys). Some chemical compounds can react adversely on contact with the skin and result in immediate and rather painful damage to the contacted area. For example, sodium hydroxide (NaOH) will adversely react on contact with water. Since skin can be covered with a thin layer of water in the form of sweat at any given time, contact with sodium hydroxide flakes or powder may cause a burning of the skin tissue as the chemical reacts with the water. Obviously, depending on the quantity of sweat present, the reaction can be quick or delayed (until a person begins to sweat). Less obvious are problems (reactions) that develop on contact with agents that do not result in any immediate indications or symptoms of exposure. When the exterior surface of the skin itself is affected by such exposures, the resulting disease is usually termed dermatitis or, more specific to occupational causes, industrial dermatitis. Indications of this disease are generally visible in some degree, as dermatitis literally means inflammation of the dermis. Rash, redness, itching, scaling skin are all subsequent to a dermatitis condition. While there are an unknown number of substances and conditions that can result in dermatitis, the hygienist should be aware that almost all industrial cases are caused by absorption hazards in the following categories:

1. Chemical: solvents, stimulants, sensitizers
2. Mechanical: trauma (cuts, abrasions, friction)
3. Physical (energy): heat, cold, radiation
4. Biological: bacteria, viruses, fungi, parasites
5. Botanical: plants (poison ivy, oak), wood products

The chapters in Part II of this Basic Guide explore many of these harmful agents further. As the reader begins to study these hazards, understanding how the skin works will help facilitate the recognition of such hazards in the workplace.

Indirect results of absorption are often less obvious because the effect may not be apparent for quite some time. This makes such exposures difficult to recognize and evaluate. For example, skin exposure to a substance such as 4,4 methylene dianiline (MDA), a common constituent of some epoxy resin compounds, will show no outward signs of irritation to the skin. However, after repeated exposures, the skin color near the point of contact may turn yellow or brown. A strong hepatoxin (harmful to the liver) and a carcinogen (cancer-causing agent), MDA can eventually damage the liver to a point of serious impairment.

OTHER SENSORY ORGAN EXPOSURES

In addition to the organ known as skin, there are of course other sensory organs of the body that are susceptible to damage from exposures to hazards in the workplace. Principally, the proper function of the ears and the eyes can be damaged or degraded due to occupational exposure to a variety of hazards. The effect can range from temporary damage to permanent partial or total loss of the organ's sensing abilities.

With some exceptions, the principal occupational hazards to the eyes involve those that pose an impact hazard. Solid debris such as dust, fibers, metal shavings, or other such particulate matter that become airborne projectiles during a manufacturing process can cause serious damage on impact with the eye. Similarly, liquid substances such as chemicals, some metals, cryogenic materials, and other industrial solutions can inflict damage ranging from minor irritation to corrosion and blindness if they should contact the eye. Perhaps the most insidious eye hazard involves exposure to radiation energy. For example, invisible ultraviolet light and infrared light can damage the eye before a person is aware of the exposure. Employees who work with lasers are also susceptible to eye damage should their eyes be exposed to the intensified light beam generated by most lasers.

For the most part, exposure to eye hazards in the work environment are much easier to recognize as such than hazards to human hearing. Eye hazards are usually physical in nature, meaning some physical thing (dust, mist, fibers, a gas, liquids, etc.) must be present for the hazard to exist. Even radiation, the possible exception to this hypothesis, is still a physical form of energy usually generated during some specific, known work practice and is therefore readily recognizable as a potentially harmful hazard to the eyes.

The ears, on the other hand, are susceptible to harm primarily from exposure to damaging levels of sound. Sound, or noise, is another form of energy. However, unlike the eyes, damage to the ears can often occur without their owner realizing it until long after the actual exposure. The eyes respond rather quickly to the effect of exposure, while the ears can be damaged slowly, over prolonged periods of time, with no initial indication to their owner that anything harmful has occurred.

The following is a very brief discussion of the anatomy of the ear to familiarize the reader with the hearing process, how it works, and how it can be affected/damaged by occupational exposures. A similar and equally brief discussion on the eyes will also be provided.

The Ears

The ear is an extremely complex and delicate organ. As shown in Figure 4-7, the composition of the ear can be divided into three areas: the outer ear, the middle ear, and the inner ear.

Outer Ear

The outer, or external, ear consists of two basic parts. The flap of trumpet-shaped skin and cartilage that is visible on the each side of the head is called the *pinna*. It functions to channel or focus sound waves toward the second element of the outer ear, the auditory canal. This canal is approximately $1^1/4$ inches long in the average adult and generally takes an inward, forward, and downward direction into the head. Because of this curve in the auditory canal, the pinna should be pulled upward and backward when ear protection is being inserted to straighten the canal and ensure proper placement of the protector. Modified sweat glands in the auditory canal secret a wax-like substance called cerumen that provides a sticky bactericidal coating to prevent debris from entering deeper into the canal and also keeps the outer ear healthy and free of infection. However, the hygienist should be aware that this wax can often become impacted, clogging the auditory canal and causing hearing loss or even deafness. Employees complaining of hearing loss should be checked by medical personnel for impacted cerumen. The cause of the problem may not be occupational at all, especially when the hearing loss is symptomatic to one ear only. Usually occupational exposures will affect both ears nearly equally since most people cannot selectively chose one ear over the other to hear the sounds that are occurring around them. Hearing loss in one ear only is therefore highly suspect as an occupational exposure. Also, when employees use hearing protection in the form of inserts (especially expandable foam plugs), impacted cerumen can occur quickly. Hence, medical personnel evaluating any complaint of hearing problems will generally check for impacted wax as a routine measure during examination. The tympanic membrane (or, more commonly, the eardrum) stretches across the inner end of the auditory canal and separates the outer ear from the middle ear.

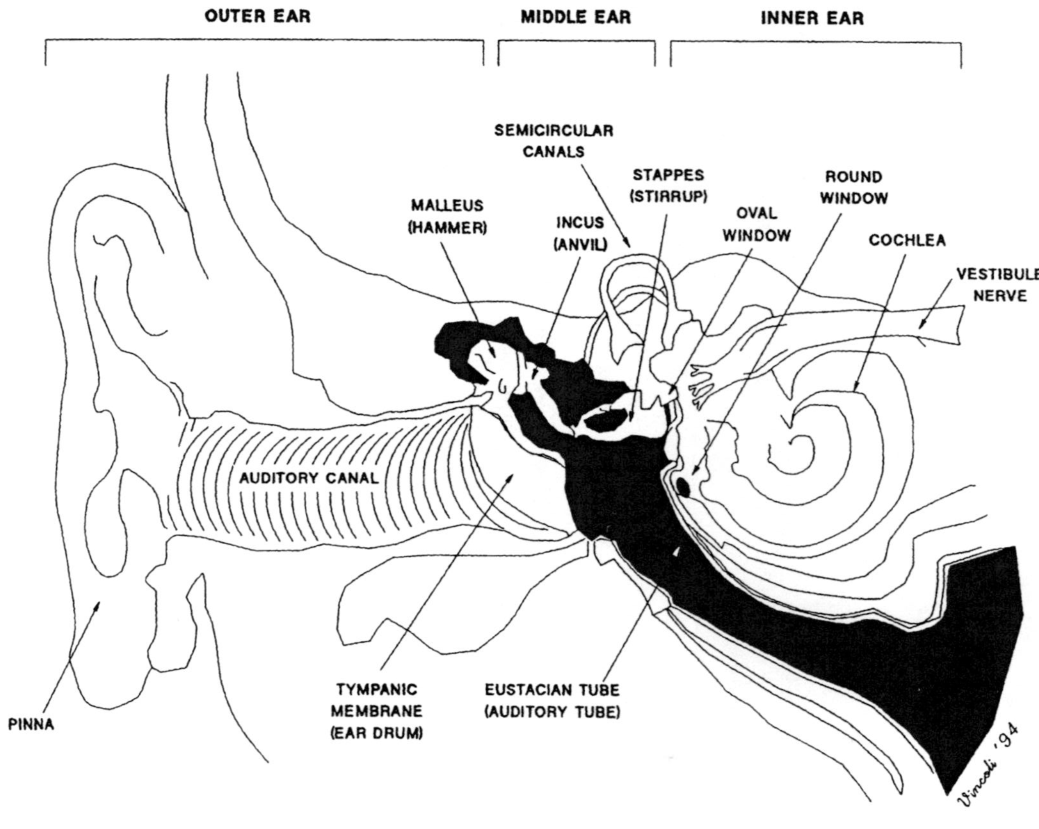

FIGURE 4-7 The composition of the ear can be divided into three areas: the outer ear, the middle ear, and the inner ear

Middle Ear

The middle ear (Figure 4-7) is also referred to as the tympanic cavity. It is a tiny cavity hollowed out of the temporal bone. It is in this small area where the three bones (or ossicles) of the inner ear are contained: the malleus, incus, and stapes. The more common names of these bones describe their shapes: hammer, anvil, and stirrup, respectively. The "handle" of the hammer is attached to the inner surface of the eardrum. The "head" of the hammer is attached to the anvil which, in turn, attaches to the stirrup. There are also several openings into the relatively tiny middle ear: one from the auditory canal (covered over by the tympanic membrane but an opening nevertheless); two into the inner ear (an "oval window" into which the stirrup fits and a "round window" that is cover by a thin membrane); and one into the Eustachian tube. The Eustachian or auditory tube is composed partly of cartilage and fibrous tissue and is lined with mucus. It extends downward, forward, and inward from the middle ear cavity to the pharynx (throat). Thus, the Eustachian tube provides the pathway by which throat infections can invade the inner ear. The useful function of the Eustachian tube is to provide equalization of pressure against the inner and outer surfaces of the tympanic membrane. It prevents membrane rupture and the discomfort that marked pressure changes can produce. When pressure changes are gradual, a person may not even realize that the Eustachian tube is doing anything of benefit. However, when pressure changes are sudden and dramatic (as in ascending or descending altitude in an aircraft), that person may experience a sudden clogging sensation in the ears that may be followed by some degree of pain and discomfort. Small children are particularly susceptible to severe problems since their Eustachian tubes have not fully developed into adult length, and pressure relief is therefore more difficult and often quite painful. Swallowing, yawning, or chewing often relieves the pressure on the ear drum because these actions cause air to spread rapidly through the open Eustachian tube. Atmospheric pressure then presses against the inner surface of the ear drum. Since atmospheric pressure is continuously exerted against the outer surface of the ear drum as well, the pressures are equalized, and the pain subsides.

Inner Ear

The inner ear, also called the labyrinth because of its complicated shape, consists of two main parts: a bony labyrinth and, inside this, a membranous labyrinth. The bony labyrinth contains the vestibule, the cochlea, and semicircular canals. The membranous labyrinth consists of the utricle and saccule (inside the vestibule), the cochlear duct (inside the cochlea) and the membranous semicircular canals (inside the bony ones). The actual organ of hearing, known as the organ of Corti, rests inside and throughout the inner chamber of the cochlear duct.

The bony walls of each component are separated from their membranous counterparts by a fluid. These components of the inner ear provide an awareness or "sense" of gravity, acceleration, and other such environmental stressors acting on the body. When problems with equilibrium occur, an employee may have trouble staying balanced while walking or standing still. While first impressions may suspect drug or alcohol abuse as the cause of such behavior, it is also quite possible that a problem exists within the employee's inner ear cavity. A medical evaluation should verify the nature of the problem. The industrial hygienist should never disregard any potential causal factor until all factual information can be obtained. The health hazard recognition process in such cases requires an understanding of the cause and effect relationships associated with illness or injury to the inner ear.

The Hearing Process

Hearing results from stimulation of the auditory area of the cerebral cortex (temporal lobe) of the brain (Figure 4-4). Before reaching this area of the brain however, sound waves must be projected through air (air conduction), bone (bone conduction), and internal fluids to stimulate nerve endings and set up impulse conduction over nerve fibers. In air conduction of sound, sound waves traveling through the air enter the external auditory canal with very little, if any, assistance from the pinna. At the inner end of the canal, these waves strike the tympanic membrane, causing it to vibrate. The vibrations from the tympanic membrane move the malleus (hammer), whose handle is attached to the membrane. When the hammer vibrates, it moves the incus (anvil) which, in turn, moves the stapes (stirrup). The stirrup moves against the oval window into which it fits precisely. At this point, fluid conduction of sound waves begins. When the stirrup moves against the oval window, pressure is exerted inward into the fluid-containing components between the bony and membranous labyrinth of the inner ear. This starts a "ripple" in the fluid (called perilymph) that is transmitted to the organ of Corti inside the cochlear duct. The wave travels onto the basilar membrane that supports the organ of Corti. From here, the sound ripple is transmitted to and through more perilymph fluid until it expends itself against the round window. Small hair cells at the base of the organ of Corti are stimulated by these ripples. They initiate impulse conduction by the cochlear nerve to the brain stem and produces the sense of hearing.

Bone conduction of sound occurs when sound waves actually travel through the bones of the skull; a route that bypasses the external and middle ear and transmits the sound directly to the inner ear.

Obviously, the process of hearing is a very complicated and quite remarkable phenomenon. The brief explanation provided here is meant only to impart a basic understanding of this process. Additional study, outside the scope of this publication, is highly recommended for the interested reader.

The Nature of Hearing Loss

Losing the ability to hear is a devastating and life-changing occurrence. While hearing can be affected by numerous factors (such as presbycusis or age-onset hearing loss), the reason for occupational hearing loss is often directly attributable to exposure to excessive unwanted sound (i.e., noise). At this point it is important for the reader to understand how hearing loss occurs (i.e., how the hearing mechanism of the ear can be damaged) rather than the mechanics of noise and sound. Chapter 7 will discuss noise and sound in greater detail.

Sound waves traveling in air generally move in a cyclic fashion. These cycles, commonly referred to as hertz (Hz), are what enter the auditory canal to be "heard" by the brain. The average human ear can detect such waves in the range of 20–20,000 Hz. As a point of reference, the pitch of human speech ranges between 300 and 4,000 Hz. When the ability to hear sounds in this range is diminished, normal communication is also hindered.

The industrial hygienist should be aware of the three categories or types of hearing loss that can be attributable to either occupational exposures, illness, or off-the-job abuses of hearing. Briefly, these three types of hearing loss are described as follows:

1. *Conductive Hearing Loss.* A person may have diminished hearing capabilities by air conduction but may show normal response when tested for bone conduction hearing. During a hearing test (audiogram), for example, a person may have depressed hearing when the earphones are worn properly (i.e., covering the opening of the auditory canal). But when the earphones are worn above the pinna, against the side of the head, bone conduction of sound directly to the inner ear may show no indication of impairment. This would indicate that the deeper structures of the inner ear are intact. The hearing problem may be caused by something as simple as impacted ear wax in the outer ear or by a more serious displacement of the middle ear bones (hammer, anvil, and stirrups) caused by some sort of trauma or shock (such as an explosion or other similar impact noise damage).

2. *Sensorineural.* When hearing difficulties occur during both air and bone conduction testing, the cause is usually attributable to problems deep within the inner ear. It can involve an impairment of the cochlea, the auditory nerve, the organ of Corti, or any combination of these mechanisms. This type of hearing loss is referred to as sensorineural.

3. *Mixed Hearing Loss.* It is, of course, quite possible that a combination of both conductive hearing loss and sensorineural hearing loss can occur

either at the same time or in succession. When this occurs, it is referred to as mixed hearing loss.

The Effects of Exposure

The effect of exposure to sound levels that can cause damage to hearing are primarily referred to as varying degrees of shift in the threshold of normal hearing. Specifically:

1. *The Temporary Threshold Shift (TTS)*. The TTS occurs after brief exposure to high-level sound. Its affects appear to be the greatest (most intense) immediately after the exposure and decreases with time if no further exposures occur. Prolonged or chronic exposures to sounds capable of producing a TTS can result in more permanent damage. In most cases, the TTS will prevent an individual from detecting low tones as well as the normal spoken voice. For example, an employee returning home after working in noisy conditions for most of an eight-hour day may find difficulty in comprehending voices over a telephone. Such employees may also wake the next day to find their television volume extremely high when the night before they had to set the volume control higher to hear it. These are strong indications of a TTS. At first, the employee may not recognize that anything is wrong. It is therefore the hygienist who must be alert to the possibility of such problems. Through surveys, inspections, and sound level monitoring with special instrumentation (see Chapter 7), recognition of noise exposure hazards can prevent a TTS condition from permanently affecting an employee.

2. *The Permanent Threshold Shift (PTS)*. The PTS is similar to the TTS except there is little if any recovery no matter how much time has passed since the exposure. The PTS basically means that the individual has permanent damage to the sound sensing mechanisms inside the ear. Since "permanent" generally means irreversible, such circumstances are not only tragic; if the damage is due to occupational exposures, PTS can involve legal ramifications as well. Suits against employers for occupational hearing loss are difficult to defend. It is more advantageous for everyone involved to prevent such an unfortunate occurrence from happening in the first place. The industrial hygienist must be able to recognize when the possibility for a TTS/PTS situation exists. Being alert to conditions in the work environment, knowing how to measure noise levels, evaluating audiograms for indication of problem areas, and educating employees to recognize when hearing problems may be suspect are all methods to control hearing loss exposures.

As a primary sensory organ, the human ear provides a quite remarkable service to its host. Without the ability to hear, other functions such as speech may also be impaired (the inability to hear one's own voice can severely affect the way words are spoken and pronounced). Indeed, hearing is one of the most precious and delicate of the human senses. While the total protection and preservation of this sensory function is beyond the control of the industrial hygienist, occupational exposures and their causes can and should be a primary focus of any industrial hygiene effort.

The Eyes

As with hearing, human sight is a most extraordinary phenomenon. Damage to the eyes due to some occupational exposure is primarily the result of impact hazards. Exposure to damaging radiation in the form of ultraviolet light is a form of impact hazard. Since the eye is by far the human organ most vulnerable to injuries in and out of the workplace, the eye is almost always at great risk. Occupational exposures to any such hazards that can cause harm or loss of sight should be of concern to anyone interested in the preservation of employee health and well-being, especially the industrial hygienist. Ensuring the provision and proper use of eye and face protection is therefore an essential element of any industrial hygiene program.

Unlike the other human systems, organs, and functions described in this chapter, it is usually quite obvious when an exposure has resulted in damage or harm to the human eye. Indications are usually (but not always) immediate, painful, and, with few exceptions, debilitating. When a person's eyes have been exposed to a hazard such as a corrosive liquid or gas, or invaded by some type of foreign body (fiber, dust, shaving, etc.), the individual typically will be unable to continue normal activities. In extreme cases, the pain may be so severe that the person may not even be able to move without assistance.

Given the specific nature of eye hazards, an understanding of the basic anatomy of the eye is not as critical as being able to recognize the eye hazards that are common to most work environments. Therefore, the following paragraphs will provide only a brief description of eye anatomy so that the reader can more fully appreciate the sense of sight and how the eye can be affected by exposure to recognizable hazards in the workplace.

The Eyeball

Approximately five-sixths of the eyeball lies recessed in the orbit that is protected by a bony socket in the skull. Only its small frontal surface is exposed. The eyeball itself is not a solid sphere. It contains a large interior cavity that is divided into two separate cavities: the anterior or frontal cavity and the posterior or rear cavity. A clear, watery substance known as *aqueous humor* fills the frontal cavity and often leaks out when the eye has been injured. The rear cavity of the eye, which begins directly behind the lens (Figure 4-8), is considerably larger than the frontal cavity. It contains a soft, gelatinous material known as *vitreous humor.* This semi-solid material helps maintain sufficient pressure within the eyeball cavity to prevent the eye from collapsing. Incidentally, an obliterated artery known as the hyaloid canal once ran through the vitreous humor to supply blood to the area between the lens and the optic disk during gestation. However, as eye formation and development were completed, the no-longer needed artery broke apart. Many people can still see pieces or the remains of this artery, commonly referred to as *floaters.* These floating pieces of artery are not on the surface of the eye but are actually trapped inside the ball itself. Floaters will usually be visible for the rest of a person's life.

The outer surface of the eyeball is composed of three layers or coats. From the outside in, they are the sclera, the choroid, and the retina (Figure 4-8). Both the sclera and the choroid consist of a front (anterior) and rear (posterior) portion. The sclera, which is made up of extremely tough, white, fibrous tissue, provides most of the eye's natural protection against impacts and invasions (by dusts, liq-

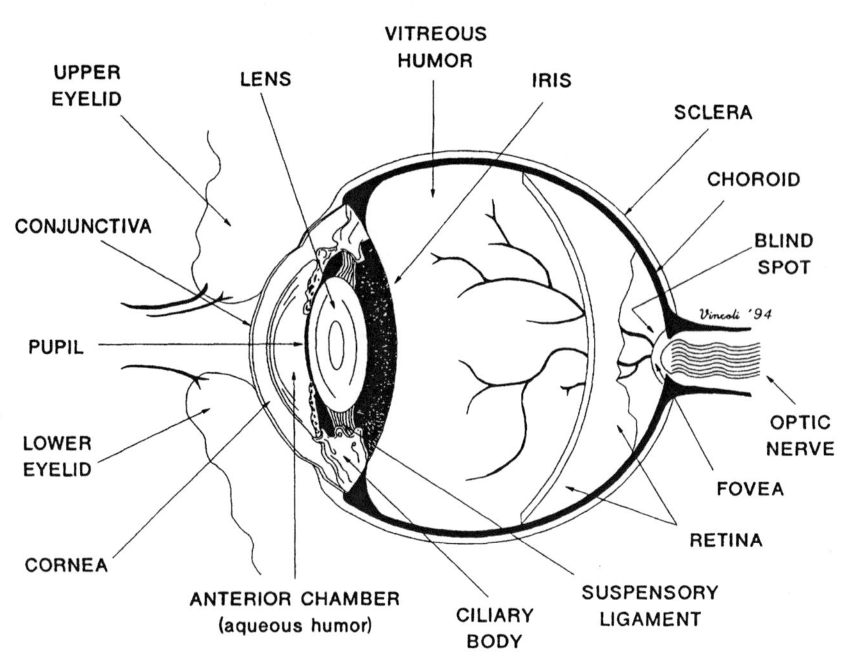

FIGURE 4-8 The major elements of the human eye

uids, etc.). The anterior end of the sclera, known as the cornea, lies over the colored part of the eye (iris). While the sclera is white and opaque, its cornea end is actually transparent and contains no blood vessels. Although the cornea is quite dense in composition, it must be transparent to allow light to pass to the eyes' light receptors. This is why the visible surface of the sclera is often referred to as the "whites" of the eyes. Certain liver disorders that elevate the bile pigment content of the blood, such as hepatitis and jaundice, can cause the skin surface and, more noticeably, the white sclera to appear yellow in color. However, occupational exposures to some chemical agents such as 4,4 methylene dianiline and epichlorohydrin (both used commonly in the manufacture of certain epoxy resin compounds) can also cause a yellow discoloration of the skin and, if exposure is long term, to the sclera as well. Successful control of occupational exposures requires an ability to recognize when individual symptoms are indicative of occupational exposures, as opposed to some non-occupational physical disorder. The hygienist should be aware of the potential signs of exposure to the chemicals used in the workplace and the possible relationship to any physical indications of exposure.

The middle or choroid coat of the eye contains a great many blood vessels and a large amount of pigment. Its anterior portion consists of three separate structures: the ciliary body, the suspensory ligament, and the iris.

The ciliary body is formed by a thickening of the choroid and fits like a collar into the area between the front of the retina and the back of the iris. Attached to the ciliary body is the suspensory ligament that actually blends with the elastic capsule known as the lens and holds it suspended in place. The iris is the colored part of the eyeball. It consists of circular and radial smooth muscle fibers arranged to form a doughnut-shaped structure. The whole that appears to be in the middle of the iris is known as the pupil.

The retina is the incomplete third layer and innermost coat of the eyeball. It is incomplete in that it has no frontal or anterior portion. It consists of mainly nervous tissue and contains three layers of neurons. The photoreceptor neurons are more commonly referred to as rods and cones because of their respective shapes. The rods and cones constitute the eye's visual receptors. They are highly specialized structures for stimulation by light rays. They differ as to number, distribution and function. The estimated number of cones is approximately 7,000,000; there are said to be somewhere between 10 and 20 times as many rods. Cones are most densely concentrated in a small depression found near the center of the retina. Their population becomes less and less dense as they move outward from this central region of the retina. Rods, on the other hand, are entirely absent from this area and increase in density toward the periphery of the retina.

The remaining two neuron layers of the retina extend back to a small circular area at the rear of the eyeball known as the optic disk. It is here where perforated fibers from the sclera emerge from the eyeball as the optic nerve. Incidentally, the optic disk is sometimes referred to as the blind spot, since light rays striking this area cannot be seen because it contains no rods or cones—only nerve fibers.

Accessory Structures

Accessory structures of the eye include the eyebrows, eyelashes, eyelids, and lacrimal apparatus.

The eyebrows and eyelashes serve a cosmetic purpose and give some protection against entrance of foreign objects into the eyes. Small glands located at the base of the eyelashes secrete a lubricating fluid. These can frequently become infected, forming a small inflammatory protuberance or growth known as a sty.

The eyelids consist mainly of voluntary muscle and skin, with a border of thick connective tissue at the free edge of each lid known as the tarsal plate. The tarsal plate is the visible ridge that is exposed when the eyelid is peeled back when removing a foreign object from the eye. Mucous membrane called the conjunctiva lines each lid and continues over the surface of the eyeball where it is modified to give transparency. Inflammation of the conjunctiva (conjunctivitis) is a fairly common infection. It is often called *pinkeye* because it produces a pinkish discoloration of the eye's surface. In the workplace, there can be a variety of substances and materials that are particularly irritating to the conjunctiva. Exposure to fiberglass fibers, for example, can lead to conjunctivitis.

The lacrimal apparatus consists of structures that secrete tears and drain them from the surface of the eyeball. The almond-shaped lacrimal glands, are located in a depression of the frontal bone at the upper end of each bony orbit. Approximately a dozen small ducts lead from each gland, draining the tears onto the conjunctiva at the upper outer corner of the eye. The lacrimal canals are small channels, one above the other that empty into the lacrimal sacs. The openings into the canal can be seen as two small dots at the inner end of the eyelids. The nasolacrimal ducts are small tubes that extend from the lacrimal sacs into the nasal cavity. All tear ducts are lined with mucous membrane, which is actually an extension of the mucosa that lines the nose. When this membrane becomes inflamed and swollen, the nasolacrimal ducts become plugged, causing tears to overflow from the eyes instead of draining into the nose as they normally do. Hence, a common cold, influenza, or even allergies are often accompanied by watering eyes. Likewise, exposure to certain irritating chemical compounds can also cause the eyes to water. In the presence of such chemicals, this may be the first indication that an exposure is taking place.

The Physiology of Vision

Having briefly described the basic anatomy of the eye and its related components, the process of vision itself can now be explained. In order for vision to occur, the following conditions must exist:

1. An image must be formed on the retina to stimulate its receptors (rods and cones)
2. The resulting nerve impulses must be conducted to the visual areas of the cerebral cortex (reference Figure 4-4)

The Retinal Image

Four distinct processes focus light rays so that they form a clear image on the retina:

1. Refraction, which literally means the deflection or bending of light rays
2. Accommodation, which provides focusing capabilities of the lens to adjust for near vision (less than 20 feet)
3. Constriction, which causes the pupil to adjust to varying levels of light intensity to ensure proper image perception
4. Convergence, which moves the two eyeballs inward so that their visual axes converge on a single object

Problems with the proper function of any of these four processes can cause problems with normal visual perceptions. For example, errors of refraction may prevent the proper bending and focusing of light rays onto the retina. Some common errors of refraction include nearsightedness (myopia), farsightedness (hypermetropia), and astigmatism. The nearsighted (myopic) eye sees distant objects as blurred images because it focuses rays from the object at a point in front of the retina. Opposite conditions exist in the farsighted (hyperoptic) eye. Astigmatism is a more complex condition in which the curvature of the cornea or of the lens is uneven, causing horizontal and vertical rays to be focused at two different points on the retina. Visual acuity, or the ability to distinguish form and outline clearly, is indicated by a fraction that compares the distance at which an individual sees an object (usually letters of definite size and shape) clearly with the distance at which the normal eye would see the same object. Thus, if a person sees clearly at 20 feet from an object that the normal eye would also see clearly at the same distance, then his or her visual acuity is said to be 20/20. When the eye sees clear images at 10 feet that the normal eye sees clearly at 20 feet, then visual acuity has dropped to 20/10. As people grow older, they become farsighted due to the lenses losing their elasticity and therefore their ability to bulge and accommodate for near vision. This condition is known as *presbyopia*.

Stimulation of the Retina

Rods are known to contain a pigmented compound known as rhodopsin (or "visual purple"). It is formed by a protein in combination with retinene, a derivative of vitamin A. Rhodopsin is highly light sensitive. Hence, when light rays strike a rod, its rhodopsin rapidly breaks down. This chemical change initiates impulse conduction by the rod. If the rod is exposed to darkness for even a short time, rhodopsin reforms and is ready to function again.

Cones also contain photosensitive chemicals. It is presumed that the cone compounds are less sensitive to light than the rhodopsin. Brighter light seems necessary for their breakdown. Cones are therefore considered the receptors responsible for daylight and color visions. Rods, on the other hand, are thought to be the receptors for night vision because their rhodopsin quickly becomes almost depleted in bright light and is slow to regenerate. This explains why seeing is difficult when first passing from a well lit area to an area of darkness. But when the rhodopsin has had time to reform, the rods again start functioning, and dark adaptation has occurred. Night blindness occurs in individuals with a marked deficiency in vitamin A (which can be a safety concern, especially for those employees required to drive during the evening hours as part of their job function).

Fibers that conduct impulses from the rods and cones reach the visual cortex of the brain (Figure 4-4) via the optic nerves and vision is perceived.

Recognizing Workplace Eye Hazards

As stated at the beginning of the section, hazards to the eyes are primarily categorized as impact or physical hazards. The most commonly encountered injury to the eye occurs as a result of blows from other objects. Blows from a blunt object cause pressure against the eye which can, in turn, result in contusion (bruising) of the iris, the lens, retina, or even the optic nerve. If the blow is sufficiently violent, the eyeball itself may rupture. Contusions can cause severe problems if not treated properly and immediately. Another common eye injury, corneal lacerations and abrasions, can occur relatively easily before the individual has time to react. Sharp objects near the eye are obvious concerns. However, if a sharp piece of foreign matter should get into the eye, rubbing and blinking may actually cause a small abrasion or laceration of the corneal tissue. Neither are very threatening to long-term eye health unless the wound becomes infected. Thermal burns are not common but can occur if the eye comes in close contact with flame or a high heat source. Such burns are extremely destructive to eye tissue and rarely heal properly. Irradiation burns, on the other hand, are more likely to occur in an average industrial setting. Damage caused by exposure to intense light can range from negligible impairment to the severe destruction of tissue. A common source of industrial radiation that can be damaging to the naked eye is the ultraviolet (UV) waveband usually generated during welding operations. Direct exposure to a welder's arc can cause inflammation of both the conjunctiva and the cornea itself. Less common a hazard but still a threat is that posed by exposure to infrared (IR) radiation energies that pass directly through the cornea to be absorbed by the lens and the retina. Although the damage mechanism is not entirely understood, direct exposure to visible light (such as that emitted by lasers) can also cause damage to the eye. Rods, cones, and the structures of the retina, lens, and cornea all appear susceptible to damage from intense, concentrated, or amplified visible light exposure.

Another impact hazard to the eye is that posed by exposure to liquids, including their mists, vapors, and gases. Many of today's industrial processes require the use of a wide variety of chemicals and their compounds. Chemicals can be toxic, corrosive (see Chapter 6), or a combination of both. On contact with the eye, the results can range from negligible or minor irritation to total and permanent loss of sight. For example, acid exposures typically cause immediate damage to the eye tissue, while the damage caused by exposure to some caustics may not initially appear too serious. Afterward, however, the eye tissue may show signs of extreme damage due to caustic exposures. This is because most acids tend to damage the surface protein that layers the eye and normally do not penetrate much further, while the alkaline properties of the caustic will soak into the tissue itself. Eye exposure to some chemicals and toxins (and even some viral agents) can present a more insidious hazard to the employee as well. In some cases, there may be no immediate or apparent indication that the exposure has caused any harm whatsoever. But, unfortunately, harmful agents can be absorbed through the eye tissue quite efficiently and enter into the bloodstream to cause a seemingly unrelated systemic problem. Depending on the specific nature and conditions of a particular chemical exposure to the eye, the resulting effects could manifest as conjunctivitis, iritis (inflammation of the iris), and perhaps even glaucoma (an increase in the pressure inside the eyeball itself).

It cannot be overstated that the human eye is perhaps the most vulnerable of the sensory organs and, therefore, is particularly susceptible to injury from impact hazards in the workplace. Knowing the delicacy of the special anatomical mechanisms that enable sight will help the hygienist to fully appreciate the necessity to protect this most valuable of human assets against occupational hazards.

SUMMARY

This chapter provided a brief description of the major human anatomical systems that are susceptible to harm or injury due to occupational exposures. The relationship between cause (the agent of harm) and effect (the result of exposure to the agent) was also discussed in summary fashion. This included information on the difference between the real and perceived effects of exposure and the overwhelming power and ability of the human psyche to convince a person that an illness exists when, in reality, there are no health problems of any concern. Often referred to as *psychosomatic* in nature, such illness in the workplace must be recognized as such if the hygienist is to properly meet the overall objectives of the industrial hygiene program. In terms of understanding how exposure to harmful agents in the work environment can result in specific harm, the relationship between dose and concentration was also briefly explored.

The anatomical systems and the way each can be harmed by occupational stressors has been the major focus of this chapter. Without a fundamental understanding of, at the very least, the respiratory system, the digestive system, the skin and other specific sensory organs (the ears and the eyes), the reader cannot fully appreciate the importance of industrial hygiene in ensuring the preservation of employee health. Being able to recognize the existence of the many hazards that may exist in the average workplace really meets only half the challenge of the industrial hygiene profession. Being able to recognize the specific/adverse health effects that can result due to exposure, will complete the process and ensure a successful industrial health effort.

II

Assessing Workplace Health Hazards

Hazards to human health can exist in a variety of forms. They can be visible or invisible. They can be solid, liquid, or gas. Some are toxic if inhaled or ingested, corrosive on contact with skin, or a combination of both. Health hazards can be obvious or insidious. They can exist when people are not aware that their health is at risk, as is the case of exposure to radiation. Obvious or insidious hazards can be associated with the physical elements of the work environment, such as illumination levels, for example. Poor lighting can cause eye strain at such subtle levels that the employee is unaware of any immediate or noticeable health risk. Other factors like temperature extremes, ventilation, vibration, noise, and so on, can all be *agents* of risk to employee health. Such hazards can cause immediate adverse effects as well. Excessive noise or vibration can result in harm ranging from temporary hearing loss or numbness in the fingers *(Raynaud's disease)* to permanent loss of hearing or nerve damage. Whatever the exposure, there is no chance of assuring a safe and healthy work environment without proper hazard recognition and evaluation.

The hygienist must, therefore, be knowledgeable of the various types of hazardous agents and/or conditions that can exist in the work environment. This part will focus on the potential existence of many of these hazards in the workplace. When the existence of such hazards is recognized, their routes of entry into the human body and their potential to cause harm understood through evaluation, the industrial hygiene program will clearly be on the right track toward exposure control.

In Part I, the three fundamental elements of the industrial hygiene process (recognition, evaluation, and control, or REC) were emphasized. With this knowledge as a foundation, Part II addresses some of the common workplace hazards that require the special attention of the industrial hygienist.

Each chapter in this part will focus on specific workplace hazards and their known effects on human health and well-being. Understanding the existence, special nature, and unique characteristics of these health hazards will facilitate the application of the industrial hygiene REC process. In other words, to be able to properly *recognize* when toxic or otherwise hazardous agents are present in the workplace, the reader must first understand these hazards.

Chapter 5 introduces the various health hazards associated with *air contaminants* and *asphyxiants.* Since it has been established that inhalation is the most direct route of exposure, the presence of air contaminants in the work environment can pose a significant health hazard concern. Likewise, certain commodities can displace the oxygen in the air or interfere with the proper function of the respiration process. This chapter will provide discussion on the various types of air contaminants and asphyxiants that can be found in the average workplace; how they affect the human system; what methods of evaluation are available; and a brief description of standard control measures that can be taken.

Chapter 6 introduces the hazards posed by the presence of *toxic chemicals* and *corrosive agents* in the workplace. Since exposure to many chemical compounds, including most *metals,* can significantly impact human health, the hygienist must also beware of these particular workplace hazards. Toxic chemicals and corrosive agents are generally hazards due to their tendency to enter the body primarily through absorption, the second most common route of entry. However, many chemicals can also present an inhalation hazard and, to a lesser extent, an ingestion hazard as well. Such hazards often have the tendency to adversely affect more than one anatomic system. Therefore, the hygienist must be especially alert to the existence of these hazards in the work environment. Even though the evaluation techniques and control methods for these hazards may be similar to those used for other types of hazards, specific attention must still be applied to the implementation of these measures.

In Chapter 7, the very specific hazards associated with exposure to various types of energy are discussed. This includes temperature extremes, excessive noise and vibration, radiation, and illumination. Such exposures tend to be insidious since detection, especially at subtle levels, can often be difficult to recognize. In fact, understanding the subtleties of exposures to hazardous energy is essential to the entire REC process.

In modern industry, there are also numerous opportunities for occupational exposure to a variety of *biological hazards.* These include, but are not limited to, *carcinogens, mutagens,* and *teratogens.* Also, there are hazards associated with exposure to *bacteria, molds,* and *fungi.* Occupational exposure to certain *pathogens* has also become a concern for the practicing hygienist. Chapter 8 provides a brief examination of the various biological hazards and the effects of exposure.

Chapter 9 presents a general discussion of the health hazards posed by *ergonomic* problems in the workplace, their evaluation, and control. This includes problems associated with *repetitive motion* and *cumulative trauma disorders.*

In general, the REC process is not always easy. While many hazardous situations or conditions may be obvious, an equal number can go unnoticed until long after they have begun to affect the health of the employees. Since it is the *human health* aspect of the work environment that is of concern to the industrial hygienist, the chapters of this part will also examine the various ways in which employee health can be adversely affected by exposure to various health hazards.

Common techniques associated with the measurement and assessment of these hazards will also be briefly examined in these chapters. Chapter 10 will conclude this Part with a discussion on the use of evaluation tools such as instrumentation and measuring/sampling devices. Understanding how exposures are measured in terms of established safe limits is a key aspect of the evaluation process.

On completion of Part II, the reader should posses the basic knowledge necessary to begin the health hazard recognition, evaluation, and control process. This knowledge, together with the information provided in Part III on supplementing the industrial hygiene effort, form the basic foundation of the industrial hygiene discipline.

5

Air Contaminants and Asphyxiants

INTRODUCTION

Since people have to breathe to survive, exposure to contaminants in the air seems like the most threatening of all occupational health hazards. While all exposures should be of concern to the industrial hygienist, the recognition of air contaminants in the workplace is of particular importance since they can threaten the health of the entire work force.

It is extremely important for the hygienist to understand the terminology associated with the description of different contaminants. Air contaminants can exist in a variety of forms that extend through the three states of matter: solid, liquid, and gas. It is common to hear the terms *fume, gas,* and *vapor* used to describe the same contaminant. Indiscriminate use of unlike terms is not only confusing and inaccurate, it can be dangerous. For example, using respiratory protection designed to protect against exposure to organic vapors will be ineffective if used around fumes. A misunderstanding of specific terms can also adversely affect your effort to comply with regulations, since most health standards are specific to the various types of contaminants.

RECOGNIZING AIR CONTAMINANTS IN THE WORKPLACE

Air contaminants can take on a variety of forms, threatening the health of the worker on a broad scale. Recognizing the existence of such pollutants is the first step in protecting the health of the employee who must work in the presence of such hazards. Airborne substances that are solid are defined as *particles* to distinguish them from liquid droplets. When there is no need to differentiate between the two, the collective term *particulate* is used. The following brief discussion of the common types of airborne contaminants found in today's workplace, as shown in Figure 5-1, is provided to help clarify the confusion that is so apparent throughout industry.

Vapors

A volatile substance that is normally in the solid or liquid state at ambient temperature and pressure will emit a gaseous contaminant that is referred to as a *vapor*. Through the process of evaporation, a liquid or even a solid changes into a vapor and subsequently mixes with and contaminates the air in the immediate area. It is the measured amount of this contamination that becomes a hazard to human health. Commonly encountered with the use of solvents, vapors can pose a particular threat to those in the area through inhalation. Depending on the air flow through the workplace, vapor infiltration into the air handling system can also carry the contaminant to employees in surrounding work areas as well. Since most vapors are heavier than air, they tend to gather low and remain near the floor. Open tanks, pits, and other such depressions in the normal ground level will act as receptacles for low-lying vapors. Hence, entry into these areas becomes particularly dangerous if solvents have been used in or around the work area.

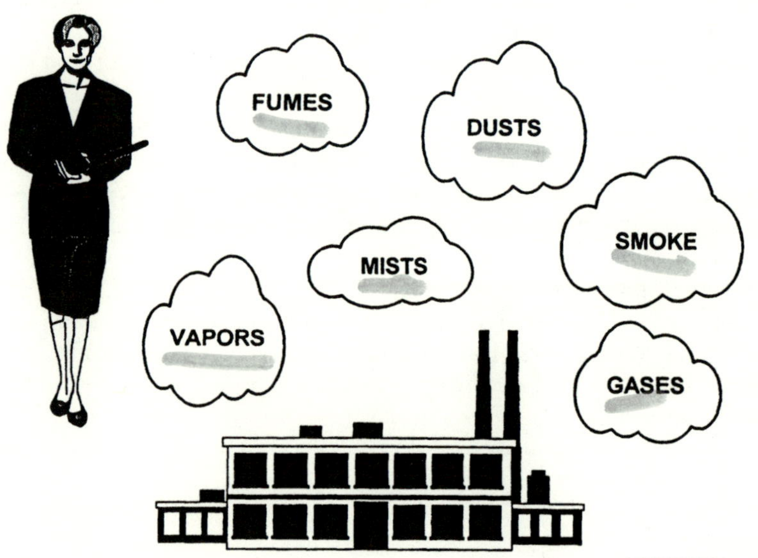

FIGURE 5-1 The industrial hygienist must consider many common types of airborne contaminants that are found in today's workplace

Usually, vapors in the air are measured in terms of parts per million (ppm) or, depending on their density, milligrams per cubic meter (mg/m^3).

Some vapors also present a physical hazard to workers. Certain vapors are extremely flammable. Excessive build-up of flammable vapors will present an extremely dangerous and highly explosive situation.

It is therefore critical to the health hazard recognition process for the industrial hygienist to be aware of work processes and tasks that produce vapors. Reviewing work procedures, examining purchase requests, inspecting the various work areas, and assessing work practices are all methods that can be used to facilitate this recognition process.

Gases

A *gas* is a thin, formless fluid capable of indefinite expansion to occupy the limits of its container. A gas can be either an element (such as argon) or a compound (such as carbon dioxide). Gases exist because their molecules are unrestricted by the cohesive forces normally found in liquids and solids. But gases are convertible by pressure (through compression) and temperature (cold) into a liquid or, eventually, a solid state. Air, a gas, is essential to sustain human life and health. The gas exchange process described in Chapter 4 (The Respiration Process), allows for the continued oxygenation of every cell in the body. However, any displacement of normal air by the presence of some other gas could significantly impair respiration. Many gases are odorless and colorless and exposure can be silent and deadly. Carbon monoxide (CO), for example, is of particular concern since, as a by-product of combustion, it can be found in many industrial work settings. Since CO diffuses approximately 19 times faster than oxygen through the alveolar membrane, it will quickly reach the bloodstream. And, since CO has an affinity for blood some 300 time greater than oxygen, it will rapidly displace oxygen in the blood. As the CO binds with hemoglobin, it forms carboxyhemoglobin, deadly to every human cell. Once again, to properly recognize the existence of gaseous health hazards, the hygienist must fully assess every possible work process where gas generation is a potential. For example, the use of gas-powered industrial trucks (i.e., forklifts) indoors can generate extremely toxic gases. In fact, internal combustion engines produce hundreds of toxic by-products that can be quickly and effectively distributed throughout the workplace by the facility's air handling system.

Gases can also be flammable or combustible, creating the physical hazards of fire and/or explosion.

Aerosols

An *aerosol* is a dispersion of solid or liquid particles in a gaseous medium (such as the atmosphere) of such microscopic size that they can remain airborne for prolonged periods of time. Aerosols are generated by fire, erosion, sublimation, condensation, and abrasion of minerals, metallurgical materials, organic materials, and other substances in construction, manufacturing, mining, agriculture, and transportation. Aerosols are classified on the basis of physical nature, particle size, and method of generation. Examples include mists, fogs, dusts, fibers, and smokes. An aerosol's ability to enter the body and the rate at which it is absorbed depends on its solubility characteristics and its particle size distribution.

Mists

Created first as a vapor, a *mist* (an aerosol) is actually a suspended liquid in droplet form that has been condensed back into the liquid form. Mists can also be generated when a liquid has been dispersed such as by splashing, foaming, or atomizing. While a mist may be considered a vapor in transition, it is a definite separate state of liquid matter. The liquid particles contained in a mist typically measure between 40 and 500 micrometers. In contrast, a fog particle is generally smaller than 40 micrometers in size.

Examples of mists encountered in the workplace include oils produced during certain machine operations, acid mists generated during electroplating or other types of chemical processing activities, or paint spray mists coming from a spray booth or similar paint spraying process. Mists are hazardous through inhalation and skin absorption. Depending on the composition of the mist liquid, results of exposure can range from contact dermatitis or inflammation of the breathing airways, to a severe systemic reaction such as hepatitis, or a debilitating chemical bronchitis.

Smoke

Aerosol particles that consist of carbon, soot, and other combustible materials less than 0.1 micrometers in size are referred to as *smoke*. The result of incomplete combustion of carbon-based materials such as coal or oil, smoke can contain both liquid droplets and microscopic solid materials. Tobacco smoke that has been exhaled into the air by a cigarette smoker contains a frightening combination of more than 4,700 moist, tarry, noxious by-products that are particularly hazardous to others in the general vicinity.

Cigarette smoke is a lung hazard to workers because it is inhaled many times during the day, and it has a marked effect on the mechanisms designed to protect the lungs (refer to Chapter 4). It irritates the mucus-lined airways and causes an increase in the level of mucus secretion. It also slows down the normally rhythmic beat of the cilia. As discussed in Chapter 4, cilia movement at a constant speed and direction is essential for the proper removal of trapped particulate matter from all areas of the respiratory system. Also, smoke will decrease the activity of the macrophages which, in turn, allows foreign materials and substances to remain in the lungs. Cigarette smoke will cause lung disease with no help from any other lung hazard. But when smokers are exposed to lung hazards at work, their protective mechanisms, already handicapped by the smoke, can enable a more rapid development of occupational lung disease. Cigarette smoke is a carcinogenic (cancer-causing) agent. It is also a powerful booster of other carcinogens such as asbestos, arsenic, chromium compounds, and many other toxic and hazardous substances that may be found in the workplace. Hence, smokers who are also exposed to such substances increase their chances of getting lung cancer. Unfortunately, in most cases it is the occupational exposure that will be blamed for the condition. Seldom will the employee's personal smoking habits be considered as a contributing (or causal) factor. While the national view of smoking in the workplace (or, for that matter, any place) continues to change, it is unlikely that people will ever completely stop smoking. As long as there are smokers who must also work with or around respiratory hazards, the hygienist must remain cognizant of the higher health hazard risk facing each smoking employee. Knowing of the magnifying effects smoking can have on those exposed to respiratory hazards will help the hygienist recognize the full potential of workplace respiratory health hazards.

Fibers

Particles whose length exceeds their width are generally referred to as *fibers*. These can be generated from minerals, such as asbestos, and human-made sources, including fiberglass. Other fibers include organic materials such as hemp and animal-borne fibers. For the purposes of general classification, some fibers are assigned a minimum size criterion. For example, asbestos particles must be at least three times longer than they are wide to be considered a fiber for occupational sampling purposes. Fiber behavior in the lungs is thought to be different from that of spherical particles.

Fumes

A *fume* is actually a small solid particle generated as the result of the condensation of a metal or plastic from a gaseous state after it has been heated. For example, welding causes the rapid volatilization of metals into gases, followed by their equally quick condensation on contact with cooler air. This creates small suspended particles of approximately 0.1 to 1.0 micrometers in diameter. Because these fumes are so very small, they can be readily inhaled by those in the surrounding areas. Inhalation of metal fumes can cause severe poisoning. Depending on the metal involved, the results of fume exposures can be quite deadly. The term *mad as a hatter,* for example, is in reference to the makers of fine felt hats during the 17th century. The process required the use of heated mercury nitrate. The fumes that resulted caused severe dementia, sometimes to the point of death.

When the surface being heated is coated with other toxic materials, such as a lead-based paint, an additional respiratory hazard is generated. In any assessment of work processes, the hygienist must be alert to all possible health hazards associated with work tasks.

It should be noted that the term *fume* is often used to describe a gas or vapor, especially by the lay person. However, gases and vapor are not fumes. In most cases, the heated material rises to form an oxide in the air. This characteristic is what differentiates the fume from a gas or vapor.

Dusts

Solid particles generated as a result of handling, crushing, grinding, rapid impact, detonation, or some other abrasive action of organic or inorganic materials (rock, metal, wood, grain) are referred to as *dusts*. These particles, which may arise from soils, bedding, or surfaces such as floors or walls, can range in size from 0.1 to 25 micrometers. Dusts smaller than five micrometers will usually remain airborne for prolonged periods of time. Hence, only those dusts that are smaller than five micrometers in size pose a particular threat on inhalation. Dusts particles this small can reach deep into the lungs and affect the entire respiratory process. Recognition of dust hazards in the workplace can be particularly difficult since particles smaller than 50 micrometers are not generally visible to the naked eye. Therefore, it should be assumed that any process generating visible dust particles is also generating an equal or greater amount of invisible dust. Since dust particles of 10 micrometers can reach the upper respiratory tract and those 5 micrometers and smaller can be deposited in the lower respiratory tract, the hygienist must ensure the proper recognition, evaluation, and subsequent control of all possible dusts hazards in the workplace.

THE HAZARDS OF ASPHYXIANTS

When people are deprived of oxygen, they will die through suffocation, as will any individual tissue that is deprived of oxygen. Any substance that interferes with the oxygenation of the tissues is referred to as an *asphyxiant*. Under normal conditions, air contains approximately 21 percent oxygen. Humans require a minimum of 19.5 percent oxygen content in air to adequately sustain all of their biological functions. When the oxygen content of the air begins to drop below 19.5 percent level, noticeable effects include drowsiness, disorientation, subtle loss of motor control, and an inability to think properly. As the oxygen content approaches 16 percent and lower, the person will likely be unconscious. Prolonged deprivation of oxygen will result in irreversible brain damage and eventually death. In technical terms, total asphyxiation will result in a condition known as *anoxia* (a complete loss of blood oxygen content), while partial asphyxiation results in a condition of low blood oxygen content known as *hypoxia*. However, if hypoxia is allowed to continue for a prolonged period, brain damage or death is still likely.

As a health hazard, asphyxiation can generally be divided into two sub-categories: *simple asphyxiation* and *chemical asphyxiation.*

Simple Asphyxiants

A simple asphyxiant is essentially an inert gas (hydrogen, nitrogen, helium, neon, argon, methane, and so on) that dilutes or displaces the atmospheric oxygen content to levels below that which are required to support human life. On inhalation of a simple asphyxiant, the blood oxygen content is insufficient to sustain normal tissue respiration, and suffocation occurs. Brain cells will begin to die with complete deprivation of oxygen for a period of three to five minutes. Because simple asphyxiants are not metabolized or changed into injurious chemicals once inside the body, they are not considered toxic or corrosive. Simple asphyxiants only prevent the blood oxygen content from reaching normal levels.

The hygienist should be alert to any tasks or processes where exposure to such hazards are a potential. An example would include any form of *confined space entry.* Understanding this health hazard begins the recognition process which, in turn, is the basis of any industrial hygiene effort.

Chemical Asphyxiants

A chemical asphyxiant causes a direct chemical action in the body to either prevent tissue respiration totally or to interfere with the transportation of oxygen from the lungs to the tissues, even though the blood itself may be well oxygenated. As discussed earlier, carbon monoxide (CO) is an odorless gas that combines with the blood's hemoglobin (carboxyhemoglobin), blocking oxygen transport to the tissues and causing death, even at very low concentrations. It should be noted that people who smoke cigarettes will normally have a marked increase in carboxyhemoglobin content. This means that their blood oxygen content is below normal levels. These individuals will most likely be effected much quicker by any subsequent exposure in the workplace to a chemical asphyxiant like carbon monoxide. Cyanide, on the other hand, is an inhibitor of cytochrome oxidase (an enzyme required for proper cell respiration). Exposure to cyanide inhibits the cellular respiration process even though the blood oxygen level may be normal. Other chemical asphyxiants, such as hydrogen sulfide, cause a paralysis of the area in the brain responsible for regulating respiration.

INDOOR AIR QUALITY

The problems associated with indoor air are not new. Benjamin Franklin once observed, "No common air from without is so unwholesome as the air within a closed room that has been often breathed and not changed." Hence, in addition to those hazards that are presented by the presence of air contaminants and asphyxiants related to identifiable work practices, there are other problems associated with indoor air pollution that may not be attributable to any specific work function or task. In some office environments, for example, the air can be more harmful to workers than the outdoor air in some of the most polluted cities (Brody, 1982). Many people do not realize that their "perpetual colds" or their consistent "hay fever" may actually be caused by the air in their own homes or at work. While there is not much that can be done to effectively evaluate the problem in a worker's home, the hygienist can do much to determine if the air at work is a health hazard. Once the cause of such problems can be isolated and identified, the hygienist can provide recommendations to eliminate or control exposures while the employee is on the job.

Indoor air quality problems have been linked to a wide variety of adverse health effects. The hygienist should be alert to complaints of headaches, respiratory problems, frequent colds and sore throats, chronic cough, skin rashes, eye irritation, lethargy, dizziness, and even memory lapses. Long-term effects include an increased risk of contracting some cancers, although substantiation for such claims are not yet conclusive. Although older workers and those who already suffer with chronic ailments such as asthma, allergies, and heart and lung diseases seem especially vulnerable, symptoms may also occur in otherwise healthy individuals.

Virtually every office building is a potential source of excessive amounts of one toxic pollutant or another. These include (but are not limited to) nitrogen dioxide, carbon monoxide, hydrocarbons, formaldehyde, radon, sulfur dioxide, and asbestos. Additional indoor air quality problems are caused by workers themselves, who use products such as hair spray, deodorants, perfumes and colognes, glues, nail polish, and (ironically) air fresheners at work. Building heating and cooling devices, if not properly maintained, can also contribute to the problem by releasing molds and bacteria into the indoor air. Carpeting, walls, furniture, and some types of treated clothing all release pollutants into the air that is breathed.

The levels of potentially hazardous substances in indoor air often exceed those allowed outside and are sometimes as great as or even greater than some permissible industrial exposures to the same substances. The Environmental Protection Agency (EPA) has concluded that the average person spends 90 percent of the time indoors. While most of that time is spent at home, the remainder is in the workplace where the hygienist can make some difference in the air that is breathed on a daily basis.

Major Indoor Pollutants

Major causes of indoor air quality (IAQ) problems can be linked to one or more of three basic hazards: air contaminants, inadequate ventilation, and poor humidity control. Table 5-1 shows the percentage of IAQ problems attributable to the various causes while Table 5-2 provides a categorized listing of the common indoor air pollutants that can be found in an average workplace. Figure 5-2 shows the relative size of these pollutants in relation to human visual perception. The following is a brief description of some of the major indoor pollutants, their sources, and their effects.

Formaldehyde

Although not as common as it used to be, this ubiquitous chemical is still a hazardous indoor air pollutant. Most commonly found in buildings constructed prior

---

TABLE 5-1   The percentage of common IAQ problems and their various causes

COMMON CAUSES OF IAQ PROBLEMS	
CAUSE	*PERCENTAGE*
Inadequate ventilation	52
Indoor contamination	17
Outdoor contamination	11
Microbial contamination	5
Building furnishings and fabrics	3
Unknown sources	12

TABLE 5-2   A categorized listing of the common indoor air pollutants found in an average workplace

IAQ CONTAMINANTS
RESPIRABLE PARTICULATES
Tobacco smoke (contains over 4,700 pollutants)
Allergens (mold, pollen, fungi, spores)
Pathogens (bacteria and viruses)
Asbestos and manufactured fibers
GASES AND VAPORS
Carbon monoxide (CO) and carbon dioxide (CO_2)
Radon (radioactive elements and their daughters)
Volatile organic compounds (VOCs)
Oxides of nitrogen (NO_x)
Formaldehyde (HCHO)
Trivalent oxygen (ozone or O_3)
Pesticides (termiticides such as aldrin and chlorodane)
Polychlorinated biphenyl (PCB)

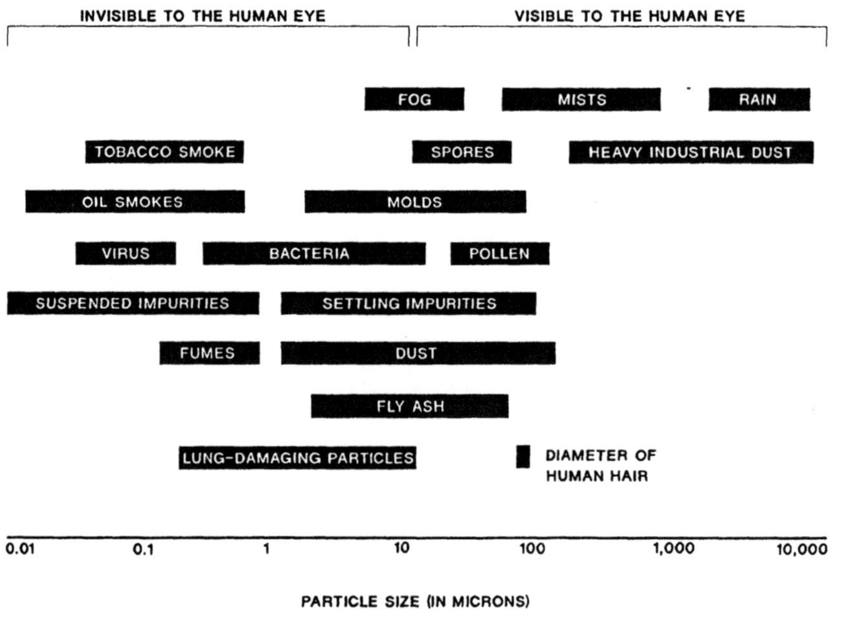

FIGURE 5-2 The relative size of common air pollutants in relation to human visual perception

to 1982, formaldehyde can be a health hazard in even the most benign office settings. It seeps out of urea formaldehyde foam insulation (not polyurethane insulation). A nearly colorless gas with a pungent suffocating odor, it is found in particle board used in wall construction, partitions, and many of the prefabricated cupboards and furniture products made in recent years. It is used in the manufacture of rugs and carpets made from synthetic fibers, in draperies, in permanent pressed clothing and linens, and in many drugs and cosmetics. Effects of exposure include eye irritation, frequent upper respiratory irritations and infections, chronic headaches, periodic memory lapses or drowsiness, and, in some older workers, chest pains and heart problems. There is also some indication that the chemical may propagate nasal cancers with prolonged exposures. Formaldehyde is also a known sensitizer. This means that people exposed over long periods of time may become sensitive to the pollutant, and each subsequent exposure will result in progressively more severe reactions.

Nitrogen Dioxide

As a major by-product of combustion, this irritating air pollutant with its pungent, acrid odor can be found in facilities where gas vehicles such as forklifts (both natural and liquid propane) are used indoors. The facility's air handling unit can effectively disperse this pollutant to other areas of the work environment. Any facility that uses gas ovens or heaters may also contain high levels of nitrogen dioxide in the breathing air. Symptoms of exposure include and increased prevalence of colds and bronchitis and a significant reduction in lung function. It has also been implicated in long-term respiratory problems and, possibly, heart disease and cancer.

Carbon Monoxide and Hydrocarbons

These chemicals are also released into the air as by-products of combustion. Carbon monoxide is an insidious pollutant (odorless and colorless) that can cause drowsiness, headache, impaired heart function, and even death in high concentrations. The hydrocarbons, such as benzopyrene (a product of incomplete combustion), cause cancer in laboratory animals and can damage the liver, respiratory system, and nerve tissues. Indoors, the main source of these pollutants are tobacco smoke, some pesticides, and any automotive or internal combustion engines used to power in-plant vehicles. People who smoke or who work next to those who do are more likely to have impaired lung function due to exposure to side-stream tobacco smoke. In fact, the hazards of passive inhalation of tobacco smoke by non-smokers has led to an increasing ban of smoking in many workplaces across the country.

Radon

This radioactive gaseous element is emitted from radium that is found naturally in soils, rocks, bricks, concrete, and even water (primarily well water) in and around many businesses and homes throughout the country. Because many business structures are well sealed for energy conservation, the radon can accumulate to levels much higher than the natural levels found outdoors. Although specific studies are not conclusive, it is probable that lung cancer can result from prolonged exposure to high levels of radon gas.

EVALUATION METHODS AND TECHNIQUES

The industrial hygiene methods and techniques commonly used to evaluate the nature and degree of air contaminants include:

1. Review of work practices and procedures
2. Surveys and assessments of the actual work environment
3. Evaluation of the material safety data sheets (MSDS) for substances that may emit gasses, vapors, mists, fumes, and any other form of airborne contaminant
4. The use of air sampling instrumentation and devices

The industrial hygienist must be capable of recognizing the level of evaluation that is required for a particular air contaminant problem. For example, a general sampling of the overall work environment may be required to determine the full extent of a suspected airborne contamination problem. In such cases, the hygienist may obtain air samples using a grid pattern approach to ensure a comprehensive evaluation of the entire area. In other instances, direct sampling the workers' immediate breathing zone may be in order to properly evaluate the level of individual exposure to particular air contaminants. The hygienist must be familiar with the types of sampling instrumentation available to facilitate the evaluation process. Of course, the substances being sampled for will determine the types of equipment to be used. The following is a brief and generic discussion of the common sampling devices that can be used by the industrial hygienist for the evaluation of air contaminants.

Sampling Devices for Gases and Vapors

There are three basic categories of sampling devices available to the industrial hygienist for determining the level or concentration of gases and/or vapors in the work environment: short-term, long-term, and passive samplers. The first type is used to establish a short-term or "snapshot" assessment of the level of air contamination at the time the sample was taken. These grab samples are adequate for real-time determination of air quality in terms of specific concentrations of a particular gas or vapor. They measure contamination levels only at the time and location they were taken and do not provide a comprehensive analysis of the true exposure risk for the overall environment. Such samples can be obtained using a syringe, a small hand-held pump, or a squeeze bulb to draw air directly through some type of detection media. The detection media may be impregnated filter paper, some type of solid absorbent material, a gas detection tube, or a glass cylinder containing treated materials that will change color in the presence of the sampled gas or vapor (colorimetric tube). The level of color change indicates a level of concentration, usually expressed in parts per million (ppm). Whatever the type of device used, the hygienist must realize that such samples will provide quick determination of exposure risk for immediate reference but will not demonstrate long-term contamination levels that may actually exist in the workplace.

When situations warrant more exacting sampling of the work environment air, continuous or long-term samples can be obtained using a variety of instrumentation and devices. These include bubblers with reactive or absorbing solutions, adsorption tubes, or, more commonly, portable battery-operated sampling pumps. Sampling pumps draw air through a sampling media at a continuous rate for a set period of time that may be eight hours or more. The sampling media include impregnated filter papers, activated charcoal filters, and silica gel adsorption tubes. Once the sampling session is complete, the collected samples can be examined under laboratory conditions. The trapped sample material is eluted from the sample media using chemicals such as methyl alcohol, ethyl alcohol, carbon disulfide, dimethyl-sulfoxide, or even water (depending on the material sampled). Once separated, the quantity of material present can be evaluated (usually using sophisticated instrumentation such as a gas chromatograph) and, based on the length of time the sample was taken, the true level of contamination in the workplace for an average eight-hour work day can be determined (extrapolated).

A third type of sampling device, indicator badges, are worn by the employees in the work area. These badges, known as *passive samplers*, either change color on exposure to a particular substance or can be subsequently analyzed in a laboratory. However, because the air volume passing through the badge is unknown, such devices are typically used to provide a qualitative check on the work environment.

Whatever the type of sampling device used, the hygienist must always ensure that the equipment has been properly calibrated. When grab samples are taken, at least one tube from the batch should be checked for efficiency. Long-term samples sent to a laboratory for analysis should always include one blank filter from the same batch as a control procedure to ensure accuracy. General practice is to submit blanks at a ratio of 10 percent of the total number of samples, with a minimum of two blanks. Instrumentation used during the sampling process should be calibrated just before the sampling session and again just after the session is complete. This procedure will establish documented assurance of the equipment's accuracy and efficiency.

Sampling Devices for Particulates

Recalling the discussion on the respiratory system from Chapter 4, the reader will remember that the respiratory tract consists of several unique components, from the nose and trachea to the tiny alveolar sacs. The effects inhaled particles have in the respiratory system depend on their final location in the tract and the body's ultimate disposition of these particles. Obviously, the deeper the penetration into the respiratory system the higher the risk of harm and illness. Based on their ability to affect the respiratory system, standards for measuring particulate concentrations can be discussed as either *respirable mass* or *total mass concentrations* (the latter being the basis on which most hygienic standards are based). Respirable dusts are those usually 10 microns in diameter or less. Those five microns or less can reach the deep components of the respiratory system. Therefore, when sampling for airborne particulate matter, the hygienist must ensure that the sampling medium used has been designed to adequately separate the matter from the air—even matter with diameters less than five microns. Devices that can make this separation include horizontal elutriators and cyclones. In either case, the hygienist should choose a filter considering the potential size of the particulate to be sampled so the respirable mass loading can be accurately measured.

Sampling for airborne particulate matter can be more complicated than it is for gases and vapors. Aside from the variations of airborne concentrations posed by time and work processes, the hygienist must also consider the variations that are characteristic of the particulate matter itself and its existence in normal airspace. In other words, the movement of airborne particulate matter through a work environment is not entirely consistent or predictable. Its movement will follow air flow patterns through the space, changing or shifting in consonance with such changes or alterations in the air flow. Hence, even under the best of circumstances, the hygienist may find that an air sample taken at a given location will only provide a rough estimation of the average particulate concentration within a few feet in diameter around the sample.

Another consideration involving the characteristics of the particulate matter being sampled concerns the selection of the sampling method. The accuracy of a sample can be severely jeopardized if the wrong sampling method was used. For instance, if a particular material happens to be soluble in water, then it should not be sampled using a collecting medium that will dissolve it. Likewise, if particles tend to stick together in clusters, and individual definition is desired for analysis, they should not be collected in deep section on a filter.

A wide variety of sampling methods and devices for particulate matter are currently available to the hygienist. These include but are not limited to the following:

Filtration

The most common method of collecting particulate matter, filtration employs a variety of filtering media such as cellulose, glass, and plastics (or any combination of these materials). Filters are used together with a filter housing that conveniently attaches to an air flow indicator and an air pump. This sampling system can be used for high volume flows (e.g., 60 cubic feet per minute) as well as for low volume personal samplers (that may be set as low as 1 liter per minute). Under analysis, the specific weight loading of the sampled material on the filter medium can be used to determine the concentration of either total mass or respirable mass particulate matter.

Impactors

Another common method used to sample for dust and aerosol particulate, impactors take advantage of sudden changes in air flow direction and the momentum of the dust airborne particles as they impact a flattened surface. Since particles travel with the direction of air flow, introducing an impactor plate into the flow can effectively trap particulate matter for later analysis. The flat surface may be dry or coated with a special adhesive material to trap the particles. Under laboratory procedure, the sampled material on each plate is weighed or otherwise analyzed by size classification to determine specific airborne concentrations that existed at the time of the test.

Impingers

Somewhat like impactors, the impinger takes advantage of air flow velocities and dust movement. However, impingers utilize a somewhat more complex method to obtain samples. Air is sampled through the impinger apparatus jet at high velocity and impinges (or strikes) onto a plate that is immersed in a liquid medium. When the particulate matter is impinged on the plate it looses its velocity and is wetted by the liquid at the same time to be trapped on the plate for later analysis.

Centrifugal Separators

When sampling for large particles, centrifugal separators (or "cyclones") are most effective. By using inertial force to separate particles according to size and mass, the centrifugal separator can collect dusts for later analysis. However, the momentum of small particles (less than one micron in diameter) is not substantial. Therefore, they are not efficiently collected using this method. In industrial hygiene sampling, the greatest utility of this technique is as a separator for non-respirable dusts.

Electrostatic Precipitators

Used most in the control of aerosols from large power plants, the electrostatic precipitator (ESP) technique has also been applied to air sampling as well. ESPs are extremely efficient for particulate matter and are available in a variety of electrode configurations. Essentially, they utilize an electrode maintained at either positive or negative potential (between 11,000 and 17,000 volts) and a second collecting electrode. This system produces a very intense electric field causing a corona discharge between the electrodes. Dust particles subjected to the corona become highly charged and are subsequently drawn to the collecting electrode, where they are deposited for analysis. Obviously, such a technique would have serious disadvantages in explosive or combustible environments. The discharge will also produce ozone that is irritating to the respiratory system. Hence, extreme caution is required whenever the ESP is used for sampling.

Elutriators

Although similar in concept to the centrifugal separator, the elutriator operates under normal gravitation conditions. Dust particles moving in air flow across a surface will continue to travel until normal gravity forces them to settle. Heavier particles will obviously settle quicker than lighter (smaller) ones. Elutriators collect the settled particles for analysis. Since heavier particles will be more abundant using this collection method, the elutriator can serve as a selective dust sampling method for separating non-respirable dusts.

Piezoelectric Sensors

This method utilizes an external vacuum to literally pull aerosols into a sample chamber. The particulate matter present in the air stream are then deposited onto a piezoelectric sensor by an electrostatic system. A change in frequency of oscillation of the sensor is used to measure the weight of the collected particles while a second sensor compensates for the effects of temperature, pressure, and other ambient air conditions that may affect the accuracy of the sample. As the frequencies from the two sensors are "mixed," the measuring sensor frequency is subtracted from the reference sensor frequency. The resulting rate of change in the output frequency is directly proportional to the mass concentration of the collected particulate matter.

There are, of course, many other air sampling methods, techniques, and procedures available to the industrial hygienist that are beyond the scope of this introductory text. Those briefly discussed here are commonly used, proven methods that should be considered whenever airborne contaminants have been recognized as a potential threat to the health of the work force.

RESPIRATORY PROTECTION PROGRAM REQUIREMENTS

Perhaps because the respiratory system is so vulnerable to attack from the wide variety of airborne contaminants and/or asphyxiants, the Occupational Safety and Health Administration (OSHA) has issued requirements aimed at respiratory protection. These requirements, located under Title 29 of the U.S. Code of Federal Regulations (CFR), Part 1910 (general industry), Subpart I (personal protective equipment), at 1910.134 (respiratory protection), focus on the use, care, and maintenance of respiratory protective devices. They also include general statements concerning respiratory protection. In terms of controlling exposures to those hazards that threaten the health of the respiratory system, OSHA has been quite specific under this Standard with regard to employer actions and responsibilities.

Actions to control exposures to occupational diseases that can be caused by breathing air contaminants should begin with attempts to eliminate or prevent such contamination. If elimination is not operationally feasible, engineering and administrative actions to control the exposure potential should be taken. According to OSHA, such actions can include enclosure of the operation (confinement or isolation), the use of general and local ventilation, or substitution of less toxic materials. Because OSHA places such importance on the implementation of engineering controls, it has issued a separate standard concerning ventilation requirements (29 CFR 1910.94).

Industrial Ventilation

Ventilation is one of the most widely used methods of controlling environmental contaminants. Two types of ventilation are used in the workplace: general (or dilution) and local exhaust. The choice of one over the other is dependent more on situation and circumstance than on any specific engineering principle or formula. The hygienist should be actively involved with the engineering community in the facility when the subject of industrial ventilation is discussed. While engineers tend to concentrate on the practical application of the methods and techniques necessary to ensure the maximum efficiency of any given system, process, or task, the hygienist must consider the industrial health aspect as well. Working together should prove mutually beneficial for all interests.

General ventilation is the simple act of exchanging external or outside air for inside air to accommodate either occupant comfort or contamination control. Comfort control requires the cooling (in summer) or the heating (in winter) of this air, as required. When used for the purpose of contamination control, general ventilation has certain limitations. For instance, if the quantity of contaminant is relatively great in comparison to the intake capacity of the ventilation system, it will not be possible to dilute the air adequately. In addition, when contaminants are within workers' breathing zone, general ventilation of the work space may not do anything to prevent exposures. Therefore, this method of industrial ventilation is best suited for those locations where the contaminants have low toxicity and their generation is consistent throughout the work day in known locations. Dilution ventilation can also be successfully used to reduce the concentrations of combustible vapors within an enclosed area to concentrations below the lower explosive limit (LEL). General ventilation has been somewhat successful in controlling exposures to organic vapors. However, when large volumes of air are required to safely dilute contaminants such as dusts or fumes, a more effective technique is to literally capture the contaminant near its source using local exhaust ventilation.

Local exhaust ventilation provides nearly complete capture of the contaminant, regardless of its rate of generation, toxicity, or physical state. This procedure requires relatively low air volumes as compared to general ventilation. The major disadvantages of local exhaust ventilation are its cost and its usually complex and site-specific design, which render them difficult to move from one location to another). The exhaust hood is usually the most visible element of a local exhaust ventilation system. The specific design of the hood will determine its air flow capacity. Even with minimum air velocity, contaminants are captured and drawn into the hood and the system's connecting eductors. The nature of the contaminant itself will determine the rate of capture, based on their individual characteristics of inertia and size (mass). Gases, vapors, and some fine dusts (below one micron), for example, require relatively low capture velocities compared to heavier particulate matter. On the other hand, large dust particles, especially those released at a high velocity (such as during grinding or brazing operations), would be difficult to capture unless directed toward the hood.

Only when it is determined that effective engineering controls are not feasible, or while they are being implemented, should appropriate respiratory protective equipment be provided and used. The intent of this order of precedence for exposure reduction is to ensure, so far as possible, that all appropriate consideration has be given to hazard risk reduction by every possible means before personnel are required to wear protective devices. In other words, if the hazard of exposure is eliminated to begin with, then respiratory protection will no longer be of concern. Unfortunately, total elimination is not always practical, feasible, or possible in the real world of industrial operations. Realizing this, OSHA first promulgated the Respiratory Protection Standard (29 CFR 1910.134) in 1971. It establishes the minimum acceptable requirements for a respiratory protection program and governs the care, use, and maintenance of respiratory protective devices (e.g., respirators, gas masks, etc.). The following provides a brief summary and analysis of these requirements and the issue of respiratory protection as a method of exposure control.

Respiratory Protection Regulations: An Overview

The approval or certification of respirators is based on regulations developed for such devices by the Bureau of Mines, U.S. Department of the Interior, during World War I. From 1919 to 1973 the Bureau performed tests and issued all approvals on respirators for coal mine use from its laboratories in Pittsburgh, Pennsylvania. On March 22, 1972, as per the Federal Coal Mine Health and Safety Act of 1969, NIOSH and the Bureau issued joint regulations for approval of respirators in the Federal Register (37 CFR 6244). Subsequently, on March 15, 1973, these regulations were amended (38 CFR 11458), and the Bureau officially transferred the testing of these devices to the NIOSH Testing and Certification Laboratory at Morgantown, West Virginia. Regulations for respirator approval are now incorporated in the Federal Regulations as Title 30 CFR Part II. All approvals under these regulations are issued jointly by the Bureau and the Institute.

Under the OSHA Respiratory Protection Standard, the employer must implement a program designed to ensure the control of personnel exposure to the hazards posed by airborne contaminants. As a minimum, the respiratory protection program must contain the following basis elements:

1. There must be written standard operating procedures governing the selection and use of respiratory protection.
2. Respirator selection must be based on the hazards to which a worker is exposed (i.e., the right respirator for the job).
3. Users of respiratory protection must be appropriately trained on the proper use, care, and maintenance of the devices as well as the known limitations of respiratory protection.
4. Procedures covering the storage, cleaning, disinfecting, and inspection of respiratory protective devices must also be established.
5. There must be appropriate surveillance of work area conditions to assess employee use of devices and to ensure that area conditions have not changed.
6. There must also be regular inspection and evaluation to determine the continued effectiveness of the program.
7. The employer must ensure that only approved or accepted respirators are selected for use in the workplace (selection guidelines are provided by the American National Standards Institute (ANSI) in Standard Z88.2).
8. The employer must also assess the medical status of any person assigned to duties where the use of respiratory protection will be required. No employee should be assigned to such task until it has been determined that the employee is physically able to perform the required work while using the respiratory protective equipment. A physician must determine what health and physical conditions are pertinent. There should also be a periodic (e.g., annual) assessment of the respirator users' medical status.

Types Of Respiratory Protective Devices

Although it may appear that an enormous variety of respiratory protective devices are available for occupational use, they can essentially be divided into two basic categories:

Atmosphere Supplying Respirators

This type of equipment will protect the wearer from harmful surrounding atmospheres that pose a threat to life and health by supplying respirable air (air mix or oxygen) from a source (liquid or compressed) that is independent of the working environment of the wearer. The two common types are the self-contained breathing apparatus (SCBA) and the supplied air apparatus.

The SCBA makes the wearer independent of any other source of air by making use of liquid or compressed air, oxygen cylinders, or oxygen generating canisters that are carried by the user. The SCBA is normally for use during emergency and rescue situations. It may be used for up to three hour periods and is often classified on the basis of this use period. Using a SCBA requires special and adequate training, since it may be relatively complex in operation. It will protect against gases, vapors, dusts, fumes, smokes, and mists in any concentration or combination that unprotected skin can endure. The SCBA will also protect the user from the hazards of asphyxiation, such as those that can be found in many confined spaces. It is the only type of respiratory equipment that provides such complete protection while allowing nearly complete freedom of movement and travel for considerable distances from respirable air. Its only serious limitation is that of time, which can vary considerably from its rated specification depending on individual breathing capacity, exertion and work level requirements, and other characteristics that can vary from user to user. The SCBA is suitable for use by fire fighters (air type only), in mines, aboard ships, when entering unventilated tanks, and nearly all types of emergency situations that may be encountered in modern industry.

The supplied air apparatus makes use of an external air supplying source (e.g., a compressor, a blower, cylinders, or a tank) of respirable air supplied directly to the user through an air supply hose connected to the apparatus worn by the user. This type of equipment provides a great level of protection against the same variety of hazards as the SCBA. However, it requires the use of filtering devices (especially when a compressor provides the breathing air) and is limited by the length of the supply hose.

Air-Purifying Respirators

These respirators are non-emergency use devices designed to afford protection by filtering some or most of the contaminant from the inhaled air. Since they rely on the presence of sufficient breathing air levels in the work area atmosphere, albeit slightly contaminated, the air purifying respirator will not protect the user in an oxygen deficient or highly contaminated atmosphere (i.e., those that are considered immediately dangerous to life and health). In such work environments, use of the air purifying respirator is largely inappropriate, creates a false sense of protection, and can result in injury, illness, and even the death of the user.

The common types of air purifying respirators are the particulate removing respirator (or dust mask), the gas/vapor (or chemical filter) respirator, and the combination gas/vapor and particulate removing respirator.

The particulate removing respirators provide minimum protection against particulate matter contaminants such as dusts and some forms of mists, fogs, smokes and sprays. They usually consist of a formed filter with a half-masked facepiece attached so that inspired air passes through the filter media. These respirators, usually designed to be disposable, simply remove contaminants from inhaled air by physical trapping and electrostatic attraction. Their period of usefulness is determined by the time required to saturate the filter to a point where breathing becomes difficult. Obviously, this minimum protection is not adequate for atmospheres containing high levels of airborne contaminants or for use in oxygen deficient areas.

The gas/vapor (chemical filter) removing respirator usually consists of a small cartridge-like filter attached directly to and carried by a half-mask facepiece. They are actually low capacity gas masks. Designed for non-emergency use, these respirators are intended for use in atmospheres that may be breathed without protection but which will produce discomfort or chronic poisoning after a prolonged or repeated exposure of several hours or days. Chemical cartridges are usually color coded for classification and easy identification. The colors correspond to a specific chemical or group of chemicals or materials for which the respirator is designed to filter. In addition, they must carry NIOSH approval along with any designated use limitations. Table 5-3 shows the color coding for chemical cartridges to be used with gas/vapor removing respirators.

TABLE 5-3 The color coding for chemical cartridges to be used with the gas/vapor removing respirators (Source: OSHA 29 CFR 1910.134)

ATMOSPHERIC CONTAMINANT	OSHA (ANSI) COLOR CODE
Acid gases and organic vapors	Yellow
Organic vapors	Black
Acid gases	White
Hydocyanic acid gas	White with 1/2 inch green stripe completely around cartridge base
Chlorine gas	White with 1/2 inch yellow stripe completely around cartridge base
Ammonia gas	Green
Carbon monoxide	Blue
Hydrocyanic acid gas and chloropicrin vapor	Yellow with 1/2 inch blue stripe completely around cartridge base
Acid gas and ammonia gas	Green with 1/2 inch white stripe completely around cartridge base
Acid gases, organic vapors, and ammonia gas	Brown
Radioactive materials (except tritium and noble gases)	Purple (magenta)
Particulate (dusts, fumes, mists, fogs, or smokes) in combination with any of the above gases or vapors	Cartridge color coded for contaminant as designated above, with 1/2 inch gray stripe completely around cartridge near its top
All of the above atmospheric contaminants	Red with 1/2 inch gray stripe completely around cartridge near its top

The combination gas/vapor and particulate removing respirator is essentially and physically the same as the gas/vapor removing respirator discussed above. However, in addition to the specific chemical cartridge attached to the half-mask facepiece, a dust pre-filter is fitted on the cartridge. The dust pre-filter provides protection against dusts, some fumes, mists, and smokes in relatively small concentrations.

OSHA regulation for respiratory protection provides great detail on the specifics of each general area outlined above. It is highly suggested that the industrial hygienist, who may have specific concerns regarding respiratory protective devices, consult both the OSHA regulation located at 29 CFR 1910.134 (respiratory protection), and the ANSI Standard Z88.2-1980 (Standard Practice for Respiratory Protection).

SUMMARY

It is a generally accepted fact that airborne contaminants will exist in the average workplace. The industrial hygienist must recognize when and how they are created. By understanding the work involved, knowing the processes required, and being familiar with the materials used in the workplace, the hygienist should be able to recognize when such hazards are likely to exist. Dusts, fumes, aerosols, gases, and vapors are all types of air contaminants that can threaten the health of the worker. Knowledge of these will help the hygienist determine how such hazards can exist.

By assessing workplace processes, the hygienist should be able to recognize when the hazards associated with asphyxiation are likely to occur. Identification of the various work tasks and their associated materials during the industrial hygiene assessment will facilitate the overall hazard recognition process. Asphyxiation is literally a silent killer and, therefore, quick and accurate recognition of this hazard, as well as the proper understanding of the conditions that are conducive to oxygen deficient environments, are particularly critical to the prevention of injury, illness, and/or death.

Another silent danger in some work environments is the existence of insidious airborne contaminants that can degrade indoor air quality to levels that are unhealthy or even dangerous for people who must work indoors. The hygienist must be sensitive to the potential for such hazards and be able to properly assess indoor work environments for adverse conditions that might affect human health.

A wide variety of sampling methods and techniques are currently available to ensure the proper evaluation of recognized hazards. Once the nature of a particular air contaminant problem is understood through evaluation and assessment of samples, the hygienist can recommend appropriate actions to control exposure potential in the workplace. The general control methods discussed in Chapter 3 under General Principles of Industrial Hygiene (also reference Figure 3-2) can be employed here to ensure that the basic elements of hazard exposure control have been properly considered. In addition, the hygienist should be familiar with the general principles of ventilation as a control measure, as well as the basic requirements of respiratory protection under OSHA law.

6

Toxic Chemicals and Corrosive Agents

INTRODUCTION

The use of chemical substances and compounds in virtually every line of work is one of the most prolific hazards found in industry today. In fact, humans come into contact with a wide variety of chemicals and/or their compounds every day, both at home and at work. In modern society, the use of chemicals has become necessary for nearly every aspect of human existence. Vehicles would not run without chemical fuels, dry cleaning would not be possible without chemicals, and food preservation and processing requires the use of hundreds of different chemical compounds. In the work environment, an even greater reliance on the use of chemicals can be seen, regardless of the nature of the work itself. Indeed, many of the manufactured products that most people simply take for granted would not be available without the use of chemicals.

In terms of hazard exposure, many chemicals can be harmful to human health in a variety of ways. Understanding their existence, their uses, and their capacity to cause harm is essential to the industrial hygiene process. While chemicals may be an essential element of modern human existence, exposure and subsequent illness are not necessary by-products of their use. The industrial hygienist must play a key role in ensuring that such exposures are kept to an absolute minimum.

RECOGNIZING TOXIC CHEMICALS

The hazardous properties of chemical substances can be divided in two basic categories: *toxicity* (or health hazard) and *flammability and explosivity.* This discussion will focus on the first category. The term *toxic* is used readily in society today to describe virtually anything that will cause harm or damage to human health. The news media in particular seem to love this word. Its very sound conjures up images of painful suffering and death resulting from exposure to some kind of poison. In terms of accuracy, it is safe to say that the word *toxic* is an overused generalization of a cause-and-effect relationship between chemical exposure and results that are not clearly understood by those who rely on this term to describe a specific hazard to human health. Of over 9 million compounds that have been identified or synthesized, toxicity data have been documented for approximately 100,000 substances—only a small fraction (Patnaik, 1992). Most of the available information is based on animal studies. Human toxicity data are known for a relatively small number of substances. Hence, the hygienist is cautioned that the use of the term *toxic* to describe the hazardous nature of a chemical may not be totally accurate in the majority of the cases encountered in today's work environment.

With regard to chemical exposures, *toxic* refers to a harmful effect on some biological structure or its function and the condition under which this effect occurs. A toxic substance can be a liquid (hydrochloric acid, for example), a solid (lead), or a gas (ammonia). The relative *toxicity* of a chemical is a measure of the degree of harmful effect that arises subsequent to the exposure. Toxicity is also contingent on conditions of exposure (such as dose or concentration), route of entry, duration of exposure, the nature of the substance, and the individual attributes (size, weight, and physical condition) of the person exposed. Toxicity can be further expanded to include *corrosivity* (or irritant action); *mutagenicity* and *carcinogenic* potential; and *teratogenicity.* These characteristics will be the subjects of separate discussions later in this chapter.

Determining Exposure Levels

Human exposure to health hazards such as toxic chemicals are particularly difficult to evaluate, especially under conditions of low-level exposures. Toxic effects can be *localized* or *systemic.* Localized toxicity occurs when a chemical agent affects only the body part or area that has been exposed. This exposure may be through direct contact, inhalation, ingestion, or possibly through injection. Examples include contact dermatitis (direct contact), bronchitis (inhalation), esophagitis (ingestion), or an atrophic reaction to a localized area of the body resulting from an injection of a toxic material. Systemic toxicity occurs when a substance enters the bloodstream and has a harmful effect on tissues or functions throughout the body. Toxic substances can enter the bloodstream through the skin, lungs, or stomach.

To recognize when exposures are considered toxic, the hygienist must first realize that *all* chemicals can be toxic. While a small amount of most chemicals is not generally harmful to people, larger *doses* or *concentrations* can prove fatal. As discussed in Chapter 4 under Fundamental Principles of Cause and Effect, chemicals can be toxic by ingestion, absorption (which are both measured as *dose*), or inhalation (which is measured as *concentration*). In some specific instances of exposure, chemicals are also toxic by injection. In an effort to standardize the scientific understanding of chemical toxicity, researchers have developed methods to establish specific levels of exposure based on the average population of a given biological test species. For inhalation exposures (the most common and most direct route), the measurement attempts to identify the *concentration* of any given toxic chemical that, if exceeded, may have severe detrimental effects up to and including death of the exposed organism. These measurements are expressed as a *lethal concentration* (LC) of the toxic chemical. Typically, the measurement is provided as LC_{50}. This means that the amount stated is the concentration of a toxic chemical in the air (or other suitable vehicle) that, when administered to test animals (rats, mice, rabbits, etc.) for a defined time period, will kill 50 percent of the exposed population. The LC_{50} is expressed in parts per million (ppm) for that time period (Baker, 1985). When absorption or ingestion are the routes of exposure, the measurement is referred to as the *lethal dose* (LD) and the value expressed as LD_{50} in milligrams per kilogram (mg/kg) of body weight (or some similar mass ratio). As an example, Figure 6-1 shows a graphic representation of the LC_{50} and LD_{50} for the chemical 1,4 dioxane. This example shows an LC_{50} for an inhalation exposure to rats at 13,000 ppm for a two-hour period and an LD_{50} oral (ingestion) exposure to mice of 5,700 mg/kg. It is important to note that any value represented by an LC_{50} or LD_{50} does not mean that human exposures will be equally tolerant. Also, these values do not imply that the remaining 50 percent of the test population stayed perfectly healthy. In most cases, nearly all the test animals become ill. Many eventually die long after the conclusion of the test. For the industrial hygienist, the LC_{50} and LD_{50} methods should be considered only as measurements of the relative toxicity of a given chemical substance. They are not, by any means, conclusive or accurate in terms of human exposures. These measurements are merely a means of comparison, a basis of understanding, and a method of appreciating chemical toxicity. For instance, in the case of 1,4 dioxane, the toxicity is relatively low in test animals by all routes of exposure, as is evidenced in the above example. However, in humans the toxicity of this compound is known to be quite severe. The target organs are the kidneys, liver, lungs, skin, and eyes. Exposure to its vapors through inhalation (as well as absorption through the skin or ingestion) can cause poisoning. Symptoms to recognize exposures include drowsiness, headache, respiratory distress, nausea, and vomiting (Patnaik, 1992). Because of its toxic effects on humans, the Federal OSHA has established a permissible exposure limit (PEL) of 100 ppm. The ACGIH (1992), however, has established an even lower Time Weighted Average (TWA) Threshold Limit Value (TLV) of 25 ppm. The hygienist should, therefore, become familiar with all exposure data that may exist regarding any toxic chemical used in the facility. Knowing about

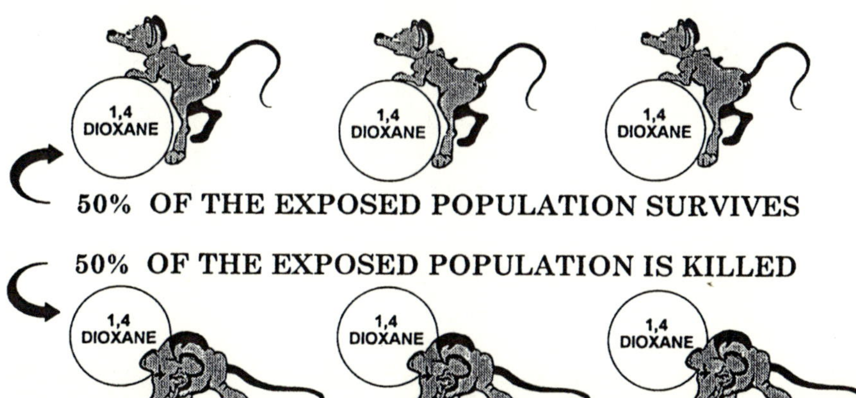

FIGURE 6-1 A graphic representation of the LC$_{50}$ and LD$_{50}$ for the chemical 1,4 dioxane. This concept of *lethal concentration* and *lethal dose* is used as a means of establishing or understanding the risks of human exposure.

the toxic nature of the chemicals in use at a facility is the beginning of the hazard recognition process. Chapter 3 introduced the general methods of evaluation and control available to the industrial hygienist (reference Chapter 3, General Principles of Industrial Hygiene, and Figure 3-2). These principles can be successfully employed in those work situations where exposure to toxic chemicals has been recognized as either a real or potential health hazard concern.

Carcinogens, Teratogens, and Mutagens

It has been established that the use of many toxic chemical substances found in today's workplace can result in harmful health effects if exposure is not properly controlled. Perhaps the most bizarre of any known health effects resulting from chemical exposures are those classified as *carcinogenic, teratogenic,* and *mutagenic*. The industrial hygienist should be concerned with the accurate identification of such substances so that proper control measures can be implemented. Exposure to the toxins described below is an extremely serious matter and uncontrolled occupational exposures must therefore be avoided (Lisella, 1994).

Carcinogens

Any substance known to cause cancer in humans and animals is a *carcinogen*. Examples are vinyl chloride and aromatic hydrocarbons such as benzene. It is generally accepted that the term "carcinogen" refers to all chemicals in which there is some evidence of carcinogenicity in experimental animals, even if the test data are inconclusive. Carcinogens represent a broad range of organic and inorganic chemicals, hormones, immunosuppressants, and solid-state materials. A substance is considered carcinogenic if properly designed studies indicate carcinogenesis in:

1. humans
2. at least two different animal species
3. one animal species in at least two separate studies
4. one animal species in addition to multiple in-vitro tests

Two stages of carcinogenesis have been identified. The initiation stage occurs after exposure of cells or tissue to a dose of a carcinogenic substance. The promotion stage which is the process whereby the altered cells proliferate.

Mutagens

A chemical substances that has the ability to produce a change (mutation) in the genetic composition of the DNA in a cell is called a mutagen. The change is capable of being passed on to succeeding generations. Such changes can be brought about by exposure to ionizing radiation (see Chapter 7) and to chemicals such as nitrogen and sulfur mustards, epoxides, and methylsulfonate.

Teratogens

Any substance that is capable of producing physical defects in an unborn fetus is a *teratogen*. These changes can result in fetal death, a high rate of embryonic mortality, or in the birth of physically defective offspring. While the scientific community has identified numerous problems associated with occupational exposure to teratogens, it is important to mention that there are currently no standards that attempt to regulate or even address exposure requirements related to the effects on an unborn fetus. It is true that existing PELs limit worker exposure to these substances, but there is very little information on their effects following exposures that are below the PEL. As a health professional, the hygienist must be concerned with all potential health effects of a workplace exposure (i.e., those to the worker and those that may affect a fetus). In most cases, lack of substantive scientific data may require a strict conservative approach to ensure optimum worker protection.

An industrial hygiene assessment that fails to recognize the existence of these hazards can place employees' health in serious jeopardy.

THE HAZARDS OF CORROSIVE AGENTS

In the discussion of toxic chemicals, it was established that some chemicals are toxic by inhalation, ingestion, and/or absorption through the skin. The toxic affects of these chemicals can be recognized by the adverse reaction in one or more of the human systems (respiratory, digestive, circulatory, etc.). Just as toxic chemicals can enter the body to harm these systems, so can corrosive agents (which include chemicals).

Basically, a corrosive agent is anything that reacts with any surface and causes that surface to deteriorate. In terms of human exposure, contact with corrosive agents can result in the local destruction of tissue, damage to cell structures, interference with biological functions (breathing, blood flow), and even the partial or total loss of the areas subjected to the exposure. The substance reacts chemically with the tissues at the point of contact. Corrosive agents can be liquid, solid, or gas. Corrosive agents, by their very nature, are also toxic. Inhalation of corrosive gaseous substances can be particularly harmful to the airways of the respiratory system. Chlorine, for example, is a pungent suffocating gas, highly irritating to the mucous membranes, that can cause pulmonary edema, lung injury, and even respiratory arrest. Sodium hydroxide (caustic soda), a solid material, is corrosive on contact with skin, eyes, and the respiratory airways. The hygienist must be especially cognizant of substances in the company's inventory that are corrosive on human contact. Through an industrial hygiene assessment, the hygienist should identify all corrosive agents, what they are used for, how they are used/handled, and what precautionary measures are in place. Frequent review of process specifications and procedures, purchase orders, and the company' MSDS library will facilitate the hazard recognition process. Recognizing the hazardous characteristics of a particular substance used in the workplace is essential to the success of the industrial hygiene effort.

HAZARDOUS PROPERTIES OF METALS

The principal hazard associated with various metals and metal compounds concerns the generation of airborne metallic fumes during some type of heating process (such as welding). The resulting health effect is referred to as metal fume fever. However, other metal ions can be extremely toxic to humans without having to first be heated and then condensed into an airborne fume contaminant. Lead dusts and lead fumes can present severe hazards to those who have been exposed (through inhalation or ingestion). Lithium metal powder, on exposure to the moisture in the skin produces a highly corrosive hydroxide. Contact with the skin or eyes will therefore result in a burn of relatively severe intensity. While many metals are essential nutrients found in the food chain, some serve no beneficial purpose once they enter the body. In fact, metals such as lead and mercury are highly toxic even at low levels.

To aid in the recognition metal health hazards in the workplace, Table 6-1 has been developed. It provides a sample listing of the more common metals found in industry today along with some of their uses. Table 6-2 describes the major health effects of exposure to these various metals. To recognize when such hazards exist in the workplace, the industrial hygienist should become familiar with the various industrial uses of the many different metals listed on Tables 6-1 and 6-2.

Metal Fume Fever

Metal fume fever is generally an acute (rapid onset) illness resulting from a brief exposure to the superheated fumes of metal oxides such as copper, zinc, magnesium, nickel, mercury, and numerous others. Health hazard recognition begins with an assessment of work processes to identify any tasks or functions involving high-heat operations and metals. Work areas where welding, smelting, or furnace work is conducted are prime candidates for health hazard assessment. For example, in welding, several metals and coatings (from the surfaces being joined or repaired to the welding rod or electrode being used) are heated to high temperatures. The various metals often vaporize, react with the oxygen in the air, and form a fume consisting of fine particles of metal oxides. The fume has a blue-gray appearance and often hangs in the air and drifts into patterns, similar to the conduct of cigarette smoke. Smelting involves heating ores, usually to separate and refine the metals present in the ore (gold, silver, etc.). A variety of toxic fumes and gases may be produced during this process as well.

It is not always possible to determine the degree of hazard posed by the individual constituents of fumes, since several hazards are almost invariably generated simultaneously. Welding and smelting fumes typically contain the oxides of several different metallic elements. At the same time, a mixture of gases such as carbon dioxide, ozone, and nitrogen oxides, are frequently formed. In copper smelting, for example, metal fumes, sulfur dioxide, and arsenic trioxide may all be generated. None are beneficial to the worker. The hygienist must be cognizant of such tasks performed in the workplace to properly recognize the potential health hazards associated with heated metals. Techniques and methods for evaluation will be discussed in subsequent chapters of this part.

Other Metal Hazards

As stated in the previous paragraph, metals do not have to be heated to present a health hazard to the worker. Inhalation or ingestion of metal dusts or metal-con-

TABLE 6-1   A sample listing of the more common metals found in industry today, with some of their uses (Patnaik, 1992)

NAME	COMMON INDUSTRIAL USES
Aluminum	As a fine powder used in manufacture of explosives and fireworks and used in photography; also found in aluminum plants and used as an alloy with other metals. Wide applications throughout industry.
Antimony	Used to make alloys such as Babbit metal, white metal, and hard lead; in bullets and fireworks; and for coating metals.
Arsenic	Used to harden metals such as copper and lead and as a doping agent in solid-state products of silicon and germanium. Its salts are used in making herbicides and rodenticides, and in semiconductors and pyrotechnics.
Barium	Used in electronic tubes and as a carrier for radium. Its salts are found in paints, ceramics, and lubricating oils. Employed in analytical work, has many other uses.
Beryllium	Used as an alloy with copper, as a neutron moderator in nuclear reactors, and in radio tubes.
Cadmium	Used in electroplating, in nickel-cadmium storage batteries, as a coating for other metals, in bearing and low-melting alloys, as control rods in nuclear reactors, in dyeing and printing textiles, in TV phosphors, in pigments and enamels, and in semiconductors and solar cells.
Calcium	Used as a deoxidizer for copper, steel, and beryllium in metallurgy, to harden lead, and in the making of certain alloys.
Chromium	Used to manufacture its alloys, such as chrome-steel and chrome-nickel-steel; in chrome plating of other metals; in painting; and for tanning leather.
Copper	Used to make utensils, electrical conductors, and alloys such as bronze and brass.
Lead	A major component in many alloys such as bronze and solder; used in tank linings, piping, and building construction; in the manufacture of paint pigments, tetraethyllead, and many organic and inorganic compounds; in storage batteries; and in ceramics.
Lithium	Used to make certain alloys, in the manufacture of lithium salts, and in the production of vacuum tubes.
Potassium	Used in the manufacture of reactive potassium salts, in organic synthesis, and as a heat exchange fluid when allied with sodium.
Magnesium	Used in the manufacture of alloys, optical mirrors, and precision instruments; in pyrotechnics; as a deoxidizing and desulfurizing agent in metallurgy; in signal lights, flash bulbs, and dry batteries.
Mercury	Used in mercury arc and fluorescent lamps; in thermometers, barometers, and hydrometers; to extract gold and silver from ores; in some switching mechanisms; and as amalgams with many metals in dentistry and other industries.
Nickel	Used in various alloys such as German silver, Monel, and nickel-chrome; for coins; in storage batteries; in spark plugs; and as a hydrogenation catalyst.
Selenium	Used to manufacture colored gases; in photocells and semiconductors; as a rectifier in radio and TV sets; and as a vulcanizing agent in the manufacture of rubber.
Silver	Used in jewelry and ornaments; in photography and x-ray films; electroplating; dental alloys; high-capacity batteries; printed circuits; coins; and mirrors.
Sodium	Used in the manufacture of many highly reactive sodium compounds, in the production of synthetic rubber, and in making sodium lamp and photovoltaic cells.
Thallium	Found in photovoltaic cells; used in semiconductor research, in low-range gas thermometers, and alloyed with many metals. Its salts are used as rodent poisons.
Titanium	Added to steel and aluminum to enhance their tensile strength and acid resistance; alloyed with copper and iron in titanium bronze.
Zinc	Used as an alloy in brass, bronze, and German silver; as a protective coating to prevent corrosion of other metals; for galvanizing sheet iron; in gold extraction; in making utensils; in dry cell batteries; and as a reducing agent in organic synthesis.

TABLE 6-2   The major health effects of exposure to the metals of Table 6–1 (Patnaik, 1992)
(continues)

NAME	HEALTH HAZARD RISK AND EXPOSURE SYMPTOMS
Aluminum	Repeated inhalation of high concentrations may result in aluminosis or lung fibrosis. Accumulations of aluminum particles in the kidney may occur, especially among dialysis patients. Exposure symptoms include impaired memory, dementia, ataxia, and convulsions. Accumulation in the brain has been suggested as a factor in Alzheimer's disease.
Antimony	Low-order toxicity. Symptoms of acute poisoning include weight loss, hair loss, and congestion of the heart, liver, and kidney. Routes of entry are primarily inhalation of its dusts or fumes, and skin absorption. The metal is more toxic than its salts.
Arsenic	A severe human poison (yellow arsenic). All arsenic compounds are toxic. Routes of entry are ingestion and inhalation, which result in absorption into the body. Symptoms include fever, gastrointestinal disturbances, irritation of the respiratory tract, ulceration of the nasal septum, and dermatitis. Chronic exposure can produce pigmentation of the skin, peripheral neuropathy, and degeneration of the liver and kidneys. It is carcinogenic in humans. Ingestion may cause tumors in the liver, blood, and lungs.
Barium	Inhalation of dusts can cause irritation of the nose and upper respiratory tract. All soluble salts of barium are acute poisons. Barium ion is toxic to muscle. Ingestion can cause severe hypokalemia.
Beryllium	Highly toxic. Inhalation of its dusts can cause irritation of the eyes, skin, and respiratory system. Other symptoms are weakness, fatigue, and weight loss. The main acute effect from single exposures is pneumonitis, while chronic exposures can cause pulmonary disease. A carcinogenic that is thought to cause lung cancer.
Cadmium	Toxic by inhalation or ingestion. Acute symptoms are nausea, vomiting, diarrhea, headache, abdominal pain, muscular ache, salivation, and shock. Inhalation of its fumes or dusts can cause cough, tightness of chest, respiratory distress, congestion of lungs, and bronchopneumonia. It is a poison that is accumulated in the liver and kidneys. It is also thought to cause certain diseases in the bone. Absorption through the digestive system is low.
Calcium	Contact of dusts with skin or eyes can cause burns. Reacts with skin moisture to form caustic hydroxide, which may cause severe burn. Fumes are strong irritants to the eyes, skin, and mucous membranes.
Chromium	Hexavalent chromium salts are prime health hazards. Symptoms are skin ulcerations, dermatitis, perforation of the nasal septa, and kidney damage. Water soluble hexavalent salts (sodium, potassium, etc.) are absorbed into the bloodstream through inhalation. Exposure to some chromite dusts may cause lung disease, including pneumoconious and pulmonary fibrosis.
Copper	Inhalation of dusts, fumes, and mists can cause irritation of the eyes, mucous membranes, and nasal septa perforation. Other symptoms are cough, dry throat, muscle ache, chills, and metal fever. Skin contact can result in dermatitis.
Lead	Toxic routes of exposure are food, water, and air. Acute symptoms include ataxia, repeated vomiting, headache, stupor, hallucinations, tremors, convulsions, and coma. Chronic exposure can cause weight loss, central nervous system effects, anemia, and kidney damage. Chronic lead poisoning affects the central and peripheral nervous systems, causing restlessness, irritability, and memory loss. Dust inhalation can cause gastritis and liver changes. It is accumulated in the bones and teeth where it is stored and released. Also inhibits enzymatic activity.
Lithium	Can react with moisture in skin to produce corrosive hydroxide that will cause skin and eye burns. Fumes are irritating to the skin, eyes, and mucous. Ingestion can cause kidney injury.

TABLE 6-2   (continued)

NAME	*HEALTH HAZARD RISK AND EXPOSURE SYMPTOMS*
Potassium	Reacts with skin moisture to cause severe burns. Contact with the solid metal can produce skin and eye burns. Fumes are highly irritating to the skin, eyes, and mucous.
Magnesium	Dust inhalation is irritating to the eyes and mucous membranes. May react with water in bronchial passage to form caustic magnesium hydroxide, which can damage the lungs. The fumes can cause metal fever.
Mercury	Highly toxic in most forms. A severe inhalation hazard that can cause bronchitis, pheumonitis, coughing, chest pain, respiratory distress, salivation, and diarrhea. Central nervous system effects include tremor, insomnia, depression, and irritability. Inhalation of vapors can damage the liver, kidneys, lungs, and brain. Some compounds highly toxic by ingestion can result in inflammation of the mouth, loose teeth, salivation, muscle tremors, jerky gait, depression, irritability, and nervousness.
Nickel	Ingestion can cause hyperglycemia, depression of the central nervous system, myocardial weakness, and kidney damage. Skin contact can cause dermatitis and "nickel itch" (a chronic eczema). Inhalation of the dusts may produce irritation of the nose and respiratory tract that can lead to lung and sinus tumors with a latent period of up to 25 years.
Selenium	Toxic to the respiratory tract, liver, kidneys, blood, skin, and eyes. Symptoms include headache, fever, chill, sore throat, and bronchitis. There may also be a garlic-like odor in the breath and sweat. Chronic exposure can cause loss of hair, teeth, and nails; depression; nervousness; giddiness; blurred vision; and a metallic taste.
Silver	Silver dusts are irritating to the eyes, nose, and respiratory tract. Toxicity is low by all routes of exposure. However, prolonged skin contact with fine particles or inhalation of dusts can result in argyria (a blue-gray discoloration of the skin). Argyria may occur in the conjunctiva of the eye, gum tissue, or nasal septum and may indicate silver accumulation in the body.
Sodium	Highly corrosive, causing skin burn on contact. Fumes are highly irritating to the skin, eyes, and mucous membranes.
Thallium	Symptoms are nausea, vomiting, diarrhea, polyneuritis, convulsion, and coma. Chronic toxicity can lead to liver and kidney damage, deafness, and loss of vision. Other symptoms include reddening of the skin, abdominal pain, polyneuritis, loss of hair, pain in legs, and, occasionally, cataracts.
Titanium	Inhalation of powders can cause coughing, irritation of the respiratory tract, and dypsnea. May also cause tumors in blood.
Zinc	Exposure to dusts causes irritation, cough, sweating, and dypsnea. Inhalation of fumes can cause weakness, dryness of throat, chills, aching, fever, nausea, and vomiting. Its salts can produce metal fume fever.

taining particulates can also be highly toxic to humans. Some metals, such as mercury, are readily absorbed through the skin and can enter the bloodstream very rapidly. Lead, another heavy metal that has no nutritional value, can be a particular nuisance to the industrial hygienist. It is a common metal found throughout the environment in paints, air, soil, household dusts, food, certain types of pottery, porcelain, pewter, and even in drinking water.

In fact, lead in drinking water, although rarely the sole cause of lead poisoning, can significantly increase a person's total lead exposure. When water stands in leaded pipes or plumbing systems containing lead for several hours or more, the lead may actually dissolve into the drinking water. This means that the first water drawn from the tap after a relatively long period of disuse can contain high levels of lead. In the workplace, the hygienist must be concerned about the possibility of contaminated water (or any other potential for lead poisoning in the facility). During the industrial hygiene assessment process, such insidious sources of potential health hazards associated with exposure to toxic metals should not go unrecognized.

HAZARD COMMUNICATION AND CONTROL MEASURES

By 1986, the prolific use of hazardous and toxic chemicals in the workplace had reached such dramatic levels that OSHA promulgated the Hazard Communication Standard. In an effort to ensure employee training, an awareness of the hazards with which employees must work, and the precautionary measures to be taken while using these materials, the Standard set the minimum acceptable requirements for a hazard communication program. More commonly referred to as the "right to know law," the Standard is based on the premise that workers have the fundamental right to know about all the hazards associated with their employment. These include actual and potential exposures to hazards that can occur during the normal course of their assigned duties or in the case of a foreseeable emergency.

Located at 29 CFR 1910.1200, the expressed purpose of the Standard is to ensure that the hazards of all chemicals produced or imported are evaluated and that information concerning these hazards is properly transmitted to the users (i.e., employer and employees) of these chemicals. The vehicle for ensuring this transfer of information is the employer's written Hazard Communication Program (HCP). Principal elements of the HCP include chemical container labeling requirements and other forms of warning, the use of the material safety data sheet (MSDS) to become familiar with a chemical's hazardous properties, and very specific employee training requirements.

It is important to note that the OSHA Hazard Communication Standard is concerned with information exchange only and does not, per say, attempt to regulate the use of chemicals. Basically, this Standard requires the implementation and/or consideration of the following minimum requirements:

Hazard Determination

Chemical manufacturers and/or importers must evaluate the chemicals they produce or import to determine if the chemicals are hazardous. This determination must be based on all available data, including known information and test and/or research findings, that establish the nature and degree of any health or physical hazards that may result from exposure or use of the chemical. The hazard determination requirement extends to mixtures of chemicals as well. The data resulting from this determination is then used as the basis for developing the information to be used by employers in the implementation of a hazard communication program.

Written Hazard Communication Program

The Standard requires each employer to develop, implement, and maintain in the workplace a written hazard communication program (HCP) that, at the very least, describes how the employer intends to comply with the OSHA Standard, especially with regard to labeling requirements, material safety data sheets, and employee training. The HCP must also include a complete list of all the hazardous chemicals that are known to be present in the workplace and the methods the employer will utilize to inform employees of the hazards of non-routine tasks as well as the hazards associated with chemicals contained in unlabeled pipes that may be located in the workplace.

Multi-employer Workplaces

When the employees of more than one employer are located at the same work site (construction contractors, for example), the provisions in the HCP must also specify how the required information exchange will occur between employers and employees.

Labeling and Other Warnings

The Standard requires the chemical manufacturer, importer, or distributor to properly label, tag, or otherwise mark each container of hazardous chemicals leaving the premises with at least the following information:

- The identity of the hazardous chemical(s)
- Appropriate hazard warnings
- Name and address of the chemical's manufacturer, importer, or other responsible party

Once a hazardous chemical is received into a workplace from a manufacturer, importer, or distributor, the employer of that workplace must ensure that containers are labeled, marked, or tagged with at least the following information before allowing those containers into inventories that can be accessed by employees:

- Identity of the hazardous chemical
- Appropriate health and physical hazard warnings, including information on required personal protective equipment

The Standard specifically requires labels or other forms of warning to be legible, in English, and prominently displayed on the container, or readily available in the work area during each work shift. Other forms of warning may include signs, placards, process sheets, batch tickets, or even operating procedures, so long as these alternative methods properly identify the applicable containers, are readily accessible by all employees in the area during each work shift, and convey the hazard warning information required by the Standard. Also, defacing or marking on a manufacturer's label is a violation of the Standard, unless the container is immediately marked with subsequent labeling that provides the required warning information.

Material Safety Data Sheets

Perhaps the best source of information the industrial hygienist can use regarding any chemical hazards is the MSDS. Under Federal Law, chemical manufacturers and importers must either obtain or develop an MSDS for each hazardous chemical they produce or import. Likewise, employers must have a copy of the MSDS for each hazardous chemical that they bring into the workplace. The MSDS must be in English and must contain at least the following critical information:

- The identity of the chemical, as used on its label
- The common and chemical name of the substance
- Physical and chemical characteristics (such as vapor pressure, flash point, etc.)
- Physical hazards (reactivity, fire and explosion potentials, etc.)
- Health hazards, including signs and symptoms of exposure and any medical conditions that are generally recognized as being aggravated on exposure to the chemical
- The OSHA permissible exposure limit (PEL), the ACGIH threshold limit value (TLV), and any other known exposure limits
- Known information on the carcinogenic properties of the hazardous chemical
- Safe handling procedures and precautions, including any procedures for the safe clean-up of spills and leaks
- Emergency and first-aid procedures
- The date the MSDS was prepared or revised
- The name, address, and emergency phone number of the chemical manufacturer, importer, or distributor who can provide additional information on the chemical

Figure 6-2 shows a sample MSDS. Although OSHA has not mandated any specific format, the information summarized above must be on the MSDS as a mini-

MATERIAL SAFETY DATA SHEET

IDENTITY *(As Used on Label and List)*	Blank spaces are not permitted. If any item is not applicable, or no information is available, the space must be marked to indicate that.

Section I – General Information

Manufacturer's Name	Emergency Telephone Number
Address *(Number, Street, City, State, and Zip Code)*	Information Telephone Number
	Date Prepared
	Signature of Preparer *(optional)*

Section II – Hazardous Ingredients/Identity Information

Hazardous Components (Specific Identity; Common Name(s))	OSHA PEL	ACGIH TLV	Other Limits Recommended	% *(optional)*

Section III – Physical/Chemical Characteristics

Boiling Point		Specific Gravity (H2O = 1)	
Vapor Pressure (mm Hg)		Melting Point	
Vapor Density (AIR = 1)		Evaporation Rate (Butyl Acetate = 1)	
Solubility in Water			
Appearance and Odor			

Section IV – Fire and Explosion Hazard Data

Flash Point (Method Used)		Flammable Limits	LEL	UEL
Extinguishing Media				
Special Fire Fighting Procedues				
Unusual Fire and Explosion Hazards				

FIGURE 6-2 A sample material safety data sheet (MSDS) (continues)

Section V – Reactivity Data

Stability	Stable		Conditions to Avoid
	Unstable		

Incompatibility *(Materials to Avoid)*

Hazardous Decomposition or Byproducts

Hazardous Polymerization	May Occur		Conditions to Avoid
	Will Not Occur		

Section VI – Health Hazard Data

Route(s) of Entry:	Inhalation?	Skin?	Ingestion?

Health Hazards *(Acute and Chronic)*

Carcinogenicity:	NTP?	IARC Monographs?	OSHA Regulated?

Medical Conditions
Generally Aggravated by Exposure

Emergency and First Aid Procedures

Section VII – Precautions for Safe Handling and Use

Steps to Be Taken in Case Material is Released or Spilled

Waste Disposal Method

Precautions to Be Taken in Handling and Storing

Other Precautions

Section VIII – Control Measures

Respiratory Protection *(Specify Type)*

Ventilation	Local Exhaust		Special
	Mechanical *(General)*		Other

Protective Gloves	Eye Protection

Other Protective Clothing or Equipment

Work/Hygienic Practices

Page 2

FIGURE 6-2 (continued)

mum. Unfortunately, because there is no requirement for standardization of appearance, it is rare that MSDS forms from different manufacturers look alike. This can make it somewhat difficult to locate specific information from one MSDS to the next, especially in times of emergency. It is, therefore, highly recommended that the hygienist become familiar with the information on each MSDS before the need for such knowledge becomes critical.

Employee Information and Training

The law requires employers to provide instruction training to their employees on the hazardous chemicals located in the workplace. This requirement is a key provision of the Standard since it establishes the basis of the information exchange on which the Standard has been based. This training must be provided at the time the employees are initially assigned to work with the chemical and whenever a new hazard is introduced into the work area. Specifically, employees must be informed of:

- The requirements of the Hazard Communication Standard
- Any operations in their work area where hazardous chemicals are present
- The location and availability of the written HCP, including the required list of chemicals and their MSDSs
- Methods and observations used by the employer to detect the presence of hazardous chemicals in the workplace (such as monitoring, visual appearance, odors, etc.)
- The physical and health hazards of the chemicals used in the workplace
- The measures employees can take to protect themselves from these hazards, including specific procedures the employer has implemented to protect employees from exposure (written work practices, emergency procedures, personal protective equipment use, etc.)
- The details of the employer's HCP, including an explanation of the labeling system and the MSDS, and how employees can obtain and use the appropriate hazard information

As a method of controlling exposures to hazardous chemicals, the HCP provides quite specific direction to employers. In all likelihood, the industrial hygienist will become intimately involved in the development and implementation of the employer's HCP. The industrial hygienist should, therefore, become familiar with all the requirements of this OSHA Standard. Because of their unique qualifications and background, as well as their overall responsibilities in the area of employee health, hygienists may be best suited for ensuring that the critical tasks required under the OSHA Hazard Communication Standard are properly implemented in the workplace.

THE TOXIC SUBSTANCES CONTROL ACT (TSCA)

In addition to the OSHA Hazard Communication Standard, the industrial hygienist should be aware of another law that focuses on the use of toxic chemical substances. In 1976, the U.S. Congress enacted the Toxic Substances Control Act, or TSCA (normally pronounced "tosca"), which is administered and enforced by the Environmental Protection Agency (EPA). A detailed discussion of all this Act's provisions is beyond the scope of this text. For a more comprehensive discussion, the reader is referred to a companion book in this series, the *Basic Guide to Environmental Compliance* (Van Nostrand Reinhold, 1993). However, one specific aspect of this Law is of particular concern to the practicing industrial hygienist. Section 8(c) of TSCA requires manufacturers, processors, and users (if such use can be determined as processing) of chemical substances and mixtures to keep records of any allegations of significant adverse reactions to health or the environment believed to have been caused by a specific substance or mixture. Since "health" is a specific concern under TSCA 8(c), the hygienist may likely become involved in the recording of these allegations. An 8(c) allegation can be made by any person, including an employee, union, citizen activist, plant neighbor, or another company on behalf of its employees. It is important to note that any such allegations need not be supported by scientific or medical evidence for the purpose of TSCA. It must only establish a link between adverse health effect and a chemical substance or mixture, a particular company's product or process, or an abnormal experience by persons or the environment. It is also important to realize that TSCA does not require the recording of known human health effects under Section 8(c). This means that any known or previously recognized human health effects of exposure to a substance, such as those typically listed in the Health Hazards section of the MSDS, do not have to be recorded. In other words, if it is already known that exposure to the vapors of a particular solvent causes depression of the central nervous system, such information is not new, and no TSCA 8(c) allegation is necessary. However, if a person alleges that exposure to those same vapors caused hair loss or some other previously unknown affect for which no documentation exists, an 8(c) allegation would be warranted. The purpose is to standardize a means by which new exposure data (whether suspect or genuine) can be collected. No submittal requirements are currently specified under the law. The employer must simply record the data and make it available on request to the EPA or any approved state agency. These records must be kept for a minimum of thirty years following the allegation.

SUMMARY

It is an indisputable fact that the widespread use of chemical agents and compounds throughout industry poses a significant health hazard risk. Because the use of chemicals has become so critical to modern industrial processes, the hygienist must be familiar with the risks to human health that can result from exposure to the hazardous properties of these chemical agents. Understanding chemical hazards is the first step toward recognition of such hazards in the work environment. Proper hazard evaluation requires the ability to first identify the nature of these hazards so appropriate control measures can be instituted.

This chapter provided basic information concerning the hazards associated with chemical toxicity. The methods of measuring the levels of exposure were also discussed. Specifically, *dose* (referring to ingestion or absorption exposures) and *concentration* (referring to inhalation exposures) are scientifically derived standards that can be used by the hygienist to understand the hazards posed by exposure to specific chemical substances. When evaluated along with other available information such as the ACGIH Threshold Limit Value (TLV) and/or the OSHA-established Permissible Exposure Limit (PEL) for any given chemical, the hygienist can more accurately assess the hazardous nature of the subject chemical.

This chapter also explored the hazards of corrosive chemical substances and how human contact can result in extremely serious and adverse health effects. This discussion was followed by a brief analysis of the hazards associated with exposure to metals and their compounds. The hygienist must be aware of the great potential for illness, injury, and even death that can occur as a result of the improper, uneducated, or careless use of toxic and corrosive chemical agents and their associated compounds. In an effort to standardize the approach to this issue, the Occupational Safety and Health Administration (OSHA) has promulgated the Hazard Communication Standard. Employers are responsible, under the law, to develop specific methods and techniques to ensure that their employees are trained and aware of the hazards associated with their assigned duties. More commonly known as the "right to know" law, the Standard is based on the premise that all workers must know of any chemical hazards, the risks of exposure to these hazards, and the actions their employers are taking to prevent or control exposures. It is highly probable that the industrial hygienist will become intimately involved in the development and implementation of the Hazard Communication Program (HCP) in the workplace. Hence, in terms of exposure control, the HCP can be a critical element of the overall industrial hygiene program. In fact, there is evidence that OSHA continues to consider its Hazard Communication Standard to be of extreme importance. Since it became effective for employers in May of 1986, violations of the Standard are consistently among those most frequently cited by OSHA during their investigations and inspections of work sites across the country. It is clear that OSHA does not take lightly the hazards of working with or around toxic chemicals.

Another Federal Regulation pertaining to chemical exposures that should be of concern to the industrial hygienist is the Toxic Substances Control Act (TSCA), as administered and enforced by the Environmental Protection Agency (EPA). Specifically, Section 8(c) of the Act requires the recording of any alleged health effects that resulted from exposure to a chemical substance if such effects have not been previously identified or documented. The primary intent is to create a historical record of significant adverse reactions and to provide a way to identify previously unknown chemical health hazards. The hygienist should become somewhat familiar with these requirements since, in all likelihood, the hygienist's job would become involved with the evaluation of any such allegations made under TSCA 8(c).

7

The Hazards of Exposure to Energy

INTRODUCTION

Energy is an essential element of life. This less-than-profound observation is more an expression of human existence than a statement of scientific oddity. Simply stated, all living things draw their energy from the sun. Without this energy, life as we know it would not exist. The food that is eaten to sustain life is possible only because of the sun's energy. Energy can exist as heat, light, sound/vibration, electromagnetic waves, and a variety other equally essential forms. Energy and, more specifically, human exposure to energy, are not entirely a bad things. However, when such exposures exceed the limits of human tolerance, energy can be the cause of serious and even life-threatening effects.

This chapter will explore the hazards associated with exposure to some common energy forms that may be encountered in most work environments. In terms of hazard *recognition,* occupational exposure to energy can be one of the most subtle and insidious hazards the hygienist may confront. For instance, the dangers of high *noise levels, temperature extremes, poor illumination, vibration,* and *radiation* exposures are not readily visible to the worker (or the hygienist). But such hazards nevertheless exist and must therefore be evaluated for their health-threatening potential and controlled as required. This chapter will also provide information on the methods and techniques to ensure proper hazard evaluation and will offer some suggestions on common control measures that can be used to reduce or eliminate the health hazards that are subsequent to exposures to energy. Since exposures to energy can be difficult to recognize, special concentration shall be placed on this first critical step in the industrial hygiene process.

It must be understood, however, that the information provided here is basic in presentation, scope, and analysis. Understanding the characteristics of energy in its various forms can be extremely complex, at best. Additional study beyond that which can be presented in this limited volume is highly recommended.

TEMPERATURE EXTREMES

Up to this point, the information discussed in this Part has focused primarily on common workplace hazards that are somewhat substantive. That is, the hazards to the worker presented by air contaminants, toxic chemicals, corrosive agents, and metals are recognizable because, when they exist, they have substance. Even asphyxiation can be attributed to the action of specific recognizable agents. In every case, the action of these specific agents have a specific effect on human health. Recognition of these agents or their subsequent effect on health is the first step in the industrial hygiene process.

There are also health hazards that cannot be attributed to any specific substance but are threatening to human health nevertheless. The agents of these hazards, however, can still be recognized through assessment of work processes. Exposure to *temperature extremes* is an example of one such hazard. While the agent of this hazard, *energy,* is not a specific substance per se, the results of exposure can still be extremely serious, even deadly. Unfortunately, a less-than-adequate assessment effort under these circumstances may permit some such hazards to affect the health of the worker before recognition finally occurs. This *after-the-fact* recognition can result in serious and quite detrimental consequences to the employee.

Exposure to Heat

Aside from the obvious *physical hazards* associated with working around heat or hot surfaces (e.g., burns, scalding, etc.), there are also less-than-obvious *health* effects that can result from exposure to hot work environments. These effects, often termed *heat disorders,* are generally common in work processes that involve the use of elevated temperatures as part of the process itself, such as furnace work, or jobs requiring the use of ovens. In terms of regulatory requirements, the hygienist should note that no specific standards currently exist to regulate occupational exposures to heat. However, in the absence of a specific standard, the employer is still required to ensure a safe and healthy work environment under the General Duty Clause of the OSHAct.

Heat disorders may also be encountered in areas where higher temperatures are likely due to the location of the job and not specific to the job task itself, such as outdoor road work or other such activities that involve direct and relatively strenuous work in open areas.

The *heat stress* an individual is subjected to is a measure of the environmental and physical factors that combine as the total heat load imposed on the body during the performance of work. Factors that exist in the work environment that contribute to heat stress include the air itself (temperature and movement or circulation), radiant heat exchange, and water vapor pressure. At the same time, the level of physical exertion by the worker during the job or task contributes to the total heat stress of the job itself. As work is performed, energy is expended, and metabolic heat is produced inside the body. Combined with the environmental factors occurring outside the body, the intensity of total heat stress at any given time during the job can be relatively high and subsequently dangerous to the health of the worker. If a high level of heat stress is allowed to continue, *heat strain* will likely follow. In general, the body will begin to strain in response to the conditions of heat stress. When this strain is particularly intense or excessive for the exposed individual, a feeling of malaise or physical discomfort or distress may precede the onset of a *heat disorder.*

It is important at this point for the hygienist to realize that such health hazards can exist. While recognition may be somewhat more difficult than with other health hazards, recognition is still essential to ensuring the health of employees. Therefore, during the assessment process, the hygienist should pay particular attention to those jobs or tasks that may require brief, intermittent, or long-term exposure to high temperatures.

Evaluating Exposures to Heat

Human body temperature is regulated by a tiny control center in the brain known as the *hypothalamus.* When a worker's blood temperature rises above a standard 98.6° F, the hypothalamus sends chemical messages that prompt the heart to pump more blood and dilate the blood vessels, especially the tiny capillaries located near the upper surface of the skin (recall the discussion in Chapter 4 on the skin). More blood flows through these surface vessels so that excess body heat can drain into the cooler atmosphere through the process of *heat conduction.* At the same time, water diffuses through the pores of the skin. When conditions are right, the water evaporates before it can be seen and the skin appears to remain dry. This process is known as *insensible perspiration.*

If this level of control is insufficient to cool the blood, the hypothalamus signals the sweat glands to pour out larger amounts of water and heat in what is known as *sensible perspiration* or sweating. When perspiration evaporates, heat energy is needed to change the liquid to vapor. This heat comes from the skin and is the reason that sweating helps to cool the body. Uses of industrial creams, lotions, or other materials that may actually be intended to protect the skin from exposure to chemical agents may clog the sensitive pores in the skin and limit or prohibit the sweating process. The hygienist should be alert to such products and practices, especially in hot work environments.

The regulation of body temperature by the hypothalamus works reasonably well, as long as the air temperature is considerably lower than normal skin temperature (that tends to be several degrees lower than the body temperature) and the humidity is not so high that sweat is unable to evaporate rapidly. However, when the air temperature is near or above normal body temperature, there is no conductive cooling of the blood. When the relative humidity rises above 60 percent, even sweating becomes a poor cooling device. At 75 percent humidity, sweating has no cooling effect at all.

The hygienist should also know that even during a fairly warm (but not hot) and relatively dry day, workers can still experience a heat illness by overworking their muscles and overheating their body. Such activity still taxes the body's cooling mechanism. Profuse sweating and perspiration may result in a large loss of body salts, disrupting the body's electrolyte balance because the heart must work harder to rid the body of excess heat. If allowed to continue, the following heat-induced illnesses may result, explained here in order of severity:

Heat Fatigue

Also known as asthenia (or weakness), heat fatigue is usually the first indication of the onset of more serious heat illness. It is characterized by symptoms such as fatigue (hence the name), headache, mental and physical inefficiency, poor appetite, heavy sweating, high pulse rate, and shallow breathing. These result from exposure to excessively hot, humid conditions. If allowed to continue, the individual will begin to experience the effects of more serious heat disorders. The worker should be removed to a cool, dry place and, if not contrary to any physicians orders, be given salt in liquid form (not tablets) in small doses to begin replacing that which has been lost. However, it should be noted that administering salt can, in some cases, make matters worse. This is why a physician should be consulted, or medical treatment sought, especially in severe cases.

Heat Cramps

The most likely victim of these sudden incapacitating pains in the abdomen or extremities is the worker who is actually in good physical condition but overexerts during particularly hot atmospheric conditions such as a heat wave. The cramps are caused by excessive sweating and loss of salts from the blood and tissues. They are best treated by drinking a salty liquid. Firm pressure on the cramped muscle may also provide some relief, but massage is not recommended. The worker should be allowed a full 24 hours rest before resuming any hot, strenuous work.

Heat Exhaustion

This disorder may be experienced by those working at length under hot atmospheric conditions, coupled with overexposure to heat and humidity and overexertion. The hygienist should be able to recognize the difference between heat exhaustion and the far more serious *heat stroke* discussed below. In heat exhaustion, the worker sweats profusely, feels weak and dizzy, and may faint or vomit. The skin is pale and may feel cold and clammy, and the body temperature is normal or below. It is extremely important that the worker be removed to a cool, dry location. Treatment calls for rest and a sponge bath with cool water. If there is no vomiting, salty liquids should be administered. Although the worker may feel well rather quickly, it is dangerous to return to any activity. Older workers, or those with heart conditions, may require transportation to a hospital for further treatment or observation.

Heat Stroke

This life-threatening condition can result from extreme overexertion or from circulatory impairment caused by illness, old age, or the use of certain drugs. In contrast to heat exhaustion, the worker will actually stop sweating and feel very feverish. The body temperature will soar (often rising to 106° F or higher). The pulse will pound, and unconsciousness may result. The skin at first will appear flush and then may become ashen or purple in color. Minutes count in this syndrome, as the body's thermo-regulatory system has completely shut down. Body temperature must be reduced as rapidly as possible. Expert medical attention is required immediately.

The industrial hygienist should be cognizant of the various types of heat-induced disorders. Recognizing when conditions are right for such problems is the first step in preventing their occurrence. Methods of determining whether physically harmful temperature conditions exist focus on the measurement of *humidity, dew point, temperature,* and a combination of *"bulb"* temperatures that indicate precise conditions of exposure (at the time the measurements are taken).

Humidity, that is the amount of water vapor present within a given space of air, is measured as the *relative humidity* (RH). This is an expression of the amount of actual water vapor in the air relative the total amount it could contain under the same conditions of temperature and pressure. Water vapor pressure is a key factor effecting the proper exchange of heat between the body and the atmosphere through evaporation. Measuring the relative humidity tells the hygienist whether environmental factors might be restrictive to those working under such conditions.

The *dew point* is that temperature at which the air, subjected to cooling temperatures, becomes saturated and moisture begins to form on exposed surfaces. Dew point is also important because it is a positive indication of existing air moisture content. Together with the RH, it tells the hygienist specific information about the air's capacity to accept heat and moisture from the worker's body during heat exchange. A high atmospheric temperature and a high relative humidity, coupled with a low dew point temperature, will cause some difficulty for the person working under such conditions.

A simple thermometer, also referred to as a *dry bulb thermometer* (DBT), is most frequently used to obtain a reading of the atmospheric temperature. Whereas, the *wet bulb temperature* (WBT) is measured by a thermometer whose bulb has been covered with a wetted wick. This method shields the bulb from the effects of radiant heat. When the bulb is exposed to a current of rapidly moving air, the wetted wick begins to dry due to the evaporation. When water evaporates from the wick, the subsequent heat loss cools the bulb to register the wet bulb temperature. The combined readings of dry bulb and wet bulb temperature are used to calculate the percent relative humidity. This measurement is effective in determining moisture evaporation rates to be expected under similar working conditions. The wet bulb temperature will help the hygienist forecast whether heat exchange problems might exist for the worker.

However, the exact rate of heat transfer depends on the temperatures and radiant heat characteristics of the surfaces involved and not specifically on the temperature of the air. The *globe temperature* (GBT), therefore, is used in combination with the bulb temperatures for more accurate reading. Specifically, radiant heat is a form of electromagnetic radiation. It cannot, however, be measured in the air. To measure the effects of radiant heat on temperature, the globe thermometer (Figure 7-1) is typically used. The globe is usually a hollow sphere whose surface is painted a flat black. A dry bulb thermometer is inserted into the globe and suspended at the approximate midpoint. As the surface of the globe is heated by the surrounding radiant heat, the temperature inside the globe is also heated, thus affecting the insulated thermometer. After a minimum of 20 minutes, the increase in temperature over the dry bulb reading in the same area is a measure of the radiant heat. As shown in Figure 7-1, a simple apparatus can be used to effectively measure the dry bulb, the globe, and the wet bulb temperatures at the same time. The resulting temperatures can be combined in a standard formula to determine the *wet bulb globe temperature* (WBGT) index, which can be used to evaluate the conditions of heat stress, as follows:

Indoors: $WBGT = 0.7(WBT) + 0.3(GBT)$
Outdoors: $WBGT = 0.7(WBT) + 0.2(GBT) + 0.1(DBT)$

Exposure to Cold

Human beings have evolved as warm-blooded, semitropical animals. As such, they must maintain their body heat. The body is better equipped to cope with hot atmospheric temperature than with cold. However, numerous job tasks and occupations require workers to perform in cold and very cold environments. Without proper protection, such work can be extremely hazardous to human health. The

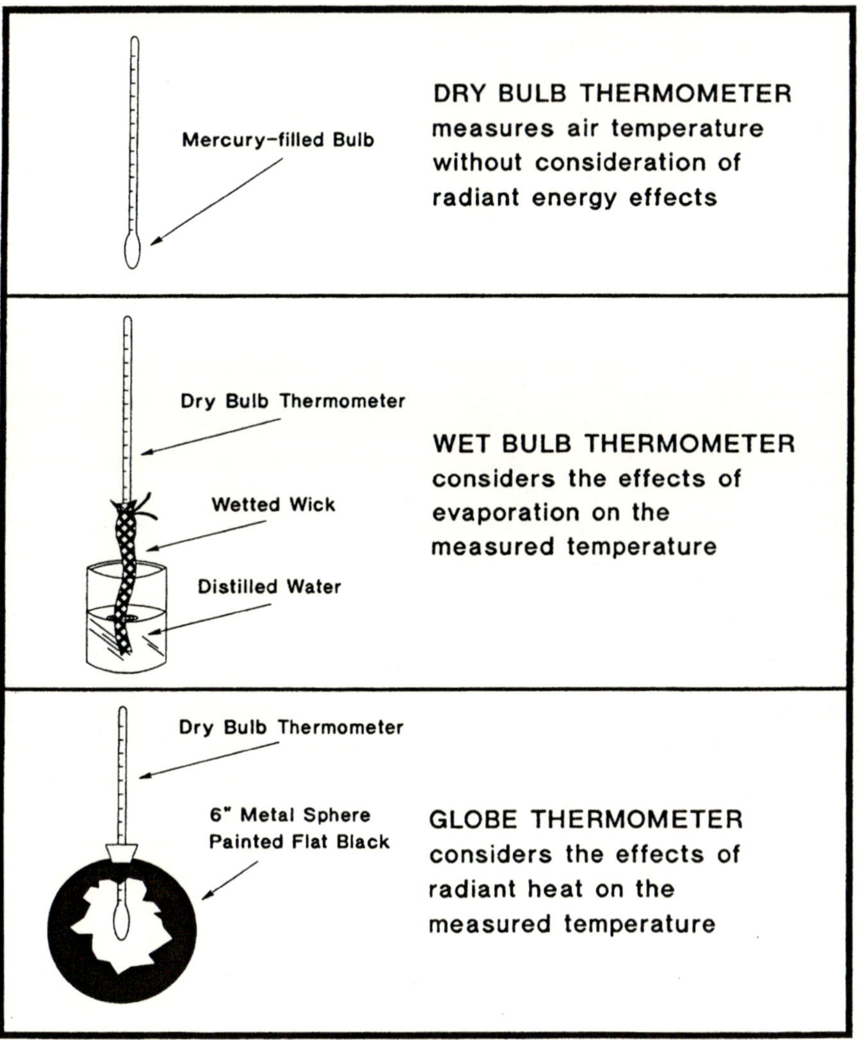

FIGURE 7-1 The various types of thermometers used to aid in the determination of existing atmospheric temperature extremes

hygienist must recognize when exposures to cold are likely to occur, which job functions are potential candidates for cold work, and what means (if any) are currently provided to protect against exposure to low temperature. Recognizing the health hazards of cold work environments are not always obvious. For instance, many cases of *hypothermia* (loss of body heat and decrease temperature due to extensive exposure to cold) occur in air temperatures between 30–50° F, which, by most standards, would not be considered extremely cold. However, when one takes into account additional environmental factors (e.g., wind chill) and individual attributes (e.g., poor physical fitness coupled with an inappropriate level of protection provided by clothing), the potential for cold-induced illness or injury increases substantially.

In addition to any outdoor work in cold climates, workers in meat packing plants and other cold storage facilities may be subjected to cold hazards under normal working conditions. Butchers, for example, must frequently perform their work in cold freezers or cutting rooms. Engineers and technicians who handle super-cold commodities such as liquid oxygen are not only exposed to the *health hazards* of a cold work environment, they are also subjected to the *physical hazards* associated with cryogenic liquids. In fact, exposure to cold environments can result in both injury to specific body parts or systems, as in the case of *frostbite*, as well as general illnesses such as the hypothermia, which can be fatal.

Evaluating Exposures to Cold

Working in cold environments can expose employees to hazardous conditions that are sometimes—but not always—much more easily recognized than those that occur in hotter environments. While thermometers measure the air temperature, they do not register the additional chilling effects the wind can have—the so-called *wind chill factor.* The wind effectively removes the warming layer of air that normally surrounds the body. A mere 5 mph wind can carry away eight times more body heat than still air. Simply stated, the wind chill factor measures the increase in cooling power of moving air. There is no differentiation made between the wind moving the air or the body moving through still air (i.e., there is still a wind chill effect). At 20° F in a 20 mph wind, the cooling effect is equivalent to calm air at –10° F (Figure 7-2).

When employees must work in cold, windy conditions, they should be advised to wear layers of relatively light, loose clothing as opposed to one thick, heavy item. A film of trapped air rests between each layer that, when heated by the body, will act as excellent insulation against the cold and wind. Wetness increases the loss of body heat since, unlike air, water is a excellent conductor of heat. Wet skin or clothing will subsequently cause the body to loose heat much more rapidly. Even at 50° F, if wet, a person can suffer ill effects of cold.

It is important to realize that body heat is more readily lost from those body parts that have the most surface area compared to total mass—namely, the hands and feet. It is true that the hands and feet are capable of withstanding a drop of 30 to 40 degrees below normal body temperature for relatively short periods of time without sustaining any noticeable damage. However, if the internal body temperature (the *core temperature*) should drop by just 10 or 20 degrees, the result may

WIND-CHILL FACTOR CHART	ACTUAL TEMPERATURE READING (in degrees Fahrenheit)							
	50	40	30	20	10	0	-10	-20
ESTIMATED WIND SPEED (mph)	EQUIVALENT WIND-CHILL TEMPERATURE							
Calm	50	40	30	20	10	0	-10	-20
5	48	37	27	16	6	-5	-25	-26
10	40	28	16	4	-9	-24	-33	-46
15	36	22	9	-5	-18	-32	-45	-58
20	32	18	4	-10	-25	-39	-53	-67
25	30	16	0	-15	-29	-44	-59	-74
30	28	13	-2	-18	-33	-48	-63	-79
35	27	11	-4	-20	-35	-51	-67	-82
40	26	10	-6	-21	-37	-53	-69	-85

FIGURE 7-2 The effects of "wind-chill" on actual temperatures

be fatal. This places the lethal deep core body temperature at approximately 78° F. This condition can occur where a worker is immersed in cold water (diving operations), is exposed to cool, high winds, is in a state of physical exhaustion, or has insufficient food. Obviously, any combination of these conditions would only expedite the body chilling process. The body tries to preserve its internal heat by constricting surface blood vessels and by burning more body fuel to increase heat production. Incidentally, contrary to popular belief, alcohol consumption expands blood vessels and actually contributes to heat loss. Even when the body (including the hands and feet) is well protected from the cold, as much as 90 percent of its heat can still be lost through a person's head. Appropriate head protection, therefore, should never be overlooked when working in cold conditions. Also, pink/flush checks, ears, nose, etc., do not indicate warmth, as is sometimes wrongly assumed. Rather, in cold temperatures they result from a slowdown of metabolism in skin cells in response to the cold. The skin looks pink because very little oxygen is being removed from the blood and red, oxygenated blood is moving through the skin's surface veins.

The following information pertains to cold-stress disorders that can be encountered in the work environment under certain exposure conditions. The hygienist should be extremely observant for the signs and indications of a cold exposure illness to ensure that employees are protected from such hazards.

Shivering

An involuntary contraction of the muscles to generate additional internal body heat is usually the first sign that the body's core temperature has begun to drop in a condition known as *hypothermia*. Shivering should not be taken lightly. Employees should be removed to areas that are warm, dry, and stable (no wind) as soon as possible. Wet clothing must be discarded immediately. Warm liquids and quick-energy foods (chocolate, for example) can be given to a conscious victim in this early stage of hypothermia. The hygienist should watch for indications of impending hypothermia at cold work locations, especially when strenuous work or heavy exertion is required. Employees who become overheated in cold climates may succumb to cold stress quicker because perspiration carries away body heat much more quickly. When cold air is inhaled directly into the lungs through the mouth, bypassing the warming effects of the nasal cavity (as is typically the case during strenuous activities), the person may overwhelm the body's ability to warm itself.

Frostbite

In frostbite, fluid between the cells actually freezes, and the cells themselves become dehydrated. Frostbite can be superficial, involving only the skin, or it can be deep, extending well below the skin surface to affect muscle and other deep tissues. In either case, the resulting cell damage may cause death of the tissue. In some cases, the damage is so severe that a digit (finger, toe, foot, etc.) may have to be amputated. The warning signs of frostbite include a feeling of extreme cold and pain in the affected area, usually the fingers, toes, ears, or nose. This may be followed by a burning sensation as the damage progresses. The frostbitten body part then becomes hard and numb, and the skin takes on a white to yellow-white or molted blue-white color. Once frostbitten, the tissue may remain permanently sensitive to cold. It is important to note that the fingers, toes, nose tips, ears, and cheeks are particularly vulnerable to frostbite, even when the rest of the body remains warm or even overheated, because blood supply to these areas is relatively poor.

Employees who must work around super-cold liquids (cryogenics) such as liquid nitrogen are particularly susceptible to site-contact frostbite. Extreme caution must be practiced when working with such commodities. The hygienist must assess the jobs where such exposures are possible and ensure the proper level of precaution and personal protective equipment (PPE) has been provided and is being properly used by all involved workers.

NOISE AND VIBRATION

As with exposure to temperature extremes, noise and vibration are examples of other non-substantive health hazards that can be found in many work environments. Energy, once again, is the agent of harm. In this case, *sonic* or *sound energy* is the culprit. While the term *sound* is generally applied to that form of energy that produces an audible sensation, *vibration* is used to describe a feeling. While there is no real difference between the sonic and vibratory forms of sound energy, the subsequent effects of exposure to each can be discussed in separate terms. The results of exposure to either can be illness or injury, depending on the level and degree of exposure. Excessive exposure to sound will have detrimental effects on hearing while excessive exposure to vibration can harm other systems and/or organs within the body. The primary objective of this section is to provide the information necessary for the hygienist to properly recognize when, how, and under what conditions such hazards exist.

Understanding Noise and Sound

Recognition of the hazard to human hearing posed by sound requires an understanding of the difference between what is referred to as *sound* and what is described as *noise*. Basically, the difference between the two terms has more to do with perception than science. What may be considered noise to one person may be music to another. In the industrial arena, noise has been described as sound that is unwanted by a listener because it is unpleasant, difficult to listen to, or provides no useful information. However, there are no specific attributes of noise that can distinguish it from sounds that are, in fact, wanted. Hence, the existence of noise as opposed to sound in any given situation is more a matter of human reaction and is therefore quite subjective. Since noise usually interferes with sounds that are wanted, it is distracting and can be harmful under certain conditions.

The conditions that can make noise harmful to human hearing are generally discussed in terms of sound *frequency* (that is related to *pitch*). Sound is produced by vibratory motion, in some medium such as air or water, in the form of a *sound wave*. To propagate from the vibrating source through the medium, the medium must have mass (to provide inertia) and be elastic. Air has both properties and is, therefore, quite suited to the propagation of sound waves. In air, sound is usually described in terms of variations of pressure above and below the ambient air pressure. These pressure *oscillations* are commonly called *sound pressure*. They are generated by a vibrating surface or a turbulent fluid flow causing the formation of alternating high and low pressure areas that propagate from the source of the vibration as sound. Sound, then, can be defined as simple oscillations in pressure in a medium with elasticity and mass. Sound may also be defined as the auditory sensation evoked by the oscillations in pressure (ANSI S1.23-1976/R1983).

The *rate* at which the pressure oscillations are produced, known as the *frequency*, is expressed in units referred to as *hertz* (Hz). One hertz is equal to one cycle (or full oscillation) per second. The distance a sound wave travels in one cycle is known as the *wavelength* (Figure 7-3). Wavelength is a very important property of sound. High-frequency sounds have very short wavelengths, while low-frequency sounds have very long wavelengths. To the human ear, frequency is perceived as pitch and is represented as a range of hearing. The frequency range of the human ear varies considerably from one person to the next. A young person, for example, with normal hearing capability should be able to perceive sounds with frequencies between 20 and 20,000 Hz at moderate sound pressures (AIHA, 1986). With increasing age, the upper frequency limit tends to decrease due to a condition known as *presbycusis* (age-onset hearing loss).

Loudness, another human subjective response to sound, varies as sound pressure and sound intensity (or sound "power") vary. Sounds that are perceived as

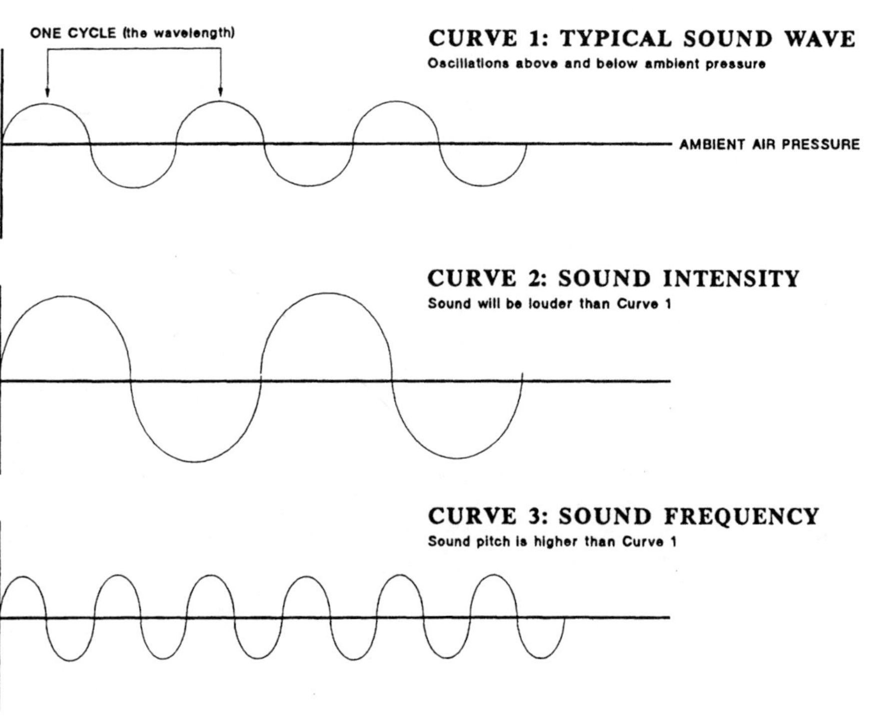

FIGURE 7-3 The characteristics of sound wave progression

SPL ~ sound pressure levels   +/- levels of SPL

loud may also be referred to by workers as noisy, distracting, irritating, annoying, or any combination of terms. During the hazard recognition process, the hygienist should be particularly alert to any such descriptive terminology from workers. Recognizing noise and vibration hazards in the workplace is often difficult since the health effects (as described in the next chapter) are generally insidious, subtle, and difficult to quantify.

In an effort to standardize the hazard recognition process, regulatory agencies have established permissible noise exposure limits that, if exceeded, require implementation of specific requirements that are often referred to as a *hearing conservation* program.

Evaluation Methods and Techniques

The principal measurement of sound level is the *bel,* a unit of measure used in electrical and communications engineering. A *decibel* (or dB) is simply one tenth of a bel and is considered to be the minimum difference in perceptible loudness of any given sound. The evaluation of sound is a measurement of its pressure level relative to a certain reference measure or quantity and is expressed as a weight on a standardized scale. The scale, an internationally recognized sound weighting network that has been developed to evaluate sound levels, consists of three networks. The A-network is said to approximate equal loudness curves at low sound pressure levels. The B-network measures medium sound pressures and the C-network was designed for high levels. While each network is designed to respond to some frequencies more than others, the A-weighted sound level measurement has become most popular for evaluation of sound and noise. The A-scale, as it is often called, appears to provide a relatively accurate assessment of industrial broadband noises that indicates the injurious effects of such noise on human hearing. Since it measures slow response to noise, the A-scale is said to most closely approximate human hearing. Measurements on the A-scale, symbolized as *dBA,* are the basis of OSHA's hearing conservation requirements.

Various types of *sound level meters* (SLMs) are available for monitoring and analyzing noise. The most commonly used device is the *dosimeter.* The dosimeter performs two functions. Its microphone is placed on the employee in the "hearing zone" while the remainder of the instrument automatically computes the noise measures. Dosimeters are usually battery operated and have evolved from simple devices computing single-number exposures to highly sophisticated (and expensive) monitors that can compute and store comprehensive data on the sound levels encountered by the subject during the entire sampling episode. Most are set to monitor exposures based on A-scale criteria that help to simplify the tasks required for compliance with OSHA's hearing conservation requirements.

Hearing Conservation Program Requirements

Under OSHA's Occupational Noise Exposure Standard (29 CFR 1910.95), employers are required to implement and administer a hearing conservation program whenever employee noise exposures equal or exceed an 8-hour time-weighted-average (TWA) of 85 dBA or, equivalently, a dose of 50 percent. The Standard also refers to the 8-hour TWA of 85 dBA or a dose of 50 percent as the *action level.* This means that employer action to prevent occupational exposures to noise must begin when these noise levels are reached. The Standard establishes the minimum actions to be taken for an acceptable hearing conservation program. However, as with all requirements that pertain to employee safety and health, the industrial hygienist must evaluate the adequacy of these minimum requirements with regard to the specific exposure problems encountered in a particular facility. More often than not, the practicing hygienist must implement even tighter controls to ensure a proper level of protection.

In general, the OSHA Standard requires employers to implement a *monitoring program* as the first action once information indicates that employee exposure to noise has reached or exceeded the action level. Monitoring requires physical sampling using calibrated noise detection devices and instrumentation to identify those areas where noise emission is reaching the maximum allowable levels for specific periods of time, as shown in Table 7-1. Employees exposed to such levels must be identified for inclusion in the hearing conservation program and provided with adequate hearing protectors.

TABLE 7-1 Maximum permissible noise exposure levels based on time (Source: OSHA 29 CFR 1910.95)

PERMISSIBLE NOISE EXPOSURES	
Duration per Day in Hours	*Sound Level in dBA*
8.0	90
6.0	92
4.0	95
3.0	97
2.0	100
1.5	102
1.0	105
0.5	110
0.25 or less	115
Note: Exposure to impulsive or impact noise should not exceed 140 dBA peak sound pressure level.	

Once the high noise areas have been identified, the Standard requires employers to notify each employee who is exposed to noise levels that are at or above the action level. Additionally, employers must establish an audiometric testing program and offer, free of charge, audiometric testing to exposed employees.

Specifically, within six months of an employee's first exposure to noise at or above the action level, a *baseline audiogram* must be established against which all future audiograms can be compared. It should be noted that an audiogram is usually administered to employees who have not been exposed to high noise levels for the preceding 14 hours. This is meant to ensure a true baseline measurement of the worker's hearing capabilities. At least annually thereafter, the employer must obtain new audiograms from each affected employee. Employees must be made aware of the results of their audiograms, especially when a *standard threshold shift* (STS) has been observed. They must also be informed as to the actions the employer intends to take to ensure the continued protection of hearing in the workplace. An STS is defined as a change in hearing threshold relative to the baseline audiogram of an average of 10 dBA or more at 2,000, 3,000, and 4,000 Hz in either ear. It should be noted that the evaluator, usually an occupation physician, nurse, or specially trained technician, must consider the possible effects of presbycusis (age on-set hearing loss) when interpreting test results.

In addition to establishing the written hearing conservation program and the audiometric testing requirements, the Standard also requires employers to provide *hearing protectors* at no cost to all employees exposed to noise at or above the action level. To ensure that this protection is properly worn, the industrial hygienist may have to conduct spot inspections or surveys of the work area. Also, hearing protection should be provided to any employee who has not yet had the baseline audiogram or to any who have experienced an STS. There are a variety of choices in hearing protection available to industry today. While the Standard does require employers to offer their personnel a variety of hearing protection device, the industrial hygienist should first evaluate these devices to ensure their suitability to the job and the hazard. By reviewing manufacturer-supplied data on each device's *attenuation* (noise reduction) capabilities, along with the specific need as measured in the facility or work environment, the hygienist can effectively select only those types of protectors best suited for applicable requirements. As a minimum, hearing protectors must attenuate employee exposure at least to an 8-hour TWA of 90 dBA (for employees who have already experienced an STS, the protector must attenuate to 85 dBA during the 8-hour TWA).

Employees provided with hearing protection must be properly trained regarding the purpose of hearing protectors (advantages, disadvantages) and instructed on their use, care, and maintenance. Training must also be provided on the effects of noise on hearing and on the purpose of the audiometric testing program. This training should be updated as required and repeated annually for each employee involved in the hearing conservation program. Employees must be assured access to any information and records that are applicable to the hearing conservation program, including training materials and results of their own audiometric tests.

In all likelihood, the industrial hygienist (and/or possibly the occupational health nurse, if the company has one on staff) will be responsible for ensuring the proper implementation of the hearing conservation program. This task will also include an appropriate level of *record keeping* (exposure monitoring data, audiograms, hearing protector specifications, and the like).

Whatever the specific hearing conservation requirements for any given workplace, the hygienist should become very familiar with the requirements for a *minimum acceptable program* as mandated in OSHA's Occupational Noise Exposure Standard. It should also be noted that the minimum requirements established by this and every OSHA Standard often must be exceeded to ensure adequate employee safety and health. Compliance with the minimum requirements simply may not always be enough to meet this objective. Intelligent and thoughtful consideration of all possible contingencies when employee safety and health is at issue will often mandate more stringent hazard controls than those prescribed by the regulatory agencies.

RADIATION

Radiation is also a means of propagating energy. Essentially, the two types of radiation that can be encountered in the work environment are *ionizing radiation* and *nonionizing radiation*. Ionizing radiation is described as any electromagnetic or particulate radiation capable of producing ions, directly or indirectly, during its passage through matter. Conversely, nonionizing radiation is simply any electromagnetic radiation that does not cause ionization.

To understand these concepts further requires a basic knowledge of the terminology used. *Ionization* is the physical separation of a normally electrically neutral atom or molecule into its electrically charged components. This term is also used to describe the degree or extent to that this separation occurs. Ionization results in the physical removal of a negatively charged electron from the atom or molecule, either directly or indirectly, leaving a positively charged ion. The separated electron and ion are referred to as an *ion pair*. Electromagnetic radiation refers to a traveling wave-like motion resulting from changing electric or magnetic fields. When the relative lengths of the waves are short, they are capable of producing ionization. Examples include x-rays and gamma rays. When the wavelengths are relatively long, they are nonionizing. Examples include radio waves, radar, ultraviolet lasers, microwaves, and other radio-frequency (rf) radiation. Figure 7-4 shows a graphic representation of the *electromagnetic spectrum* and where the various types of ionizing and nonionizing radiation exist relative to visible light. *Particulate radiation* (also known as *corpuscular* radiation) concerns the presence of alpha or beta particles, electrons, positrons, neutrons, or heavy particles. Such particles are generated or emitted from radioactive materials and can penetrate, in varying degrees, solid materials (including skin). As the particles pass through a material, ionization of that material occurs, resulting in damage on the molecular level. Airborne radioactive particulates present a partic-

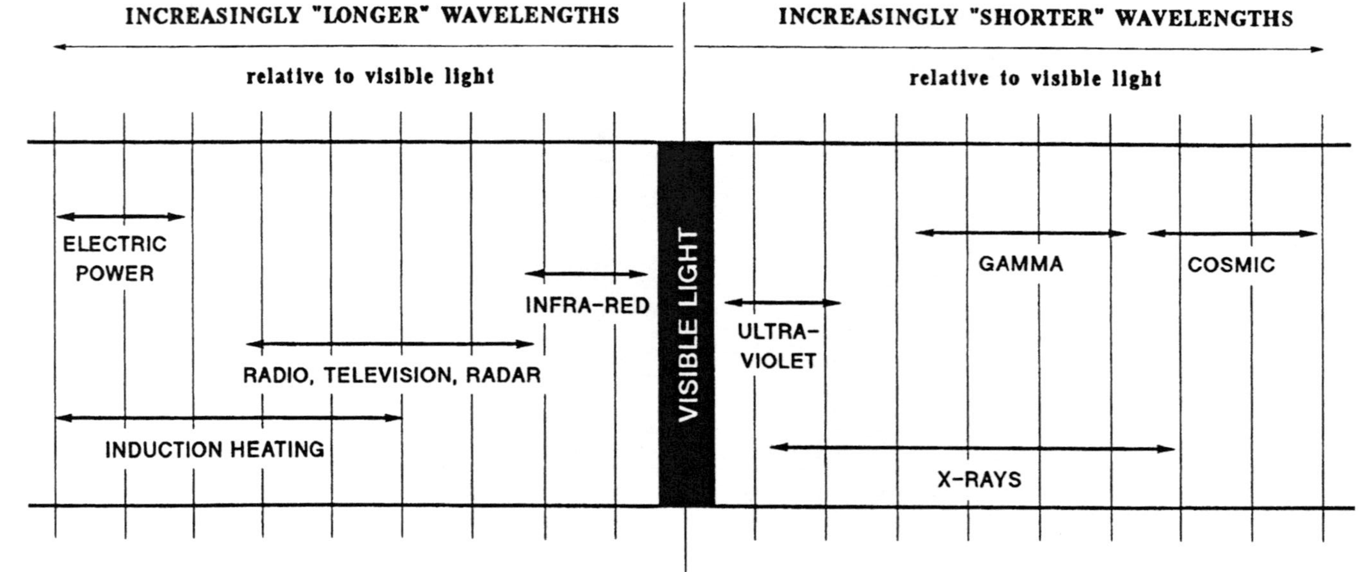

FIGURE 7-4 A graphic representation of the electromagnetic spectrum and the location of the various types of ionizing and nonionizing radiation relative to visible light

ularly dangerous health hazard. Once such particles enter the body and have access to the soft inner tissues and organs, the resulting effect can be extremely deadly.

Recognition of radiation hazards in the workplace is a relatively easy process, since most organizations know where, when, and how such energies are used in their facility. Also, various state and federal agencies have mandated specific requirements associated with the use and control of radiation sources. Therefore, the industrial hygienist should be somewhat familiar with the work processes in the facility that involve the use or generation of radiation hazards. X-ray machines are generally not difficult to recognize. Similarly, the uses of particulate radiation are usually quite specific to a work task or procedure.

The professional discipline that is concerned with the study of radiation is known as _health physics,_ and the specialist in this field is the _health physicist._ However, in many cases the responsibility for radiation safety will often fall on the industrial hygienist or, in the absence of a qualified hygienist, the occupational safety professional. Whatever the particular responsibilities may be in any given organizational structure, the primary requirements, practices, and principles of radiation safety remain the same: the control of radiation exposures to personnel. The advanced study of radiation introduces extremely complex and somewhat abstract concepts that are beyond the scope of this text. From the basic industrial hygiene perspective, however, the introductory materials provided here should facilitate the initial recognition process. To accomplish this objective requires effective _understanding of radiation hazards,_ efficient _accident prevention practices,_ and consistent application of _control measures._ Quite often, just _knowing_ that a particular radiation hazard exists may be enough to prevent occupational illnesses resulting from unprotected exposures.

Understanding Radiation Hazards

As previously established, radiation is yet another form of energy. Light is a visible form of radiation. Invisible forms of radiation, such as microwaves, exist and can cause harm on exposure. All forms of radiation energy exist in a phenomenon referred to as the _electromagnetic spectrum_ (Figure 7-4). The electromagnetic spectrum covers a broad range of energies that are usually classified in terms of their frequency or wavelength. Included in the spectrum are x-rays, gamma rays, ultraviolet radiation, visible light, infrared (heat), microwaves, and radio waves. For the purpose of this basic discussion, understanding the complex and somewhat arbitrary relationship that exists between each wavelength in the spectrum is not critical. However, it should be noted that the _shorter_ the wavelength, the more ionizing the radiation becomes (e.g., x-rays, gamma rays). Conversely, the _longer_ wavelengths are nonionizing sources of radiation (e.g., radio frequencies, infrared). Also, from the industrial hygiene perspective, it should be remembered that electromagnetic radiation can create certain biological reactions in humans or animals once it is *absorbed* by living tissue. This discussion will focus first on ionizing radiation followed by a brief evaluation of the hazards posed by nonionizing radiation.

Ionizing Radiation

In the simplest of terms, all matter consists of atoms. Each atom consists of two basic components. These are the core or nucleus and its orbiting, negatively charged electrons. The nucleus is composed of positively charged particles (*protons*) and neutral particles (*neutrons*), as shown in Figure 7-5.

Radioactivity, then, refers to the spontaneous emission of very fast atomic particles and/or rays by a nucleus. This can occur naturally or as a result of the bombardment of neutrons by another particle. The principal concern for the industrial hygienist is the harmful effect of exposure to radiation. Basically, human cells are constantly in a state of change. After their creation, they perform specific biologically necessary functions as they grow and multiply, until they finally die and are replaced by new cells so the process can continue. With exposure to a radioactive

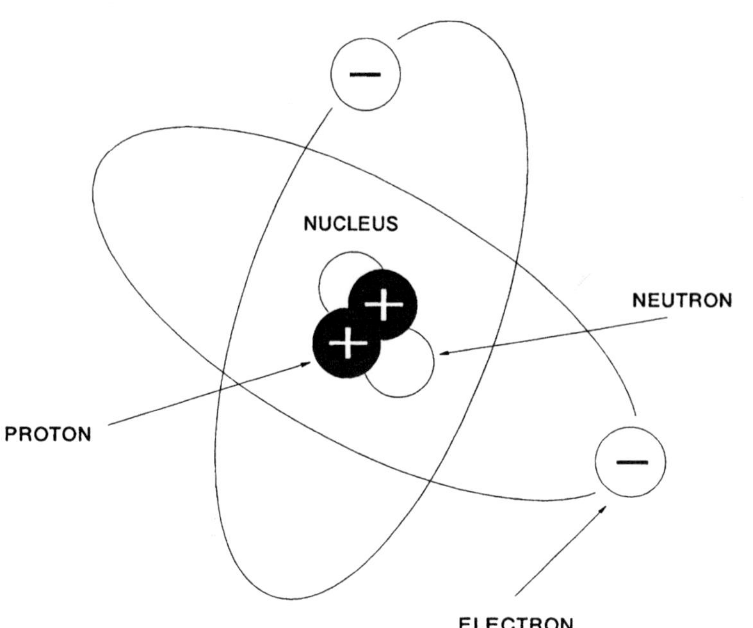

FIGURE 7-5 The nucleus of an atom is composed of positively charged particles (*protons*) and neutral particles (*neutrons*).

material, the cell may react in a variety of ways, depending on the *dose* received and when in the cell's life cycle the exposure occurred. The results range from little or no noticeable effect to changes in biochemical reactions (inability to reproduce new cells), the production of diseases, physical changes (mutations) in the cell structure, and cell death. These changes are essentially due to the ionization process, which violently strips away electrons from atoms and interferes with their chemical bond with other atoms. After ionization, atoms will normally attempt to recombine. In complex living cells, this may not always be successful, and a mutated cell may be created in the process. Then, as the mutated cell divides and reproduces, additional damaged/mutated cells are created, and overall human health is threatened.

Radioactive materials spontaneously emit one or more types of radiation. The three major types of ionizing radioactivity of concern to the industrial hygienist are *alpha* particles (expressed as α), *beta* (expressed as β), and *gamma* (expressed as γ). Figure 7-6 shows the relative hazard of each, as described below, in terms of their ability to penetrate barriers (a determining factor for exposure risk).

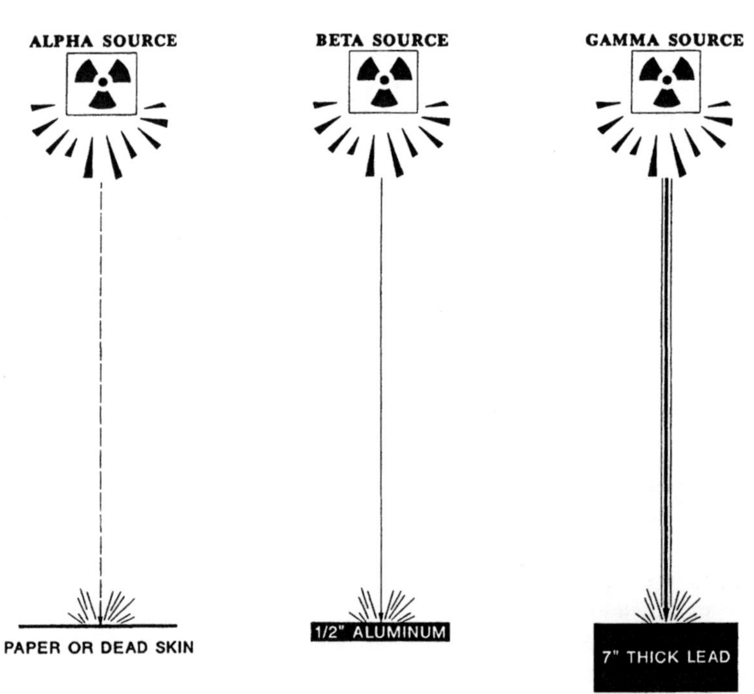

FIGURE 7-6 The relative penetrating hazard of alpha, beta, and gamma radiation

Alpha Radiation. In terms of hazard risk, alpha radiation is often thought to pose the least threat. While this assumption may be generally true, alpha particles can be extremely hazardous under certain conditions. Alpha particles actually originate in the nucleus of a radioactive atom. They are emitted during a process of radioactive decay known as *disintegration.* Since the radioactivity of such atoms will continue to disintegrate over time, the amount of particle emission is known as the *activity.* As alpha particles are emitted, they produce dense ionization along their path of travel through a material, including air. Although this characteristic makes alpha particles a serious hazard, it also tends to slow the speed of travel dramatically. In fact, the range of travel in unobstructed air is only approximately four inches. Hence, a thin barrier such as a piece of paper or the layer of dead skin that surrounds the human body can effectively prevent further travel or penetration of the alpha particle. On the other hand, if alpha particles are inhaled or ingested, they can be extremely destructive to the soft tissue of internal organs such as the kidneys, liver, and lungs. Also, since some alpha emitters are chemically similar to calcium, they can assimilate into the bone and remain there for long periods of time—perhaps years. As they continue the process of disintegration, the particles emitted will continue to damage internal tissues and organs.

Beta Particles. As a radioactive atom disintegrates, it ejects electrons from its nuclei. Once ejected, these electrons are known as beta particles and carry a negative charge of one unit. This leaves the disintegrating atom in a positively charged state (as opposed to its original neutral state), which changes the atom into another element. When beta particles ejected from a nuclei hold a positive charge, they are called *positrons.* If a positron should combine with an electron, its subsequent annihilation will produce gamma radiation. The travel distance in air for a beta particle depends on its emission source and can range from between 6 inches to 60 feet. They are even smaller than the tiny alpha particles and may travel at velocities close to the speed of light. When a beta particle is caused to slow down or stop, side-stream x-radiation (known as *bremsstrahlung radiation*) is often produced. The principal external threat to health posed by exposure to beta particles is caused by their ability to penetrate the skin to depths up to 0.5 inches. Skin burns can result due to excessive exposures. Internally, these particles are extremely hazardous.

Gamma Radiation. Gamma rays are considered potentially harmful at all levels of exposure. A gamma ray comes from an exited or unstable nucleus of the atom. Similar to x-radiation in that it is also ionizing and electromagnetic, gamma radiation can penetrate into vital tissues and organs. Depending on its wavelength, it can also be absorbed through the skin. Table 7-2 shows the results of gamma radiation exposure on the various listed body systems and organs.

TABLE 7-2   Results of gamma radiation exposure on various body systems and organs

BODY SYSTEM OR ORGAN	EFFECTS OF EXPOSURE
Lungs	Alveoli degeneration, cancer
Blood (including bone marrow)	Anemias, leukemias, cancer
Eyes	Cataracts
Skin	Erythema, dermatitis, cancer, hair loss
Reproductive system	Decreased sperm and egg-cell production, sterility, genetic changes
Lymphatic system	Spleen damage, lymphocytopenia

Units of Exposure Measurement

It has been established that ionizing radiation can take the form of both electromagnetic radiation (gamma, x-ray) and particulate radiation (alpha, beta), which differ drastically in terms of origin, emission, and general characteristics. The standard units of measure that have been developed to assess exposures are therefore based on *absorbed energy* and not on radiation type. The intent is to standardize the measuring process using the *dose equivalent* concept so that different degrees of effect resulting from exposure to the differing forms of radiation could all be related to a common unit of measure.

The primary unit of measure is referred to as a *Rem* (roentgen equivalent man). It is derived based on the multiplying effect of two other units, the *rad* (radiation absorbed dose) and the *RBE* (relative biological effectiveness). The rad is a measure of the dose of any ionizing radiation to body tissue in terms of energy absorbed per unit of mass of body tissue. One rad is the dose corresponding to the absorption of 100 ergs per gram of tissue (1 millirad, or "mrad," equals 0.001 rad). The RBE is a predetermined value for the various forms of radiation where gamma, x-ray, and beta are equal to an RBE of 1 and alpha is equal to an RBE value of 10. The RBE represents the relative effectiveness of the same absorbed dose of two ionizing radiations in producing a measurable biological response. Hence, a Rem is a *dose unit* equal to the dose (in rads) multiplied by the appropriate RBE value for a given radiation. It is represented by the simple formula *Rem = rad × RBE*.

Regulatory agencies such as OSHA have established limits based on the Rem unit for occupational exposures to ionizing radiation. In its Ionizing Radiation Standard (29 CFR 1910.96), OSHA has mandated that no employee shall be exposed to any form of ionizing radiation during any one calendar year in excess 5 Rems (whole body dose plus accumulated total occupational dose).

Effects of Exposure to Ionizing Radiation

The major factor affecting radiation injury is the dose rate, or the time span over which a certain exposure is received. Since the body will recover somewhat between successive exposures to radiation, an acute *(rapid)* exposure will have more drastic physiological effects than the same total exposure distributed over a long period of time *(chronic)*. While the exact effects of an acute radiation exposure cannot be accurately predicted, factors that should always be considered include the following:

1. *The specific amount of radiation absorbed by living tissue.* The larger the amount, the greater the biological damage.
2. *Depth of penetration.* Some types of radiation, such as gamma and x-rays, have a great penetrating ability and can thus affect certain vital organs while other types of radiation may cause only surface damage unless they are ingested or inhaled.
3. *Area of application.* As in the case of thermal burns, the quantity of body area exposed is an important consideration. For example, an exposure of several thousand rad to a localized section of the body is often used in radiation therapy treatment. The same exposure, however, to the whole body would be fatal.
4. *The age of the exposed individual.* Young people seem more *radiosensitive* than mature people, perhaps because of the more rapid cellular reproduction occurring regularly within the younger body. For this reason, the U.S. Nuclear Regulatory Commission (NRC) as well as OSHA will not permit anyone under the age of 18 to be occupationally exposed to ionizing radiation.
5. *Sex.* Male reproductive organs and tissues are more sensitive than are the female reproductive tissues. This is because the testicles are located externally, where they can be more directly subjected to radiation exposures, while the ovaries are effectively shielded from radiation deep within the abdomen. Also, sperm cells, unlike ova, multiply rapidly during sexual maturity and are thus more radiosensitive during this phase of development.
6. *Nutritional and general health status.* A healthy, well-nourished individual can generally withstand a greater radiation exposure than a sickly, undernourished person.

Because of the number of variables, the exact outcome of an individual case of radiation exposure cannot be accurately predicted. However, in terms of both acute and chronic exposures, the industrial hygienist should be aware of certain signs, indications, and symptoms of *radiation sickness.*

There are three basic categories generally discussed when considering the results of exposure to radiation. In the *sub-clinical category,* no obvious disease can be attributed to exposures below 100 Rem. There is normally no appreciable degree of sickness or incapacitation associated with these levels. While nausea may occur in a small percentage of the exposed population, it is uncertain whether this is due to the actual exposure or to other unexplored psychological factors. No therapy or hospitalization is usually required, although there may be some minor detectable blood changes.

As the 100 Rem level is exceeded, the *hematopoietic category* is entered, and people will begin to show definite symptoms of acute radiation sickness. As its name implies, the effects of exposures in this category are primarily on blood and blood-forming tissues and organs (such as bone marrow). In fact, in exposures between 100 and 1,000 Rem, the most significant damage is to organs like the bone marrow. At these exposures, people begin to die, and the upper limit of 1,000 Rem is fatal in approximately 80 to 90 percent of cases. Symptoms can include nausea and vomiting, malaise and fatigue, fever, blood changes and possible bleeding. The nausea and vomiting are not persistent, and the onset usually correlates with dose; that is, the larger the exposure, the sooner the onset of nausea and vomiting. In this category, the white blood cell count will rise during the first 24 hours following the exposure but then decline. The lowest point of white blood cell count usually occurs between four and six weeks following exposure. During this time, secondary infection is likely to occur. If the victim survives the radiation exposure and subsequent illness, the white blood count may return to normal within several months. When death does occur, it is usually several weeks after the exposure. Although the outcome of any given exposure cannot be accurately predicted, the LD_{50} for man is believed to be in the range between 400 and 600 Rem.

In the range between 1,000 and 5,000 Rem, the *lethal category,* the most characteristic symptoms are the result of serious damage to the gastrointestinal tract (refer to Chapter 4's discussion on the digestive system). Nausea and vomiting are prominent, and malaise, fatigue, and diarrhea are severe and increasing. The lining of the gastrointestinal tract may be sloughed. The crypt cells in the intestinal tract lose their lubricating capacity, and the villi are left unprotected and will eventually cease to function. This is a serious and extremely dangerous condition, as imminent failure of the intestinal lining itself is now probable. In this category, blood changes are also more pronounced, but the victim usually dies before external symptoms occur. Death is probable within one to two weeks.

In extremely high exposures over 5,000 Rem, the most drastic and immediate damage is to the central nervous system. Nausea, vomiting, and explosive diarrhea will appear within minutes. Symptoms of shock may also be prominent and convulsions are likely. Death is expected to occur within a day or two after exposure. Evidence of the serious nature of such exposures followed the nuclear disaster in the former Soviet Union, Republic of Ukraine. In April of 1986, the Chernobyl nuclear power facility experienced a core meltdown in one of its reactors. Those unfortunate individuals who responded first to the disaster (mostly fire fighters and plant personnel) died within days following exposures that were calculated to be well in excess of 5,000 Rem.

Accident Prevention Practices and Control Measures

In general terms, accident prevention measures for ionizing radiation should be site specific for each factory, plant, or facility. For instance, precautionary measures and procedures would differ dramatically between a health care facility and a nuclear power plant. However, in an effort to establish the minimum acceptable requirements for an effective radiation safety program, OSHA has established the Ionizing Radiation Standard at 29 CFR 1910.96. Within this regulation are generic precautionary guidelines that can be used regardless of the industry or application to ensure the preservation of employee health. In summary, the OSHA Standard requires at least the following precautionary measures:

1. Surveys of work areas must be conducted to evaluate the radiation hazards that exist as a result of the production, use, release, disposal, or presence of radioactive materials or other sources of radiation (x-ray machines, for example).
2. Personal monitoring equipment must be provided at no cost to employees who are likely to be exposed to a specified dose of ionizing radiation during their assigned duties. Such equipment may include film badges, pocket dosimeters, or film rings.
3. Caution signs, labels and/or signals must be used to identify radiation areas, rooms, containers, and other material storage locations. The standard ANSI symbol, as shown in Figure 7-7, must be used on all signs to indicate the presence of radioactive materials. Facilities shall also be equipped with warning signals (horns) if a leak of large quantities of radioactive materials that would warrant the immediate evacuation of personnel is possible.

FIGURE 7-7 The standard ANSI symbol for radiation, as adapted by OSHA, that must be used on all signs to indicate the presence of radioactive materials (Source: OSHA 29 CFR 1910.96)

4. Instructional training must be provided to all individuals who work in or frequent any portion of a radiation area. Training should include instruction on the safety problems associated with exposure to such materials and the precautionary measures (including the use of protective devices and equipment) to minimize exposure. Employees must also be informed that they have the right of access to any reports on radiation exposure monitoring that the employer may have performed in their work areas.
5. Storage and waste disposal procedures for radiation materials must be in accordance with acceptable methods as prescribed by applicable regulatory agencies, both state and federal.

In addition to the accident prevention measures summarized above, the Standard also provides information concerning the notification requirements following radiation incidents, including exposures. These requirements address method of notification (telephone, telegraph, written), as well as specific exposure criteria that will mandate notification if these levels are exceeded.

Methods of Detection and Sampling

A variety of methods are available to the industrial hygienist to detect and sample levels of ionizing radiation for subsequent evaluation. All methods of detection rely on the ionization characteristics of radiation. Measurement of source activity is determined based on the rate at which ionization occurs within the measuring instrument. Differences between devices lie only in the *medium* in which the ionization takes place inside the instrument and in the way the measured activity is reported for evaluation. For example, *personnel exposure monitors* are designed to measure the total cumulative radiation exposure in terms of *absorbed dose.* Whereas, field survey instrumentation measure exposure in terms of a *dose rate over time,* such as Roentgens per day or milliroentgens per hour, while *laboratory* sampling and evaluation devices primarily measure radiation in terms of *radioactivity* (curie, nanocurie).

In general, the types of equipment described below are commonly used to monitor radiation exposures.

Film Badges

The most basic type of monitoring device is the *film badge.* Worn on the clothing, it can monitor for gamma and x-ray radiation as well as beta and alpha radiation. It is constructed of a small piece of photographic film encased in an opaque cover with a metal back support. This methodology is available as a badge that can be pinned to clothing or as a ring. On exposure, the silver atoms in the film react with the radiation activity in a similar way that light rays will expose the film.

When the film is later developed, the amount of film exposure matched against a control badge that was not exposed helps determine the amount of radiation exposure witnessed by the badge. Since the developed film is as final as a photograph, it provides a permanent record of exposure.

Thermoluminescent Dosimeters

The thermoluminescent dosimeter (TLD) ("thermo" meaning heat, and "luminescence" meaning light) is similar to the film badge in that it, too, can be worn as a ring or badge, but it is thought to provide much more meaningful results. It is used primarily for the measurement of x-ray, gamma, and beta radiation. The principle of operation is simple. Crystalline solids and minerals, such as lithium fluoride, often emit light when exposed to ionizing radiation. Light is emitted from these materials when trapped electrons are released through a heating process. TLDs are almost completely unaffected by most environmental conditions such as shock, pressure, exposure to chemical vapors, and atmospheric changes (humidity, temperature).

Pocket Dosimeters

Designed to measure x-ray and gamma radiation (and perhaps beta), the direct reading portable pocket dosimeter provides excellent/quick information on exposure dose. Normally a small pen-shaped device that clips onto a shirt pocket, this dosimeter consists of a quartz fiber, a scale, a lens (for observing the movement of the fiber across the scale) and an ionization chamber. The fiber is electrostatically charged. On exposure to radiation, air in the chamber becomes ionized and the quartz fiber moves across the scale in direct proportion to the amount of radiation present in the chamber. This method provides a direct (and immediate) read-out on radiation dose to the user.

Ionization Chambers

If a variable source of direct voltage is made across an enclosed volume of gas (air, for example) an ionization chamber is formed. When exposed to radiation, the air in the chamber becomes conductive and the rate of radiation exposure can then be quantified. Used primarily for x-ray, gamma and beta radiation measurements, it can also be used for alpha radiation if configured properly.

Geiger-Mueller Counters

The Geiger-Mueller (GM) counter utilizes an ionization chamber that is usually a glass tube filled typically with argon that produces ion pairs when irradiated. To count efficiently, the instruments have self-quenching tubes filled with 10 to 25 percent ethylene or iso-butane to quench the ionization avalanche started by exposure to ionization radiation. The quenching gases are decomposed by the radiation and, therefore, have a limited useful life. GM instruments are extremely sensitive for measuring low intensities of beta, gamma, and x-radiation.

Scintillation Instruments

Scintillation instruments rely on the light produced when ionizing radiation interacts with phosphor or crystal capable of producing this light. The resulting *scintillations* are then photo-multiplied and amplified. They can be used to detect either alpha, beta, gamma, or x-rays. These detectors are extremely sensitive (especially to gamma radiation). While scintillation instruments are useful at both high and low dose rates, experience has shown that they appear to have their greatest utility in measuring very low dose rates.

Nonionizing Radiation

Those radiation energies that exist in the longer wavelengths (see Figure 7-4) of the electromagnetic spectrum are not capable of producing ionization and are therefore referred to as *nonionizing radiation*. However, these energies are still of concern to the industrial hygienist since unprotected exposures can cause effects in biological systems. The mechanisms associated with such effects are extremely complex and beyond the scope of this publication. For the purpose of this chapter, the discussion on nonionizing radiation shall be divided into the categories of *invisible light* and *visible light*.

Invisible Light

In the simplest terms, exposure to the wavelengths that produce *infrared* (IR) and *ultraviolet* (UV) light, like those responsible for *radio frequency* (RF), *television* (TV), and *microwaves* (MW), can result in damage to skin and, in some instances, the underlying muscle tissues. The harmful effects are the result of the propensity of energies at these wavelengths to excite atoms to the point where their orbiting electrons become extremely unstable. Although electron ejection does not occur, the incident energies colliding violently with atoms can harm body tissue on the molecular level. The resulting dermatological and/or ocular health effects are of principal concern to the industrial hygienist.

Ultraviolet (UV) Radiation. It has long been known that unprotected, prolonged exposures to UV light can result in skin tanning and burning, pigmentations, and skin cancer. Similarly, when the eyes are subjected to UV light, conjunctivitis, keratitis, regressive corneal lesions, photosensitivity, and possible cataract formation can result. In the workplace, UV exposure is likely around electric welding operations, blueprinting, laundry marking, advertising, photoengraving, and sterilizing (food, water). There are also numerous uses in the medical profession.

The exact biological effect of exposures to UV radiation is dependent on the intensity of the exposure and the distribution of the energy from the generating source. Ultraviolet radiation will be absorbed by the epidermis layer of the skin; the shorter the wavelength the more energy absorbed. In the eye, most of the radiation will be absorbed by the cornea.

Infrared (IR) Radiation. Industrial applications for IR radiation include any process where surface heating is required such as drying (paints, varnishes, inks, adhesives), heat-shrinking (plastics, textiles), metal pipe fitting and forging, dehydrating (paper, leather, meats/foods), or spot application of heat for various desired effects. The principal hazard to health is created by the capacity of IR radiation to produce localized, high-intensity heat concentrations on exposed surfaces. Burning of skin and (less commonly) eyes are, therefore, of principal concern to the hygienist.

Radio Waves and Microwaves. These type of energies are normally emitted from antennas associated with radio and television transmissions and radar. However, as most everyone is aware, microwave energies are also commonly used in cooking applications and in the medical profession. As with other forms of nonionizing radiation previously discussed, these energies can also be absorbed on contact with a target surface. Health effects result from the ability of the energies in these wavelengths to induce electric and magnetic fields within biological systems. The results of exposure can range from little or no noticeable thermal effects to heating and subsequent damage of the affected tissues. While data on the effects of exposure to radio frequency (RF) transmission is inconclusive, there has been a tendency in recent years to suggest a possible link between blood disorders such as leukemia with long-term exposure to RF radiation. The principal effect of exposures to microwave radiation is the heating of exposed tissues. Low-level microwave radiation will normally be absorbed by the skin and will result in surface heating and possibly burning. Higher-level microwave energies can penetrate deeper into the underlying tissue to cause sever localized damage. Intense exposures may actually cause body temperature to rise to dangerous levels.

Visible Light

Illumination. Illumination is obviously one of the more visible factors of the work environment that tends to be of concern to most employees. Except for laser light, involuntary defensive mechanisms will prevent intense light from entering and damaging the eye. In most cases, the hygienist will find that employees are seldom concerned with too much light. Problems usually arise when there is too little light for them to properly perform their job duties. While there is little evidence that poor lighting will result in any permanent eye defects, it is a fact that performance levels improve and accidents rates decrease when the work environment is comfortable and has sufficient illumination for easy visual perception. Also, because working in poor lighting conditions can be particularly strenuous on the eye, proper lighting is also essential to avoid fatigue, headaches, and (in some people) nervousness.

Lasers. An acronym for *light amplification by stimulated emission of radiation* (LASER), the laser is a device that can produce a single frequency wavelength (or a narrow band of wavelengths) at high intensity in the optical region. Nearly all modern lasers present some level of eye hazard on direct exposure to the beam. Subsequently, the hygienist must ensure the proper selection of filtering lenses for use by laser operators. The specific lens type will depend on the wavelengths involved in the laser application and the optical density needed to prevent damage to the retina of the eye. The ocular hazard created by the laser is due to its intense and focused beam. Because it can pinpoint its energy onto a very fine and precise area, even the low energy level of 0.1 watt can cause severe damage to the soft tissues of the eye. In contrast, an ordinary 75 watt light bulb that disperses its energy in all directions presents no appreciable hazard to the eye on exposure. There are a variety of laser types currently used in industrial operations. By far the most common is the *ruby laser* (also known as the *ruby crystal* laser). The ruby laser beam is generated when a single crystal of aluminum oxide doped with chromium impurities is exposed to an intense light from a xenon flashtube. This results in a burst of intense red light emanating in a beam. The *gas* laser utilizes a combination of concave mirrors and reflectors to force a laser action in a gaseous discharge of helium and neon. In manufacturing operations, a principal advantage of the laser is its ability to apply an extremely high flux of energy to the surface of a workpiece, as compared with traditional heat sources such as flame, torches, electric arcs, and plasma jets. Lasers are usually placed in two categories for manufacturing operations. The first, *light-duty lasers,* range from a few tens of watts to a few hundred watts. Typical applications include cutting and drilling ceramic substances in the electronics industry, drilling gems (for example, rubies in watch making), and cutting light-gauge metals and cloth, plastics, wood, and a variety of other materials. The second category, *heavy-duty lasers,* range from a few kilowatts to a few tens of kilowatts. Typical applications include pipeline welding, automobile part welding, surface heat treating of engine and other parts, with the number of applications expanding as experience with this laser type is gained.

For the most part, the hygienist should ensure that employees who could possibly be exposed to nonionizing radiation be provided with the proper personal protective equipment (eye, face, and skin protectors) to ensure the effective reduction of any exposure risk. Employees should also be suitably trained in the hazards associated with nonionizing radiation and the precautionary measures that should be taken to ensure maximum safety. Although OSHA has promulgated a standard entitled Nonionizing Radiation (29 CFR 1910.97), it falls drastically short of the specific and stringent requirements that have been established for ionizing radiation. This is an excellent example of how some established compliance criteria must often be exceeded to ensure appropriate protection of employee safety and health.

Units of Measure

The fundamental unit of luminous intensity is the *candle*. The rate at which light flows, often referred to as the luminous *flux*, is measured in *lumens*. One lumen is equaled to the flux (or flow of light) projected on one square foot of a sphere, one foot in radius, radiating equally in all directions from a light source of one candle positioned at the center of the sphere. The *foot-candle,* therefore, is the unit of illumination resulting from this relationship and is an expression of the *lumens per square foot* that are created by a light source (the term *lux* refers to lumens per square meter). *Footlamberts* is a unit used to measure brightness, which is an expression of light reflection from a surface. Recommended minimum light levels, expressed in foot-candles for general activities, are listed in Table 7-3. The factors that affect light quality are brightness, glare, distribution and diffusion, and color.

Although it may seem that it would be obvious if light levels are not adequate, this is not always the case. Workers who complain of headaches, nervousness, eye strain, fatigue, or other subjective symptoms of ill health may be the victims of poor lighting in their work environments. Therefore, along with the normal assessment of the many other possible agents of harm that can result in such

TABLE 7-3   Recommended minimum light levels in foot-candles for general activities

LOCATION OR TYPE OF WORK	MINIMUM RECOMMENDED ILLUMINATION IN FOOT-CANDLES
General office and office equipment	60
Drafting or drawing rooms	100
Cafeteria, general eating area	30
Aisleways, corridors, stairways	20
Manufacturing workstations	30
Inspection—ordinary	50
Inspection—fine or difficult	100
Machine or bench work	50
Tool crib or rooms	25
Laboratories—general	50
Washrooms	20
Medical department—treatments	75
Medical department—examinations	100
Locker rooms	15

symptoms, the hygienist should never rule out poor illumination as a potential culprit. Inexpensive meters used to test illumination are widely available from a number of safety and health equipment suppliers and distributors. These instruments are normally direct-reading, self-calibrating devices that require little formal training to use. Even for the most basic of industrial hygiene programs, the purchase and use of such a tool is highly recommended.

SUMMARY

This chapter focused on the extremely complex hazards associated with exposures to the various forms of energy. These energies, which exist as heat, light, sound/vibration, electromagnetic waves, and a variety of others, are essential to life in measured concentrations. However, human exposure to excessive quantities or *doses* of these energies can be the cause of serious and even life-threatening ill effects.

Exposure to temperature *extremes* (both hot and cold) can seriously jeopardize human health. High temperatures tax the human body's capability to cool itself and maintain a healthy internal core temperature. When combined with strenuous work activities, excessive temperatures can lead to heat illness, up to an including heat stroke which can be fatal. Low temperatures present equally disturbing health hazard risks. The effects of the wind, in a phenomenon known as *wind chill,* further increase the effects of cold temperature exposures. These include *hypothermia* and the more severe *frostbite*. The hygienist must be aware of such hazards and the methods used to evaluate and control exposures.

This chapter also explored the hazards associated with exposure to energy in the form of noise and vibration. In terms of hazard recognition, occupational exposure to noise can be one of the most subtle and insidious hazards the hygienist may confront. Except in highly obvious cases (i.e., extremely loud or noisy environments) exposures are not readily apparent to the worker (or the hygienist). But such hazards exist and must be evaluated for their health-threatening potential and controlled, as required. Information on regulatory requirements under OSHA's hearing conservation program was also presented in terms of exposure evaluation and control. This included monitoring methods and techniques and employee training requirements.

A discussion on radiation hazards, in both ionizing and nonionizing forms, concluded this chapter. To understand how radiation exposures can affect human health, the reader was introduced to the *electromagnetic spectrum* and the various *wavelengths* in which radiation exists. The discussion on ionizing radiation focused on the three major types of *alpha, beta,* and *gamma* ray radiation with a related discussion on x-rays. A detailed discussion of the biological effects of exposure and the control measures, as highlighted in OSHA's Ionizing Radiation Standard, were also presented. The hazards associated with nonionizing radiation are of equal concern to the industrial hygienist. In particular, this chapter briefly discussed *ultraviolet light, infrared light, microwaves and radio waves,* and *visible light.* Although the capacity for these forms of energy to cause harm may not be as great relative to sources of ionizing radiation, the hygienist must still take proactive measures to ensure control over exposure to nonionizing radiation.

Since exposures to these various forms of energy can often be difficult to recognize, this chapter also paid particularly close attention on understanding energy. Although basic in scope, the information provided here should offer an excellent primer for practitioners who must assess radiation hazards in their facilities.

8

Occupational Exposure to Biological Hazards

INTRODUCTION

Perhaps the most insidious of all health risks, *biological hazard* refers to any plant or animal that can pose a potential health risk to humans (or animals). This includes plant or animal by-products. Occupational exposure to *biological hazards* has become an issue of such great concern that OSHA was moved to promulgate a specific standard to address occupational exposure to bloodborne pathogens.

On initial consideration, it may be assumed that occupational exposure to such agents is restricted to very specific occupations, such as health care professionals. However, in reality, there is potential for exposure to biological hazards in numerous occupations in other industries, albeit not as great. In other words, while most biological hazards, or *biohazards* for short, can and do threaten specific occupational groups, others (such as the common cold virus) can adversely affect human health in literally any line of work. Biohazards can be infectious, toxic, and/or capable of producing systemic effects such as tumors and cancers in humans and animals. Exposure can occur through inhalation, ingestion, absorption, or injection. Although not discussed at great length in other chapters of this text, *injection* does become a particularly critical concern as a route of exposure for biohazards, especially in the health care professions and in laboratory/research settings.

While creative names such as *legionnaires' disease, sick building syndrome,* and *acquired immune deficiency syndrome* (AIDS) have been used to identify some specific biohazards, the basic hazardous nature of each is its ability to *infect* exposed individuals, resulting in ill-health or even death. These *infectious agents* can be bacterial, viral, parasitic, or fungal. When such agents are present in the workplace, it becomes a function of the employer's occupational health programs to prevent exposures. Ideally, this responsibility should be shared jointly by the industrial hygienist, the industrial physician, and the occupational health nurse. Each should have a specific role in identifying, evaluating, and preventing exposures to biohazards. However, since many organizations may not have all of this expertise on staff, it will most likely be the industrial hygienist who first recognizes the potential for exposure to biological hazards Therefore, the industrial hygienist should become familiar with the various biohazards and how to recognize when occupational exposures are likely.

This chapter will provide the fundamental information necessary to understand these hazards so that the industrial hygienist will be able to recognize when biohazards are present and what minimum actions should be taken to evaluate and control exposures.

RECOGNIZING BIOLOGICAL HAZARDS

One of the first steps toward recognition of biological hazards in the work environment is to conduct an assessment or survey of the organization to identify those occupations and work practices most likely to come into contact with these agents. This approach will help facilitate the industrial hygiene task since most (but not all) potential for biohazard exposure can be linked to certain types of jobs or functions. Once this has been accomplished, the hygienist can then broaden the scope of the workplace assessment further to include other, less obvious exposure potential.

To properly perform this assessment, however, the hygienist must first possess a basic understanding of the various types of biohazards, how these can come to exist in the workplace, and the effects of exposure. Following is a brief description of the infectious biohazard agents most commonly encountered in industry today.

Bacteria

These infectious agents are microscopic, unicellular organisms that replicate by dividing into two parts (*fission*). Depending on the species and cultural conditions, bacteria occur as individual cells or in clumps or chains of sister cells. Bacteria are only visible at the lower limits of resolution of the optical microscope. Their average length ranges from 2 to 5 micrometers, although some are as small as 0.2 micrometer or as large as 100 micrometers in length. Bacteria are classified, albeit somewhat arbitrarily, by a descriptive array of features, one of the most common being *shape*. For example, rod-shaped bacteria are known as *bacilli* (singular, bacillus). They are characterized by small, whip-like structures known as *flagella* which allow them to become *motile* (capable of spontaneous movement). Some bacilli have oval, egg-shaped, or spherical bodies in their cells, known as *spores*. Under adverse conditions (temperature and pressure), dehydration, and in the presence of disinfectants, the bacteria may die, but the spores may live on. These spores can then germinate when conditions once again become favorable and form new bacteria cells. Some are so resistant that they can withstand boiling and freezing temperatures and prolonged desiccation.

A second type of bacteria, *cocci* (singular *coccus*), are spherical or ovoid in shape. The individual bacteria cells of this group may occur singly *(micrococcus)*, in chains *(streptococcus)*, in pairs *(diplococcus)*, in irregular bunches *(staphylococcus)*, or in the form of cubical packets *(sarcina)*. Cocci do not form spores and are usually non-motile.

Another group of bacteria are those shaped as curved or bent rods. Of these, the genus *vibro* is composed of bacteria that are comma-shaped; and the genus *spirillum* consists of those that are twisted and spiral in form. All members of this group are motile, but none form spores. However, some of these bacteria form a gelatinous capsule or covering that is believed to protect them from adverse environmental conditions. Still another group of spiral-shaped bacteria is known as the *spirochetes*, one of which is the cause of syphilis.

Bacteria may also be classified in terms of their need for free atmospheric oxygen. Those that require atmospheric oxygen are known as *aerobic* (air-living); those that cannot survive in the presence of oxygen are called *anaerobic*. While bacteria are also dependent on the proper temperature for life and reproduction, the various species of bacteria may differ widely in the temperature requirements. For example, most of the disease-producing (pathogenic) bacteria thrive best at body temperatures. Others may live and multiply in much cooler environments, while still others live only in hot springs and conditions of extreme temperature. Freezing, as a rule, does not destroy bacteria but does prevent their reproduction. Conversely, high temperatures will quickly kill many bacteria. Most disease-producing organisms in milk, for example, are destroyed by raising the temperature to 143° F (61.6° C) during the pasteurization process. Most non-spore forming, pathogenic microorganisms are destroyed by boiling water. In the spore stage, some bacteria must be heated to 240° F (116° C) for quite some time to destroy the spores. Bacteria may also be killed using chemicals known as *disinfectants*. A substance that prevents infection or inhibits growth of microorganisms is called an *antiseptic*. Some of the most potent disinfectants are phenol and related compounds. Free chlorine gas and the hypochlorites are also excellent disinfectants. Other dependable disinfectants include tincture of iodine (for cuts and wounds), mercury-containing compounds (mercurochrome), or a 50 to 70 percent alcohol solution. Antibiotics, which are actually produced by other living organisms, inhibit the growth of bacteria or destroy them (bactericidal). There are few known bacterial diseases that cannot be eradicated if the proper antibiotic is used early in the course of the disease. Tetanus and botulism are exceptions.

These diseases are the result of extremely potent toxins produced by bacteria, rather than symptoms caused by infections of the microorganisms themselves.

Bacterial diseases may be transmitted in a number of ways. Perhaps the greatest threat in the workplace are the common respiratory diseases that are distributed via small droplets of sputum and nasal secretions from already infected individuals. Furthermore, since many bacterial agents survive until they are dried, diseases may be transmitted through indirect contact with persons through objects that they have handled. Still other diseases are transmitted by water, milk, and foods that have become contaminated. Typhoid, cholera, and diarrhea are examples of diseases transmitted in this manner. Disease-producing bacteria usually cannot penetrate unbroken skin; hence they enter the body through wounds, abrasions, or natural openings in the body. Table 8-1 lists the most common types of bacteria and the diseases they are known to produce. While the spread of some bacterial diseases, such as the venereal diseases, is not normally a problem in the average workplace, the hygienist should still be aware of their existence to ensure proper recognition and so that workers can be educated on the dangers associated with such diseases.

In the workplace, employees with wounds, cuts, abrasions, skin problems (such as rash or dermatitis) are most susceptible to bacterial infection. Food-borne bacteria can also create problems, especially in areas where massive food preparation occurs (such as in large employee cafeterias, prisons, or during military operations).

Viruses

Viruses are considered to be the smallest infectious agents, so small that they are capable of replicating themselves within the confines of bacteria cells. The majority of these extremely small infectious particles fall within the range of approximately 0.02–0.25 micrometers and therefore can be seen directly only with the aid of an electron microscope. The smallest unit of a mature virus is known as a *virion*. It consists of a single molecule of nucleic acid enveloped by a specific number of protein molecules. Years of debate have centered around the question of whether viruses are living or non-living and, although resolution of this is now considered to be simply a matter of semantics, several fundamental differences distinguish viruses from other organisms (Considine, 1995). Viruses, unlike true microorganisms, are not cells, do not replicate by binary fission and contain a genome consisting of only one type of nucleic acid (RNA or DNA, double- or single-stranded).

Viruses are obligatory intracellular parasites and, as such, cannot replicate without the assistance of a host cell. In contrast to the bacteria cells that multiply by binary fusion, viruses multiply by synthesis of their separate components, fol-

TABLE 8-1 The most common types of bacteria and the diseases they are known to produce (Considine, 1995)

COMMON NAME	MEDICAL NAME	BACTERIA RESPONSIBLE	BODY PART INVOLVED	INCUBATION PERIOD
Septic sore throat	Streptococcus sore throat (tonsillitis)	Streptococcus (several species)	Throat and nasal membranes	3–5 days
Scarlet fever	Scarlatina	Streptococcus scarlatinae	Throat, tonsils, other tissues	3–5 days
Pneumonia	Pneumococcal pneumonia	Diploccus pneumoniae	Respiratory tract, including lungs	Varies
Spinal meningitis	Epidemic meningitis	Diploccus (neisseria intracellularis)	Respiratory tract, nervous system, sometimes blood	1–5 days
Clap	Gonorrhea	Diploccus (neisseria gonorrhoea)		
Typhoid and paratyphoid fevers	Enteric fever	Short rod (salmonella typhosa; salmonella paratyphi)	Reproductive organs / Intestine	2–8 days / 10–14 days
Bacillary dysentery	Shigellosis	Short rod (shingella dysenteriae)	Intestine	1–4 days
Whooping cough	Pertussis	Small short rod (hemophilus pertussis)	Respiratory tract	7–14 days
Bubonic plague	Pestis	Short rod (pasteurella pestis)	Blood, spleen, liver, lymph nodes	2–10 days
Rabbit fever	Tularemia	Short rod (pasteurella tularenis)	Lymph nodes, spleen, liver, kidney, lungs	1–10 days
Undulant fever	Brucellosis	Short rod (brucella abortus)	General body infection	5 days–10 weeks
Lockjaw	Tetanus	Sporeforming rod (clostridium tetani)	Nervous system	2–40 days
Gas gangrene	Gas gangrene	Sporeforming rod (clostridium perfringens)	Wounded areas	Varies
Botulism	Botulinus	Sporeforming rod (clostridium botulinum)	Nervous system	18–66 days
Tuberculosis	Tuberculosis	Irregular rod (mycobacterium tuberculosis)	Lungs, bones, other organs	Varies
Syphilis	Lues	Spiral-shaped organism (treponema pallidum)	Blood and nervous system	10–90 days
Diphtheria	Diptheria	Irregular rod (corynebacterium diptheriae)	Respiratory tract	1–7 days

lowed by assembly. Several stages are involved in viral replication, but the two most important are

1. *Attachment or adsorption:* The virus becomes attached to the host cell via specific receptors. Thus, cells lacking receptors are resistant to viral attack.
2. *Penetration:* With enveloped viruses, this step occurs when the virion's envelope fuses with the cellular membrane of the host cell. Naked virions penetrate intact through cellular plasma membrane and into the cell cytoplasm. Ordinarily, without a protein coat, viral nucleic acid is incapable of entering a cell structure. This demonstrates the importance of the protein coat in rendering a viral agent ineffective. If the protein coat is missing or inadequate, such viruses become a concern.

Once inside the cell, *virulent viruses* disrupt normal cellular synthesis, causing a shift toward viral synthesis. These viruses cause the ultimate destruction of the infected cell. In contrast, there are moderate viruses that stimulate the host DNA and may propagate viral cancers. In general, the DNA viruses multiply in the nucleus of the host cell.

Commonly known viruses (for which vaccines are available) include smallpox, measles, rubella and yellow fever. There are vaccines to protect against infection from numerous (hundreds) of known influenza viruses, which have also been moderately successful in preventing the spread of influenza. Other diseases known to be caused by viral agents include the mumps, Newcastle disease, canine distemper, rabies, herpes simplex (chicken pox, shingles, Epstein-Barr), the hepatic viruses (hepatitis A and B), and the retroviruses of which the *human immunodeficiency virus* (HIV) is most familiar.

In the workplace, exposure to viral agents may occur in occupations where animals are handled regularly, such as veterinary medicine. Workers employed in the meat packing and processing industry are even more susceptible, however, due to the great amount of exposure to animal blood, blood products, and tissues. Of course, with some viruses, the greatest potential for exposure is in the health care industry (both to workers and to patients). Occupational exposure to bloodborne pathogens such as HIV and the hepatic viruses is the subject of a specific OSHA Standard located at 19 CFR 1910.1030 and will be discussed later in this chapter.

Fungus

Any *thallophytic* plant characterized by an absence of chlorophyll is a fungus (plural fungi). In fact, accept for the absence of chlorophyll, the structure of fungi resembles that of algae. Examples of fungus include mushrooms, toadstools, smuts, molds, and mildews. There are well over 70,000 species, with widely diverse habits and characteristics. Therefore, generalizations are difficult. Fungi can grow in almost every type of environment where organic substances exist. Numerous species can be found in fresh water and in salt water. Others are adapted to life on land or in the ground. Fungi are particularly abundant in the tropics because warm humid climates tend to favor the existence of numerous species.

More than 40 species of fungi are known to produce diseases in humans. There are those fungi that attack only the skin (including hair and nails), and there are those species that will invade deeper tissues of major internal organs to produce serious systemic effects. In addition to causing diseases and discomforts of an annoying nature, such as athlete's foot, jock itch, and ringworm, certain species of fungi are involved in human diseases of a more serious nature. In North and South America and in Europe, the principle fungus infections (referred to as mycotic infections) are:

1. *Blastomycosis.* A pulmonary disease with symptoms that closely resemble tuberculosis, blastomycosis is more common in males between the ages of 20 and 50. People who work outdoors where close and frequent contact with soil run a higher risk of infection. The disease can also cause skin lesions at the site of contact with the soil sporophytes. The characteristic lesion in cutaneous blastomycosis is raised and crusted and is usually visible on the upper extremities and the face.
2. *Coccidioidomycosis.* Also known as Valley Fever, this disease is caused by a fungus found in desert soils. It is also prominent in areas of semiarid conditions characterized by hot, dry summers, mild winters, and moderate rainfall. The parasitic form of the fungus that causes this disease is a spherical cell 30 to 60 micrometers in diameter that does not bud but divides internally to produce numerous endospores 2 to 5 micrometers in diameter. The reader will recall from the discussion in Chapter 4 that particulate matter smaller than 5 micrometers can reach the lower recesses of the respiratory tract. In serious cases of infection, the disease may involve bones, joints, skin, subcutaneous tissues, and internal organs, although such cases are rare. There may be a rash on the palm of the hands and the soles of the feet.
3. *Histoplasmosis.* Inhalation of or contact with the infectious fungus spores 8 to 20 micrometers in diameter can result in ulcerative lesions in the skin and mucous membranes around the mouth or in pulmonary symptoms similar to tuberculosis. The fungus is prominent worldwide in soils, especially in those that have been enriched by chicken droppings. In fact, minor epidemics have followed the cleaning of chicken coops or aviaries.

Mycotic infections have a number of common points. They usually occur in fairly well defined geographic regions. They usually occur in soil, and they are easily aerosolized and thus easily transported or spread by air currents. Exposure by inhalation is generally considered to be the most common route of entry. The disease-causing fungus exists in its natural habitat as a mold that bears infectious spores. On airborne dissemination, the spores will enter a host and develop into a yeast-like phase that becomes the pathogen.

Many fungi may also have a poisonous nature. The toxic substances present in a fungus are products of its metabolism, not substances absorbed from its surroundings. These toxic substances vary, as do their effects on humans. Some may result in abdominal cramps, nausea, vomiting, thirst, and diarrhea. Others may act on the central nervous system, causing a lack of coordination, hallucinations, and delirium, as well as gastric disturbances.

Parasites

Any organism (often microbial, but not always) that lives on or in another organism but provides no benefit to its host is a *parasite*. While all parasites are not necessarily bad, most can cause some degree of harm to their host. Although bacteria, viruses, and even some fungi are parasitic in nature, there are also plant and animal organisms that are parasitic and can cause harm to their host organism. Infections caused by such organisms include malaria, amebic dysentery, and a variety of lesser known and less common blood and gastrointestinal afflictions. The parasitic agent responsible for such infections are known as *protozoa. Helminths* are parasites that can cause diseases such as hookworm and contact with *anthropods* (mites, chiggers) can result in dermatitis.

OCCUPATIONS AT RISK AND METHODS OF CONTROL

Having provided some basic information about the nature of biohazard agents, hygienists should be able to conduct fairly accurate assessments of their workplaces to determine those occupations or job functions most likely to involve exposures. As a point of reference, the following are examples of those professions, tasks, functions, or occupations where the potential for exposures to biohazard agents is somewhat greater than in other lines of work.

Health Care Workers

These include doctors, nurses, medical technicians, toxicologists, paramedics, phlebotomists, emergency room personnel, housekeeping and janitorial staff, laundry workers, orderlies, volunteer workers, x-ray and laboratory workers, in-house pharmacy staff, dietitians and nutritional counselors, supply department personnel, and even those involved in the disposal of biological waste products.

Heath care workers are not only at risk of contracting a diseases or infection from patients, there is also a risk that they will be exposed to biological agents carried by their fellow employees. For example, a dietitian with a gastrointestinal illness could conceivably spread the infection to fellow workers and patients in a hospital setting.

The principal control measure to preclude such exposures is a practiced and enforced policy of *personal hygiene.* Without exception, all workers in the health care industry must consistently ensure that their hands are thoroughly cleansed after contact with any patient at any time. Washroom facilities must be made readily available, kept clean, and well-stocked with soaps and other disinfecting agents. The second line of defense utilizes the principles of *disinfecting* and *sterilization.* Disinfecting will clear an area of harmful microorganisms and sterilization will actually destroy or kill these agents. A strict policy of sterilization and disinfecting must be enforced to ensure the maximum reduction of exposure risk. All equipment that has (or may have) come into contact with infectious agents must be properly disinfected and or sterilized before their next use. Equipment, such as autoclaves, must be kept in proper working order and checked periodically to ensure effectiveness. Chemicals used for sterilization are often highly toxic and may even be carcinogenic, such as ethylene oxide. So, in addition to the potential for exposure to infectious agents, workers must also be protected from exposure to these chemical commodities as well. Separation of ventilation service systems is yet another critical consideration to preclude the spread of infectious agents through the health care facility.

Finally, hospital and health care workers should be provided immunizations for diseases such as rubella, tuberculosis, and, when blood contact is probable, hepatitis B.

By implementing recommended measures and by establishing an OSHA-required (see below) bloodborne pathogens *Exposure Control Plan* (ECP) that includes employee training, health care workers should be afforded the maximum level of protection agents exposure to biohazards in the workplace. The objective is still the preservation of employee health through the recognition, evaluation, and control of hazards that threaten human health in the workplace. The industrial hygienist employed in the health care industry has the advantage of having almost unlimited medical expertise readily available for consult. With the necessary emphasis on biohazards above that which is normally required in general industry, the hygienist would do well to make use of this asset as often as possible.

Laboratory Workers

In addition to those who work in medical laboratories in hospitals and other such settings (as described above), workers in biological research laboratories are quite susceptible to exposure. Laboratories concerned with research and analysis of carcinogenic, mutagenic, and teratogenic agents, as well as those conducting experiments with biohazardous agents, have a higher than average exposure risk, since their work cannot occur without direct or indirect contact with these substances. Workers can be exposed through inhalation, absorption, ingestion, and injection. Many routine laboratory procedures that involve the intentional pulverization of tissue samples can further increase the likelihood of exposure. If improper or inadequate work practices and procedures are employed, exposure risk can become nearly uncontrollable. Personnel working in animal research facilities can be exposed to biohazards, such as viruses, that are not normally encountered in the average work setting. Special attention during the work area assessment of such environments is warranted to ensure proper recognition of all potential health hazards. Worker training should cover personal hygiene, adherence to established laboratory protocol, proper accident and incident reporting requirements, techniques for disinfecting and sterilization, waste disposal procedures, and methods for ensuring adequate protection of personnel (including the use of personal protective equipment and immunizations).

Agriculture and Farm Workers

Employees of the agriculture and farming industries are susceptible to exposure from a wide variety of occupational activities that require work directly with domesticated and non-domesticated animals and/or certain plants that may carry viruses, bacteria, and fungi, as well as bloodborne pathogens associated with many farm animal species. Infectious agents can enter the body through inhalation, ingestion, and/or absorption through the skin or mucous membranes. Direct contact with infected animal flesh, blood, or blood by-products is the principle threat to the farm worker. Spore-contaminated dusts are also of concern, since they can infect both animal and human populations. The spreading of disease following such infections is then greatly facilitated. Certain pests (such as mosquitoes, flies, fleas, chiggers, mites, spiders, hookworms, and ticks) as well as rodents or other small animals (mice, rats, moles, squirrels, rabbits, skunks, mink, and foxes) and reptiles (snakes), that are commonly found around farm settings also have a role in the spreading of disease, either directly or from one host to another.

There are numerous diseases to be concerned about when assessing exposure potential in the farm and agricultural industries, as well as in general industry. The following are examples of those most commonly encountered:

1. *Anthrax.* A highly infectious disease caused by a bacteria (bacillus anthracis). This disease is mainly serious in cattle, but it can be transmitted to

humans as well. Transmission typically occurs by way of meat and animal product processing (such as hide tanning and handling). The most common form of anthrax is cutaneous. The cutaneous lesion is often referred to as a *malignant pustule.* The anthrax bacteria forms pores in the external environment. These spores can linger in the soil and in animal products for many years. A cutaneous infection is usually the result of spore contact with a skin abrasion or wound. Antibiotics have been used quite successfully to treat this form of anthrax. The *intestinal* form of anthrax results from the ingestion of anthrax-contaminated meat (notably sausage). Inhalation of anthrax begins with high fever, malaise, and a dry cough followed by dyspnea (difficulty in breathing), cyanosis (blue skin due to lack of oxygen in the blood), hemoptysis (spitting-up of blood), and chest pain. Antibiotics have also proven successful in treating ingestion and inhalation anthrax cases. The primary control measure centers on cattle vaccinations to eliminate the concern for human exposures at the source. Other controls placed on the importing and exporting of animal products have also helped reduce the risk of exposure. Once again, worker personal hygiene, training, the use of engineering measures (ventilation) and protective equipment are all considered essential elements of an adequate anthrax exposure control program.

2. *Brucellosis.* Once commonly known as undulant fever or Malta fever, brucellosis is an infection caused by the bacteria brucella and is transmitted from domesticated animals to humans. The three main species of *brucella* are those found in cattle, hogs, and goats. In recent years, a canine brucella has also been identified. Prior to the initiation of the pasteurization process, the primary source of this disease was the ingestion of raw milk and milk products (butter, cheeses). Those now at highest risk are workers in the meat packing and livestock industries. In fact, less than 10 percent of brucellosis cases result from the ingestion of milk products (Considine, 1989). The responsible microorganism generally enters through the mucous membranes of the mouth and throat, or through breaks in the skin. The organism then reaches the lymphatics, traverses the lymph node barriers, and invades the bloodstream. Once the organism enters the bloodstream, almost any organ in the body is at risk. Lesions are most frequently seen in the spleen, liver, lymph nodes, and bone marrow; however the heart, joints, lungs, and prostate are also susceptible. Because of the long incubation period, an exposed worker may not display any symptoms for periods ranging from a few days to several months. Symptoms include chills, fever, headache, loss of weight and appetite, and myalgia (muscular pain). There may also be pain in the spinal area or even evidence of acute arthritis. Pregnancy may end in spontaneous abortion. Antibiotics have been successful as treatment. Untreated, the infection may produce complications such as bone infections (osteomyelitis) or a highly localized infection of the kidneys (nephritis).

In addition to proper hygienic practices, workers with potential exposure to this disease must take care to ensure that minor cuts and abrasions receive the proper attention as soon as possible.

3. *Dermatosis.* Many skin diseases of varying severity are caused by contact with molds or fungi. In the agriculture and farming industries, the most commonly encountered fungal dermatosis is *ringworm.* It is a highly contagious disease that can be spread by animal as well as by humans. Dogs and cats that are not bathed frequently are common sources of human infections. Ringworm may be acquired by direct contact with the infection, or it may be spread to other areas of the skin of a single individual. Objects handled by the infected person also carry the fungus. Occasional sources of infection are the backs and arms of theater chairs, hair combs and brushes. Fungi thrive on damp, warm skin, especially in areas such as the crotch and on the soles of the feet or between the toes. Ringworm of the crotch is referred to as jock itch. On the feet it is more commonly known as athlete's foot. The disease usually takes on the form of one or several raised, round sores on the skin that seem to heal in the center while the edges continue to grow outward. Occasionally, the healed centers may become reinfected, and a second ring develops and grows within the first ring. In some cases, there is no healing of the center, and the lesion continues to grow. The sores of ring worm begin as small, slightly raised areas with a reddish color. As they enlarge, they become redder, and often contain one or many blistered areas. There may be slight itching or burning sensation.

Workers at risk of contracting occupational dermatosis include farmers, animal handlers (zoo keepers), pet shop employees, hide tanners, ranchers, and employees in animal laboratories. Beyond the agricultural and farming industries, athletes and employees who work in health clubs/gymnasiums, or those who are lifeguards or work in the pool maintenance industry are also at risk of exposure. Control measures focus on preventing the existence of the environmental conditions that are conducive to fungal growth. Frequent sterilization of common use areas such as shower facilities at swimming pools and gymnasiums is essential. Employee education and training is also a critical line of defense against fungal infections.

4. *Tularemia.* This is an infectious, communicable disease carried by rodents and transmitted to humans via insect bites. The infectious agent is carried by rabbits, squirrels, skunks, mink, rats, foxes, mice, coyotes, as well as some fowl, snails, and fish. Dogs and cats, although themselves immune to the disease, may harbor the infectious organism and transmit it to humans through bites or scratches. Blood-sucking insects (usually ticks, but including mosquitoes and fleas) are the most commonly identified mode of transmission to humans. Exposure to infected animals, however, is also quite

common. Persons who handle rabbit meat and hides assume a risk of becoming infected. The incubation period for this disease is approximately three days. The disease takes three forms, depending on the site of infection. The *ulcer form* causes skin sores/ulcerations that usually develop within three days following infection with regional lymph tissue involvement. The ulcer or lesion may be a seeping (as opposed a dry), scaling sore. Fever may reach 104–105.8° F (40–41° C) accompanied by severe headache. Diseases with somewhat similar symptoms include the plague, cat-scratch disease, and infectious mononucleosis. The *pulmonary form,* which can precipitate tracheitis, bronchitis, and pneumonia, is characterized by cough and difficulty in breathing. The *intestinal infection form* can result in abdominal pain, nausea, vomiting, and diarrhea. When the entry site of the infection is not apparent, it may indicate a waterborne infection.

People who work outdoors, especially in wooded areas or on farms, should take care to avoid contact with insects, especially ticks. Using personal protective equipment such as rubber gloves and face shields while handling animal carcasses that may be infected will help decrease the risk of exposure. Hunters, farmers, ranchers, fishermen, field geologists, park officials, fish, game and wildlife workers, environmental site assessors, surveyors, property appraisers and assessors, construction and road crews, and employees in animal research laboratories are all occupations that may be at risk of exposure to the tularemia disease.

5. *Tetanus.* A serious, often life-threatening disease caused by exposure to a neurotoxin that is generated by the anaerobic bacillus clostridium tetani. It occurs in soil and in the intestines of domestic animals. Spores of the organism have also be recovered from clothing, house dust, and even the air of occupied buildings (such as hospitals). These spores are introduced into the human body by way of penetrating wounds in which deep layers of tissue are punctured and exposed. When spores multiply and spread, they generate two toxins, *tetanospasmin* and *tetanolysin.* Only the former can present a serious threat to human health. It reaches the spinal cord and the brain by way of the bloodstream and can interfere with peripheral nerve responses. The toxin interferes with normal muscle control and spasms may result. These spasms vary in accordance with the relative strengths of opposing muscles. Thus, a spasm may produce an extension of the hips or knees, flexion of the biceps, and/or a condition known as "lockjaw" (*trismus*). Early symptoms of the disease are pain associated with the muscle spasms, restlessness, stiffness (neck, back, thighs, abdomen), difficulty in opening the mouth and perhaps difficulty in swallowing. The early phase of this disease, with all or some of the above mentioned symptoms, has a duration of about 72 hours, after which a phase known as general tetanus takes place. In persons who recover from this disease, the general phase

may last about one week, during which time there are violent muscle spasms. Any sudden stimulus (bright light, loud noise) may result in reaction known as tonic seizure. Fever may rise to 101–102° F (38.3–38.9° C). Spasms of the muscles in the throat are perhaps the most dangerous since the resulting airway obstruction can compromise breathing and interfere with the respiratory process. Respiratory and cardiovascular complications may arise during the course of the disease and fractures may occur, depending on the frequency and severity of muscle spasms. After the general tetanus phase peaks, several weeks are required for full recovery. Respiratory assistance may be required for two to four weeks and the patient will most likely remain hospitalized for up to eight weeks.

Because of the serious nature of this disease, industrial health programs usually include verification of worker immunization against tetanus. As a precautionary measure, it is sometimes recommended that the human tetanus immune globulin be administered whenever a worker has received a puncture injury of any degree, especially when the employee cannot recall receiving his or her last tetanus shot.

6. *Tuberculosis (TB).* Once considered to be nearly eradicated from the general population, exposure to this disease is still a concern for health care providers and any other workplace setting where employees must work in close contact or confinement with other personnel. TB is an acute inflammation of the lungs, meninges, pericardium, kidneys, genitals, bones, or lymph system caused by invasion of one or more of these organs by mycobacterium tuberculosis. The lungs, by far, are the organ most affected by this disease. The disease is contracted via person-to-person transmission. When an infected person coughs, sneezes, shouts, or even laughs, tuberculosis-contaminated droplets are deposited into the breathing environment. Usually close, continued contact is required to spread the disease and, therefore, it tends to occur among members of the same household, school children, persons participating in car pools, and others in similar close conditions. Operators of nursing homes and workers in hospitals most be constantly aware of the tuberculosis potential. Unlike most infectious diseases, there is a relatively long latent period between infection and the first indications of any symptoms. The most common of these symptoms include low-grade fever and a dry cough. There may be anorexia (loss of appetite), weight loss, and night sweats. As the disease progresses, there will also be continued low-grade fever, although more severe fever and chills have been noted in some cases, and spitting up of blood. Initially, the infection usually begins in the deep lung components. In the alveoli, the bacilli proliferate slowly and, because no enzymes or toxins are produced, there is little initial inflammatory response. Experts place TB-infected people into two categories: those with TB infection who are without disease, and those

who are infected with disease. In the first category (*inactive TB*), people do not become sick because the bacilli are dormant in their bodies. Although they test positive for the disease, they cannot spread it to other workers. Those in the second category (*active TB*) do develop disease and usually display one or more of the symptoms noted above. Because the bacilli are active in their bodies, these people can infect those around them with the disease. If not diagnosed and treated correctly, people can sustain permanent damage to the infected organ and death. Untreated, active TB still has a death rate of approximately 50 percent in the United States (Davisson, 1994).

THE BLOODBORNE PATHOGENS STANDARD

The Occupational Safety and Health Administration issued a Bloodborne Pathogens Standard (29 CFR 1910.1030) to protect workers who may be exposed to infectious bloodborne agents during the course of their occupational duties. In the final rule, which became effective 6 March 1992, OSHA estimated the Standard will protect more than 5.6 million workers and prevent more than 200 deaths and 9,200 bloodborne infections each year.

Bloodborne pathogens are microorganisms in human blood that can cause disease in humans. They include the *hepatitis B virus* (HBV) and the *human immunodeficiency virus* (HIV) that causes *acquired immune deficiency syndrome* (AIDS), which is also commonly referred to as *acquired immunodeficiency syndrome*. The greatest bloodborne risk workers face is the threat posed by HBV. OSHA estimates that occupational exposures account for roughly 5,900 to 7,400 cases of HBV infection per year. The Standard mandates engineering controls, work practices, and the use of personal protective equipment (PPE) that, coupled with employee training, will reduce on-the-job risks for all employees exposed to blood.

While the Standard specifically applies to workers in the health care industry, it may also have application to workers in general industry as well. Since exposure to blood is a potentially fatal occurrence, the Standard covers all employees who *may be reasonably anticipated* to come into contact with human blood and other potentially infectious materials *in order to perform their jobs*. The key to understanding the applicability of the Standard lies in the words "may be reasonably anticipated" and "in order to perform their jobs." Hence, the good Samaritan who assists a co-worker with a nosebleed or a cut finger would not be covered by the Standard, since such actions are not reasonably anticipated and are not required to perform a job or task. On the other hand, if a worker has been designated by the employer as a first-aid responder, then the opposite is true; it might be reasonably anticipated that the worker will come into contact with blood or blood products to perform the job as a first-aid provider Therefore, the Standard would be applicable. First responders may include police and security officers, firefighters, and emergency medical services personnel. In addition to health care workers and designated first responders, other professions that are covered by this Standard include linen service employees who handle contaminated laundry and laboratory personnel who work with blood or other potentially infectious materials (plasma, tissues, etc.).

Bloodborne Diseases

According to OSHA, the term *blood* is meant to include human blood, human blood components, and products made from human blood. But the OSHA Standard does not limit its applicability to blood exposures alone. Because bloodborne pathogens can be present or harbored in other bodily fluids, workers exposed to other *potentially infectious materials* are also covered by the regulation. Other potentially infectious materials include body fluids such as vaginal secretions, cerebrospinal fluid, synovial fluid, pleural fluid, pericardial fluid, amniotic fluid, saliva, any body fluid visibly contaminated with blood, and all body fluids that may be mixed together where it is impossible to differentiate between fluids. Also included under the general description of other potentially infectious materials are any unattached tissues or organs (other than unbroken skin) from humans (living or dead). Also included, in terms of medical research, are any HIV-containing cell or tissue cultures, organ cultures, and HIV- or HBV-containing culture medium or other solutions, as well as blood, organs, or other tissues from experimental animals infected with HIV or HBV.

Bloodborne diseases that could be encountered on the job include non-A hepatitis, non-B hepatitis, hepatitis B and delta hepatitis, as well as syphilis, malaria, and HIV. The following provides brief summary of the most commonly encountered of these diseases:

Hepatitis

An inflammatory disease of the liver, hepatitis is commonly caused by a virus, although hundreds of drugs are also known to cause the disease. Among the known viruses that produce acute hepatitis, the HBV is considered the major infectious bloodborne hazard. In general, all or some of the early symptoms will include combinations of drowsiness, fatigue, lassitude, loss of appetite, nausea, and, most specific to the disease, dark urine. Headache and very mild fever in the absence of chills may also be present. The overall pain experience is usually mild and there may be general abdominal discomfort that tends to be aggravated by movement. Itching of the skin may occur in some cases and mild arthritis may even develop. As the disease progresses, jaundice (yellowing of the skin and sclera of the eye) may be evident due to excessive bile pigments present in the blood.

Syphilis

Syphilis is an infectious disease that can be congenital or acquired and is caused by a microorganism. While the acquired form is primarily venereal in nature, health care workers can be at risk if proper precautions are not taken when treating infected patients. The causative microorganism usually gains entry through broken or abraded skin. Hence, workers exposed to infected patients do run the risk of infection if any of the patient's blood or body fluids find a way to penetrate the worker's skin (through unprotected cuts or sores, for example). In the primary stage of disease progression, there is usually a chancre sore development at the site of invasion approximately two to four weeks after exposure. Before the chancre develops, however, the organisms often have invaded the body tissues by way of the lymph channels and the bloodstream. A fully developed chancre appears as a clean, slightly raised, hard circumscribed ulcer that exudes a thin, highly infectious secretion. Workers contacting these secretions are at high risk for contracting the disease if contact is made without protection. Symptoms during the secondary stage are often so mild that they may go unnoticed. There may be sore throat, slight fever, headache, and some enlargement of the superficial lymph nodes. Often, however, there are no symptoms of any regard, and the person may make an assumption that the disease is cured. Untreated, syphilis may result in severe cardiovascular strain and, after many years, central nervous system disorders.

Human Immunodeficiency Virus

The HIV attacks the body's immune system, causing the disease known as AIDS, or acquired immune deficiency syndrome. Currently, there is no vaccine to prevent infection. According to known data, it is theoretically possible for the AIDS virus to be found in any body fluid that contains lymphocytes. Viruses have been isolated from blood, semen, saliva, tears, and vaginal secretions of infected persons. Breast milk, urine, and feces might also contain the virus. Because the HIV will depress the immune system to a point of ineffectiveness, infected persons often contract certain *opportunistic* diseases that eventually may lead to death. An opportunistic disease is simply one that takes advantage of the opportunities presented by a non-responsive immune system to infect and develop. It is important to note that some individuals infected with the HIV do not develop AIDS, but instead contract a condition known as *AIDS-related complex* (ARC) that causes a mononucleosis-type illness characterized by fevers, diarrhea, myalgia, fatigue, and sore throat. Others infected with the HIV may not develop any symptoms at all (asymptomatic) for many, many years. In the United States, the four major routes of infection are intimate sexual contact, intravenous drug administration with contaminated needles, transfusion of blood products, and passage of the HIV from infected mothers to their offspring before, during, or after childbirth (perhaps via breast milk). While exposure via the first route should not be a concern in most work environments, workers are at risk of exposure in varying degrees through any possible contact with infected body fluids. The AIDS virus is considered quite fragile, and there is no evidence that it is transmitted through casual contact. Outside the body, the virus is readily killed by detergents, alcohols, hydrogen peroxide, phenolics, and sodium hypochlorite. High and low pH and exposure to high temperatures will inactivate or kill the virus.

Means of Transmission

Bloodborne pathogens may enter and infect the body through a variety of means, including those described below.

1. Pathogens may enter at the time of an accidental injury by a sharp object contaminated with infectious material. These objects, called "sharps," may include needles, scalpels, broken glass, exposed ends of dental wires, or anything that can pierce, puncture, or cut into skin.
2. Open cuts, nicks, and skin abrasions, even dermatitis and acne, as well as the mucous membranes of the mouth, eyes, or nose, are excellent avenues for transmittal and infection.
3. Infection may occur via indirect transmission, such as touching a contaminated object or surface transferring the infectious material to the mouth, eyes, nose, or open skin. This is particularly true of the HBV. In fact, contaminated environmental surfaces are a major mode of transmission of HBV in certain settings, particularly in hemodialysis units. Unlike the AIDS virus, HBV can survive on environmental surfaces dried and at room temperatures for at least one week. Surfaces and objects can be heavily contaminated by blood materials such as plasma or serum without any visible signs.

Specific OSHA Requirements

As mentioned at the beginning of this section, OSHA has promulgated a Standard aimed at protecting the thousands of workers occupationally exposed to bloodborne pathogens. The Standard mandates engineering and administrative (work practice) controls, as well as employee educational training to reduce the risk of such exposures, as enumerated below.

1. *Engineering Controls.* The first line of defense against occupational exposure to bloodborne pathogens is to institute proven engineering controls that isolate or remove the hazard from the workplace. These are physical or mechanical systems provided to eliminate or control the hazards at their source. Engineering controls include the mandated use of appropriate

sharps disposal containers and/or self-sheathing needles to avoid needle-stick injuries. Other controls include the regular use of biohazard safety cabinets to ensure proper separation and identification of infectious materials. The use of proven sterilization and disinfecting methods and techniques (such as autoclaves) is another example of acceptable engineering controls.

2. *Work Practice Controls.* These are administrative controls placed on workers to alter the manner in which a task is performed in an effort to reduce the likelihood of exposure. For example, by prohibiting the practice of recapping needles using a two-handed technique, the potential for an off-hand needle-stick injury is greatly reduced. In fact, the bending, shearing, or breaking of needles or other sharps must be strictly prohibited. Proper personal hygiene (especially *handwashing*) must be enforced. Handwashing prevents transfer of contamination to other parts of the body or to other surfaces. If skin or mucous membranes come in direct contact with blood or blood materials, they should be washed immediately. Also, eating, drinking, smoking, applying cosmetics or lip balm, and the handling of contact lenses must not be allowed in areas where they may be exposed to bloodborne pathogens. Good housekeeping policies should focus on preventing exposures as well. In every case, the concept of *universal precautions* must be enforced. Universal precautions are a conservative and effective approach to infection control in which all human blood and certain human body fluids are treated as if known to be infectious.

3. *Personal Protective Equipment (PPE).* The OSHA Standard requires employers to provide appropriate PPE and clothing free of charge to their employees. Workers who have direct exposure to blood and infectious materials on their jobs are at risk of contracting bloodborne infections. Wearing of gloves, gowns, masks, and eye protection can significantly reduce health risks for workers exposed to blood or other infectious materials. The key to this aspect of the Standard is that the PPE selected must be suitable for its intended use. This simply means the level of protection must fit the expected exposure. For instance, gloves may be sufficient for a laboratory technician who is drawing blood, whereas a pathologist conducting an autopsy would need considerably more protective clothing. PPE may include gloves, gowns, laboratory coats, face shields or masks, eye protection, and other such equipment. The gear must be readily accessible and available to employees in appropriate sizes. The major purpose of PPE is to *avoid contamination.* Blood and other infectious materials must not reach an employee's work clothes, street clothes, undergarments, skin, eyes, mouth, or other mucous membranes under normal conditions for the duration of exposure. In addition to providing the PPE and requiring its use, the

employer must also ensure that such clothing and equipment is cleaned, laundered and keep in proper repair.

4. *Labeling Requirements.* The Standard requires the employer to properly communicate the hazards of bloodborne pathogens to employees at risk of exposure. In this regard, warning labels must be affixed to containers of biological hazardous wastes, refrigerators and freezers containing blood or other potentially infectious materials, and other containers used to store, transport, or ship blood products. Labels must include the biohazard symbol shown in Figure 8-1. These labels must be fluorescent orange or orange-red in color with the lettering and symbols in a contrasting color (usually, but not always, black). Red bags or red containers may be substituted for labels since these are also treated as biological hazard containers. While the Standard does list several exceptions to this requirement, containers must still be labeled in some way so as to properly describe their contents. In addition to the labeling of containers, employers must also post signs at the entrance to work areas where biohazards are located. These signs must also contain the biohazard symbol, as well as the name of the infectious agent, special requirements for entering the area, and the name and telephone number of the laboratory director or other responsible person.

5. *Employee Training and Education.* Any employee who may have occupational exposure to bloodborne pathogens must be provided specific information and training, at no cost and on company time. The Standard requires this training at the time employees are initially assigned to tasks where occupational exposures are likely and at least annually thereafter as well as whenever changes to processes and procedures warrant additional training. Records of training must be maintained for a minimum of three years following the date on which the training was provided. As a mini-

FIGURE 8-1 The biohazard symbol must be used on all labels, tags, or containers to signify the actual or potential presence of a biological hazard.

mum, this training, which should be provided by a person knowledgeable in the subject of bloodborne pathogens and the company's plans to prevent exposures, must include the following:

a. Access to a copy of the OSHA Standard
b. An explanation of the epidemiology and symptoms of bloodborne diseases and an explanation of the mode of transmission of these diseases
c. An explanation of the exposure control plan for the company and the means by which employees may obtain copies of the written plan
d. An explanation of the methods that can be used to recognize whether tasks or other activities involve exposure to bloodborne pathogens and information on the use and limitations of methods that will prevent or reduce such exposures (including engineering controls, work practices, and the proper selection use, care, and maintenance of PPE)
e. Information on the hepatitis B vaccine, to include information on its efficacy, safety, method of administration, the benefits of being vaccinated, and a statement explaining that the vaccine and vaccination will be offered free of charge must also be included in the training program
f. Emergency notification procedures must be covered in the training, including the correct actions to take in emergency situations involving exposure to blood or other potentially infectious materials and the follow-up that will be made available to affected employees
g. An explanation of the signs and labels to be used by the company to help control occupational exposures to biohazards

The training sessions must also allow the opportunity for *interactive questions and answers* with the person who is conducting the training. This provision is important because it virtually eliminates the videotape-only training sessions that so many companies use to meet compliance requirements. While videotape training is acceptable as a supplement to the bloodborne pathogens training, it should not be selected as the only method of information exchange. The Standard requires additional, more definitive training for those employees working in an HIV or HBV research laboratory setting.

6. *Employee Vaccinations.* The employer must also make available the hepatitis B vaccine and vaccination series to all employees who have occupational exposure. Also, a post-exposure follow-up evaluation must be made available to any employee who has been occupationally exposed to bloodborne pathogens. These services, which must be made available at no cost to the employees, are to be offered at a time and location of convenience to the employee. It is important to note that employees may refuse the vaccine. This is their right. The employer's obligation under the law is only to

make it available. Although not required, it is highly recommended that such refusals be obtained in writing from the employee(s) to document both the refusal and the employer's attempt to comply with the requirements of the Standard should any questions arise at some point in the future.

7. *Exposure Control Plan.* Perhaps the most critical element of the Standard in terms of regulatory compliance and the control of exposures to bloodborne pathogens is the requirement for a written *exposure control plan* (ECP). Any employer with even one employee who may be occupationally exposed to bloodborne pathogens must establish a written ECP designed to eliminate or minimize exposure. The ECP must include information on the *exposure determinations* made for that workplace. These determinations are based on an evaluation of each job, task, or function to assess occupational exposure potential regardless of the use of PPE. In other words, whether PPE is used is not a consideration of exposure potential. The ECP must also contain the schedule and methods the employer will use to implement the requirements of the Standard for the workplace, including the procedures to be used during and following any exposure incidents. The ECP must be made available to all employees and must be reviewed and updated at least annually or whenever new or modified tasks are introduced that may alter the effectiveness of the exposure control measures.

In addition to the basic elements of OSHA's Bloodborne Pathogens Standard described above, other requirements directed at controlling exposures during waste disposal should also be reviewed by the hygienist to ensure complete understanding of the exposure potential that may exist in the respective work environments.

SUMMARY

Many serious and potentially life-threatening diseases can be contracted as a result of exposure to *biological hazards.* While preventing or at least controlling such exposures should never be the responsibility of the industrial hygienist alone, he or she is best positioned to *recognize* when these hazards exist. Therefore, a basic knowledge of the various *biohazard agents* has been presented in this chapter. Perhaps even more insidious than occupational exposures to energy extremes, the threat posed by exposure to a biohazard is just as real and often more harmful to human health. Biohazards exist as microorganisms in the form of *bacteria, viruses, parasites,* and *fungi.* Each can present a significant human health risk resulting in the development of certain diseases which, in turn, can result in anything from mild sickness to death.

While it is true that the majority of those workers at risk of exposure to biohazard agents are employed in the *health care industry,* a significant exposure potential also exists for workers in *research laboratories* and those in the agricultural and/or farming industries. In addition, employees who work outdoors, especially in rural, undeveloped, or wooded areas, can come into contact with a broad range of biohazard agents during the normal course of their assigned duties. Recognition of such hazards can become quite difficult due to the subtle nature of these agents, their basic invisibility to the worker, and their ability to enter the human body through the full variety of exposure routes (i.e., inhalation, absorption, ingestion, and injection).

In an effort to control exposures to the specific hazardous agents that are known to be found in human blood and blood by-products, principally the hepatitis B virus (HBV) and the human immunodeficiency virus (HIV), OSHA promulgated its Bloodborne Pathogens Standard, which is located at 29 CFR 1910.1030. This Standard mandates the minimum acceptable employer actions necessary to reduce occupational exposures to bloodborne pathogens. These actions include the establishment of a written exposure control plan (ECP), employee training, labeling requirements, record keeping, and the use of engineering controls and special work practices.

While the presence of biohazards in the workplace may be necessary in some instance, unprotected exposures are not. The hygienist should make every attempt to properly recognize and identify the potential for such exposures so that proper controls can be implemented.

9

Ergonomic Hazards

INTRODUCTION

The interface between man and the work environment has become an issue of ever increasing concern over the years. Those who attempt to understand how human performance is affected by both the dynamic and static conditions that exist in the work environment focus on the principles of *ergonomics*. Basically, the intent of ergonomics is to adjust the work environment so that work may be efficiently and safely performed by people, with a minimum amount of acquired stress. This is a clarification of the traditional definition of "fitting the person to the job and the job to the person." Since it is nearly impossible to modify people to fit a task, it makes more sense to design the task to fit the worker. The study of ergonomics has become a major element in the overall workplace safety and health process. It has clear applications in safety, accident investigation and prevention, and employee wellness and occupational health assurance. However, because it addresses job performance and the ability of workers to perform their tasks, ergonomics is now considered a much broader field of study, transcending safety and health. Ergonomics extends into other disciplines such as engineering and design, marketing and sales, product liability and reliability, and so on.

Clearly, a full and comprehensive study on the subject of ergonomics would be beyond the scope of this volume. Entire books could be (and have been) written on this subject. The reader is referred to the appendices and bibliography for further information.

This chapter will provide basic information on the fundamental principles of ergonomics and the role of the industrial hygienist in recognizing, evaluating, and controlling such problems in the workplace. Specifically, *cumulative trauma disorders* and other *repetitive motion* illnesses will be explained and discussed.

Since these have been (and continue to be) the most common results of uncorrected ergonomic health hazards, the hygienist should be aware of such problems. Other ergonomic considerations such as workstation design, human performance parameters, anthropometry, and biomechanics will also be briefly discussed here.

ERGONOMICS AND THE INDUSTRIAL HYGIENE PROCESS

The application of ergonomics extends far beyond the basic industrial hygiene process and, therefore, beyond the scope of this text. It is concerned with workstation design, task engineering and analysis, the use of scientific measurement data, understanding of the forces of inertia, gravity, energy, power and work that are involved in movement, and a variety of other technical considerations not generally encountered in the practice of *basic* industrial hygiene. In all likelihood, the average industrial hygienist will have very little involvement in the design phase of workstations, control panels, equipment, tools, and displays. Although their advice is increasingly being sought by design and manufacturing firms, the average hygienist will more likely be involved in the *assessment* of ergonomic problems that already exist in their workplace. In this regard, the hygienist can be more instrumental in effecting changes in company policy, written operational procedures and instructions, performance standards, and work practices. These are internal to the company and, therefore, more easily controlled in most cases. Some experts may disagree with this summation, and it is certain that the role of the industrial hygienist will change once regulatory agencies promulgate ergonomic standards. While it is true that some design changes or modifications can be made after the fact (i.e., after equipment purchase) to control or eliminate some ergonomic hazards, it is more probable that exposure to many such hazards will not be totally controllable, depending on individual equipment design. From a practical perspective, the basic industrial hygiene process requires hazard *recognition* first, hazard *evaluation* second, and hazard *control* third. Obviously, if ergonomic hazards are eliminated through equipment or process design by the manufacturer, then the application of these industrial hygiene principles becomes moot (i.e., there will be no hazards to recognize, evaluate, and control). The effectiveness of an industrial hygiene program, in terms of ergonomic health hazards, will be measured by the ability of the practitioner to fulfill the main objective of the industrial hygiene process. Only by developing and implementing effective control measures to either eliminate or reduce exposure to existing ergonomic hazards will the industrial hygiene program be considered a success. Therefore, this chapter will focus more on the recognition of those ergonomic hazards most likely to be encountered in the workplace, their effect on human performance, and how the risk of exposure can be effectively controlled.

OSHA has recommended that an effective ergonomic program consist of at least four basic elements: work site analysis, hazard prevention and control, medical management, and training and education. Clearly, these are highly congruous with the overall function of the industrial hygienist. *Work site analysis*, in particular, closely parallels the industrial hygiene assessment that is so critical to the success of any industrial hygiene program. *Hazard prevention and control* is the objective of industrial hygiene. The hygienist is also somewhat involved in employee *medical management* when it relates to health hazard exposures. Finally, the hygienist can also be a major contributor to the *training and education* provided to employees on the subject of workplace safety and health.

UNDERSTANDING ERGONOMICS: AN OVERVIEW

The term "ergonomics" has been derived from a combination of the Greek words *ergon,* which means "work," and *nomos,* meaning "law or rule." Hence, *ergonomics* literally means "the laws of work." More broadly, it may also be defined as the relationships between people and their working environment, including equipment, machines and tools, environmental factors (light, temperature, ventilation, etc.), facilities, vehicles, written work instructions and materials, and so on. Those who study and practice ergonomics as a profession are appropriately referred to as *ergonomists.* The study of ergonomics relates human capabilities and limitations to the physical design of the products and systems that exist in the work environment. With special focus on the areas of *performance, safety and health,* and *worker satisfaction,* ergonomists attempt to resolve the problems that can develop in the interrelationships that must exist between these three factors. The *performance* factor, for example, focuses on the enhancement or extension of a worker's abilities by improving output and reducing errors. At the same time, the *safety and health* relationship is concerned with minimizing accidents, injuries, and illnesses that may result from these human limitations. *Satisfaction* pertains to the actual design of comfortable, desirable, convenient or "user friendly" products and systems. For the purpose of this text, other frequently used terms such as *human factors engineering* or simply *human engineering* will be considered virtually synonymous with ergonomics, although some technical experts may argue the subtle differences.

It is generally accepted that the study of ergonomics developed as a result of military aviation and other such matters during and following World War II. The critical tasks associated with the flying of an aircraft, for example, required effective but simple design of equipment panels and controls to ensure the accurate presentation of information and error-free operation of the aircraft. Since that time, ergonomics has become a fairly common term, and its application continues to grow. As the use of office computers and other forms of automation have increased in the workplace, so has the frequency of injuries and illnesses resulting from occupational safety and health issues such as *cumulative trauma.* Because these afflictions can often be directly or indirectly attributed to poor workplace or workstation design, the hygienist must work closely with design engineers and facility managers to evaluate the cause and effect relationships that may exist so that effective control measures can be identified and properly implemented.

Basic Principles of Ergonomics

While it is true that the field of ergonomics is extremely broad, and the hazards posed by ergonomic health hazards are numerous, a few general principles form the foundation upon which the study of ergonomics is based. In nearly every case, consideration of the following principal factors provide a starting point for a more comprehensive ergonomic investigation.

1. *Select the appropriate human or machine for a job.* It is obvious that people and machines have quite different capabilities. Neither is usually best for all functions. For example, people are usually proficient at detecting and interpreting signals under a variety environmental stressors (noise, heat/cold) and making decisions based on information received. Machines can typically respond much faster and, in most cases, under even greater influence of negative environmental stressors. However with the use of the machine, the interpretive and decision-making features that come with human performance are lost. In some instances, such as high hazard or quick response tasks where accurate but routine spilt-second performance is critical, this may be a highly desirable condition. Under different circumstances, such as fine inspection and verification jobs, the process may suffer in the absence of the human element. Hence, one of the first tasks in ergonomics is to assess the jobs that people do versus those performed by machine to determine whether the best mix between the two has been implemented. It is during this initial evaluation that shortcomings may be identified and *recognized* as such, for what may be the first time. Subsequently, an ergonomic assessment can be a valuable tool to assist the hygienist in the hazard recognition process.
2. *Consider the people-machine interface.* As noted above, there are certain jobs best suited for machines and others that should be performed by people. However, more often than not, there will be jobs that require *both* the human and the mechanical element. When people must interface with machinery and equipment, the potential for error can increase dramatically, as can the likelihood of injury and/or damage as a result of those errors.

Because the capabilities of even the best workers are still somewhat limited, any failure to properly recognize and interpret control and response requirements from the equipment they must use can lead to such errors. These errors can be errors of *omission* (not doing something that should have been done) or errors of *commission* (doing something that should not have been done). Achieving optimum human performance under all possible operating conditions and in all types of working environments is difficult at best. By evaluating the human/machine interface in each work situation, the hygienist can identify potential problem areas associated with poor equipment design or improper consideration of operator limitations and capabilities *before* these develop into accidents and/or injuries. Expecting a person to adjust to a job, the equipment, or the working environment instead of designing the job to fit the person, may force that individual to stretch beyond the safe limits of a worker's capabilities. This is a dangerous but common occurrence that often leads to injury or illness.

3. *Understand the differences between people.* Simply stated, all people are not alike. While some differences (such as height, weight, sex, physical appearance, and even age) are obvious, other differences are more subtle. These include differences in intelligence, reaction time, strength, behavior, coordination, beliefs, morals, ethics, responses to surroundings, visual acuity, training and experience, and so on. This wide variability between workers requires thoughtful consideration during the equipment and machine design process. While no piece of equipment or machine can be designed to fit every person under every condition, certain allowances can be made to help accommodate such diversities. Even though the average industrial hygienist may not be in a position to influence the design process, there are still measures that can be taken at the work site to help reduce the risk of exposure to these hazards. Providing adjustable computer station/tables and chairs is one common example. While there is no single solution that will work in every case, ergonomics requires that these differences at least be considered and evaluated for their illness and injury causing potential.

4. *Evaluate lessons learned from past performance.* When accidents and injuries (or illnesses) occur, they are tragic. Even more tragic, however, are those that repeat themselves because nothing was done to change the situation or circumstances that led to the accidents in the first place. A fundamental objective of ergonomics, and of industrial hygiene in general, is to ensure a safe, healthy, and humane workplace. It may be necessary to implement changes in a job or process based on knowledge of the causes of an accident or an error to reduce the likelihood of its recurrence. Ignorance of the ergonomic perspective during accident investigation may result in additional injuries and illnesses which, of course, are counterproductive to the overall objective of industrial hygiene.

Anthropometry

The science of measuring the human body is referred to as *anthropometry*. There are basically two categories of anthropometric data that can be obtained from such measurements. *Static anthropometric data* refers to constant (or nearly constant) data such as an individual's standing height, sitting height, length and breadth of body segments, depth of body segments, and individual postures. *Dynamic anthropometric data* describes such features as reach, range of motion and the forces generated by hands and feet in different directions. These data are primarily used to evaluate design variables such as fit and reach. Since one size rarely or never really fits all users, anthropometrics facilitate the design process. Office furniture manufacturers, for example, have utilized known anthropometric data to design adjustable features in chairs and work surfaces (for the positioning of computer keyboards and monitors). Design in consideration of user anthropometrics is not an entirely new concept. Barbershop chairs have been adjustable for many years to accommodate the characteristics of different customers. Similarly, rear and side-view mirrors in automobiles have long been adjustable to ensure that drivers of varying sitting postures can easily use them. In the design of modern equipment panels, displays, controls, and workstations, ergonomics and the application of anthropometric data have become increasingly important considerations.

Biomechanics

Utilizing mechanical principles to resolve biological problems is referred to as *biomechanics*. With anthropometrics as a foundation, biomechanics builds on anatomical data and the study of human muscular movement *(kinesiology)* in terms of displacement, velocity, and acceleration. In this regard, biomechanics considers the forces, energy, power, and work involved in movement. For example, when a weight is lifted using the hand, the load on the hand creates an additional force at the elbow transmitted through the wrist joint. When the hand and arm are placed into motion during the lift, the additional forces of inertia and moment further increase the static load on the body parts involved. By analyzing the forces created during this biomechanical relationship, ergonomic problems associated with extreme joint flexion, unusual postures, large forces, opposing forces (i.e., those not aligned with body motion), vibration, and highly repetitive motions can be more easily understood.

CUMULATIVE TRAUMA DISORDERS

As stated previously, a number of ergonomic issues and concerns can exist in the average workplace. A frequently encountered ergonomic problem is the repetitive motion illness known as *cumulative trauma disorder* (CTD). A CTD can occur as the result of performing the same motion over and over for a long period of time. Repetitive motion can cause problems due to the resulting stress or strain imposed repeatedly over the same body parts or segments. After an undefinable period of time, muscles, joints, and tendons can be damaged. This damage occurs as a result of the cumulative affects of the constant motion, hence the term "cumulative trauma disorder." Table 9-1 highlights some of the common causes of cumulative trauma. While a CTD can occur in many parts of the body, including the wrists, elbows, and shoulders, the back and neck, the hips, knees and ankles, and virtually any other body joint combinations, the most commonly occurring CTD affects the wrists in an illness known as *carpal tunnel syndrome* (CTS). Proof of its commonalty is evidenced by the ever-rising costs associated with worker's compensation claims that identify carpal tunnel syndrome as the number-one type of CTD in our nation today.

Incidentally, from a record keeping and recording perspective, it is important to note that a CTD is considered an *illness* and not an *injury* by OSHA. According to OSHA, a work-related injury is caused by an instantaneous event (such as a fall, for example). CTDs, however, develop over time and are not caused by one specific event. They are, therefore, classified as work-related illnesses. An illness is an abnormal condition or disorder that was not caused by an instantaneous event, but by exposure to environmental factors over a period of time.

Tendonitis

The shoulder, arm, and hand contain a number of tendons. The tendons connect the muscles to the bones and every time the hand or arm moves, the tendons are intimately involved. Tendons are composed of parallel white fibrous tissue, closely bound into tightly compressed bundles. The entire tendon is surrounded

TABLE 9-1   Some of the common causes of cumulative trauma in industry

COMMON CAUSES OF CUMULATIVE TRAUMA	
Awkward posture or position	The more awkward the posture or position, *the greater the risk* (bent wrist, outstretched arms, slumped shoulders)
Excessive force	The greater the force involved, *the greater the risk*
Highly repetitive tasks	The higher number of wrist, arm, or shoulder repetition, *the greater the risk*
Poor physical condition	The weaker and stiffer the muscles, tendons, and joints are, *the greater the risk*
Direct pressure on nerves	The more direct pressure on a tool or equipment places on a nerve (such as in the palm or wrist), *the greater the risk*
Vibration	The more tools and equipment vibrate, *the greater the risk*
Cold exposure	The colder the work environment, *the greater the risk*

by a fibrous sheath and, in some instances (where a wide range of movement is required) the sheath is split to form a cavity that contains a lubricant called *synovial fluid*. With forceful or repeated motion activity, the tendons can become irritated, swollen, and inflamed in a condition known as *tendonitis*. The initial result is tenderness and soreness in the affected areas. For example, *stenosing tenosynovitis* is a tendon disorder caused by excessive repetition. With extreme repetition, synovial fluid builds up inside the wrist and causes swelling and pain. The tendon becomes irritated and rough, and the area becomes inflamed. Prolonged or continuous inflammation may lead to nerve compression and more serious damage causing numbness, tingling, and burning sensations extending beyond the initial point of trauma.

Carpal Tunnel Syndrome

Carpal tunnel syndrome (CTS) is the most common type of nerve compression, a condition in which the nerve that goes through the wrist (the median nerve) is compressed by swollen tendons. The classic symptom is tingling in the thumb, index, and middle fingers, especially at night. CTS can be caused by repetitive motions such as typing and data entry or by poor wrist posture at the keyboard. Tasks requiring repetitive and forceful use of the fingers, such as crimping, manual stapling, and other assembly functions can also lead to CTS. Any repetitive job that forces the wrist to deviate off the imaginary center line between the hand and forearm, especially those requiring ulnar deviation, can strain the tendons in the wrist that can eventually harm adjacent nerves. The basic problem is a lack of space inside the wrist. The flexor tendons and median nerve of the hand pass through eight carpal bones in the wrist, wrapped by a ligament that forms a rigid tunnel about the size of a dime (Figure 9-1). The tendons are surrounded by a sheath that contains the lubricating synovial fluid. As the fingers are moved, these tendons slide back and forth inside the small carpal tunnel. Repetitive motions require constant movement of these tendons through the sheath. The tendons may move as much as 2.5 inches when the fingers move from a fully closed (or retracted) position to a fully extended position. Continuous and repetitive movements may cause a decrease in the synovial lubricant inside the tendon sheath. The entire area then becomes inflamed, causing the tendons to swell, which further decreases the size of the already-small carpal tunnel. As the swelling becomes more severe, the median nerve will be compressed or "pinched" by the tendons. This results in the illness known as CTS. In severe cases, surgery may be required to alleviate the pressure on the nerve.

Raynaud's Syndrome

Also referred to as "white finger," this disorder is caused by the constriction of blood vessels in the hand. This can occur as a result of exposure to extreme cold,

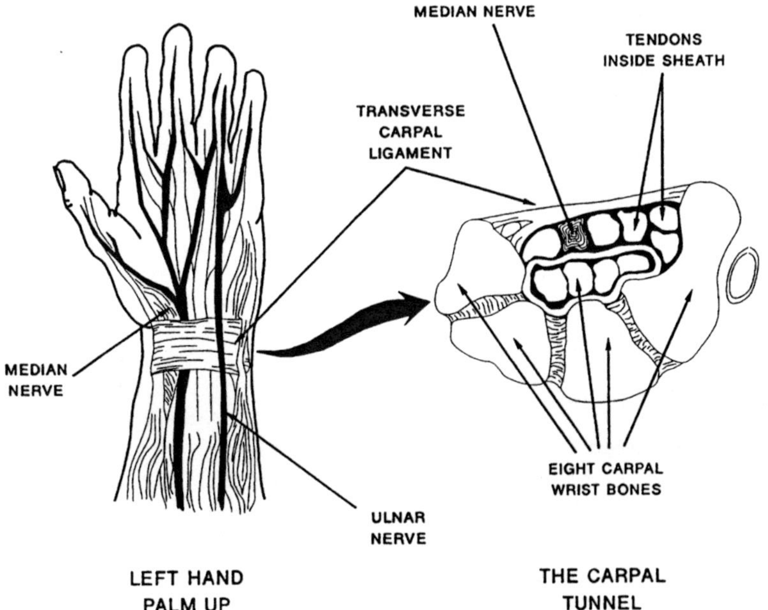

FIGURE 9-1 The flexor tendons and median nerve of the hand pass through eight carpal bones in the wrist which are wrapped by a ligament that forms the rigid, dime-sized *carpal tunnel*

vibration, or even emotional stress. Signs and symptoms include both hands becoming cold simultaneously and turning blue or white, followed by numbness and loss of manipulative control. On recovery, the hands become red, accompanied by a burning sensation. Raynaud's syndrome should be a primary concern in the design of tools and controls for machines that will induce vibration. The handles on some machines and tools (jackhammers, for example) can produce very high compression forces on the hand tissues of the palm and may lead to this CTD.

CONTROLLING ERGONOMIC RISK FACTORS

From the industrial hygiene perspective, a variety of measures can be implemented in the workplace to reduce the risk of exposure to ergonomic hazards. As stated previously, ergonomic considerations in equipment, tools, machines, or process design is the preferred way to control these exposures. However, since this is not always the case in many work environments, the hygienist must evaluate and assess the problems and recommend solutions to reduce exposure risk. Quite often, these solutions may be nothing more than the application of common sense. In other instances, the hygienist may have to come up with more innovative recommendations. Whatever the case, the primary objective remains risk reduction and control.

As indicated in Table 9-1, there are a number of conditions that can contribute to cumulative trauma in the workplace. Usually, it is not just one of these seven factors that leads to a CTD, but a combination. The more factors combined in a particular task, the greater the risk. These factors provide some guidance in developing techniques for CTD prevention. Specifically,

1. *Posture and position.* Any job or task that forces the body or any of its parts into awkward positions or postures should be evaluated to determine if alterations can be made. Tools that require bent wrists, outstretched arms, or any clumsy positioning are perfect candidates for evaluation. In some cases, straight-handled tools or tools bent to an angle that keeps the wrist straight can be used. Work surface can be tilted to produce the same straight-wrist effect. Resting wrists on sharp table corners, common during word processing operations, can eventually cause pain and discomfort and should therefore not be allowed to continue. Soft wrist pads can be used to eliminate this problem. Any job that requires long reaches can create cumulative trauma in the shoulders and back. Simply evaluating the process and relocating commonly used items closer to the worker may help resolve the problem. The safest working height is normally between the waist and shoulder. Activities above or below this *safety zone* can cause additional ergonomic problems for the worker. Neck and shoulder posture can also cause CTDs. Working in a slumped-over posture (sitting or standing) with the head and shoulders forward restricts blood flow to the arms. This, in turn, affects the tendons, nerves, and muscles of the arms.

   When evaluating workplace ergonomic problems, the hygienist must be particularly alert to employee postures and positions. These are perhaps the most difficult to control and, therefore, present one of the greatest challenges in terms of ergonomic health hazard reduction and control.

2. *Excessive force.* Whenever a task or job requires the worker to apply excessive force to work surfaces, materials, or equipment, the risk of cumulative trauma is greatly increased. It stands to reason that any force applied by the worker onto a structure or surface is also being applied to the worker's joints, tendons, and muscles. Wherever possible, tool balancing devices should be used to support the weight of heavy tools. Using fixtures or jigs to hold work products, parts, or tools will help prevent the transmission of force to the worker. If mechanical assist devices such as cranes and power tools can be made available, the worker should utilize them instead of performing the task manually. In some cases, workstation layout can be modified or rearranged to take advantage of natural body strengths and prevent sprains, strains, and cumulative trauma.

3. *Highly repetitive tasks.* Since this is one of the most common causes of CTDs, the hygienist should be particularly aware of any tasks that require repetitive motions. During task evaluation, ways to reduce the number of motions in the job should be considered. In some cases, tools can be used to perform repetitive tasks. Electric screwdrivers and power-assisted presses, for example, are preferable to their manual counterparts. Sometimes simple work station layout changes can be made to eliminate unnecessary hand and arm motions. For instance, products can be slid into their shipping boxes rather than lifted and placed into them. Even changes in work/rest schedules or alternating jobs between employees to break up the monotony of repetition can be helpful in reducing the effects of cumulative trauma.

4. *Physical fitness and condition.* Loss of strength, flexibility, and general poor conditioning of the muscles, tendons, and other soft tissue structures of the arms, shoulders, neck, and back often contribute to the development of cumulative trauma. Also, people with pre-existing medical conditions that affect these areas are at increased risk of developing a CTD at work. It is, as one might expect, extremely difficult to effect changes in this factor, since it deals exclusively with employees' personal behavior, discipline, lifestyles, and individual habits. While education, advice, and training on the merits of physical fitness can be provided, people generally will not make behavioral changes easily. It would be more effective for the hygienist to concentrate effort on actions that can help compensate for poor physical conditioning. Any worker who is assigned a new or different task or activity should be allowed a slow and gradual introduction to the job so that his or her body will have a chance to become stronger and better conditioned to the task. For example, shoulder tendonitis/bursitis often develops when overhead work is required of an employee with weak rotator cuff muscles. When the arm is lifted overhead, the rotator cuff muscles actually pull the arm bone down as the shoulder muscles lift the arm. If the rotator cuff muscles are weak, the arm bone literally rolls upward instead of down, pinching the tendon and bursa between the bones at the top of the joint. The result is an the inflammation of shoulder tendons and/or bursa. Recommending that the worker slowly become accustomed to any new tasks may help reduce the likelihood of ergonomic problems.

5. *Direct pressure on nerves.* Whenever a tool or working position places direct pressure on a nerve, there is a risk of cumulative trauma. Of particular concern is the pressure against the palm of the hand that is common with the use of many hand tools (pliers, crimpers, hammers, scissors, and so on). Numerous tool manufacturers and distributors offer a wide variety of hand tools that incorporate a bend (usually 19 degrees off-center) in the handle that is made to fit the natural curvature of the hand/palm/finger rela-

tionship. Use of such tools allows a straight wrist configuration while alleviating palm pressures. As an alternative, wrist braces can be used to ensure that the user's wrists are kept in a straight position. However, caution should be exercised if wrist braces are to be used. Depending on operational circumstances and requirements, the use of braces can be somewhat limiting and may even create a more serious hazard than that which they are trying to control.

6. *Vibration.* Vibrating tools and equipment (palm sanders, planners, jackhammers, buffers, etc.) can cause damage to the nerve in the wrist as well as other types of CTDs. When such tools are required for a job, an assessment should be made to determine if any other method(s) could be used to accomplish the desired task. If not, other techniques such as time/use limitations, alternating workers, or other such administrative actions should be considered to help reduce the potential for a vibration-induced CTD.

7. *Cold exposure.* Whenever work is performed in cold environments, certain anatomical functions are affected (see Chapter 7). The areas most susceptible to cold (the fingers, hands, and wrists) are also those that are at a high risk of cumulative trauma. When blood flow is restricted to these areas due to cold temperatures, the opportunity for CTD damage increases dramatically. Requiring the use of wrist/arm coverings, providing tools with insulated handles, and isolating workers from cold temperatures are all methods that can be used to help reduce the risk of exposure.

DEVELOPING AN ERGONOMICS PROGRAM

The industrial hygienist can play an important role in the development of an ergonomics program. However, as with any workplace safety and health initiative, it will not succeed without *commitment from top management.* Indeed, such commitment and involvement are complementary and essential elements of a sound safety and health program. Commitment from management not only provides the organizational resources that will be required to accomplish program goals, it also establishes the motivating force necessary to deal effectively with ergonomic hazards.

The second ingredient for program success is *employee involvement.* When the affected employees are involved in the ergonomic program from the beginning, the identification of existing and potential hazards will most likely be quicker, and the development and implementation of hazard abatement techniques will be more effective. For example, employees can participate in teams trained with the required ergonomic skills to identify and analyze jobs for ergonomic stress and to recommend solutions. Employees should be encouraged to report all real and perceived ergonomic concerns without fear of reprisal from management. After all, the person who best knows the problems that may be associated with any given job is probably the one who performs that function day after day. A wealth of information and time savings can be realized if the employee is allowed to participate rather than simply observe in the process.

Another essential requirement of a successful ergonomics effort is to develop and implement a *written program.* This program, which must carry the endorsement of the highest level of management, should provide an outline of the company's goals and plans for ergonomic safety and health. Obviously, the written program must be suitable to the size and complexity of workplace's operations. In this regard, the hygienist can be helpful since an assessment of the workplace can be used to identify the nature and degree of the ergonomic hazards that exist. It is essential that the written ergonomic program be communicated to all personnel since it will encompass the total workplace, regardless of the number of workers or the number of work shifts.

Finally, to ensure program adequacy and effectiveness, there should also be *regular program reviews and evaluations.* Procedures should be developed to evaluate the implementation of the ergonomics program and to monitor its progress and accomplishments. Top management should review the program regularly (annually or semiannually) to evaluate its success in meeting established goals and objectives. By reviewing trends in injury and illness rates, employee surveys, and process modifications, management can gain a better understanding of the program's success. Managers, supervisors, and employees should also review the program frequently to reevaluate goals and objectives and to discuss changes.

Program Elements

Earlier in this chapter, the four basic elements of an ergonomics program (work site analysis, hazard prevention and control, medical management, and training and education) were presented. While specific programs may incorporate additional features based on individual needs and requirements, application of these four basic elements will certainly facilitate the establishment of an effective ergonomics program. Specifically:

1. *Work site analysis.* In the performance of industrial hygiene, assessment or analysis of the workplace is critical. Work site analysis identifies existing hazards and conditions, operations that create hazards, and areas where hazards may develop. From the ergonomics perspective, this should also include close scrutiny and tracking of injury and illness records to identify patterns of traumas or strains that may indicate the development of CTDs. For example, the analysis of medical, safety, and insurance records, including the OSHA 200 Log and information compiled through medical management programs, may reveal evidence of CTD problems or trends in spe-

cific departments, process units, job titles, operations, or workstations. Use of an ergonomics checklist that includes consideration of posture, force, repetition, vibration and so on can also help facilitate the analysis process. The objectives of this analysis, then, are to recognize, evaluate, and correct or control hazards, including ergonomic hazards.

2. *Hazard prevention and control.* Once ergonomic hazards have been identified through the systematic work site analysis, the hygienist should develop measures to prevent or control these hazards. As discussed throughout this chapter, ergonomic hazards are best prevented through effective design of the work station, tools, and the job itself. Additional measures include appropriate engineering and work practice controls, the use of personal protective equipment and devices, and administrative actions to correct or control ergonomic hazards.

3. *Medical management.* Proper medical management is necessary to eliminate or reduce the risk of CTD development through the early identification of warning signs and symptoms. Plus, as a source of information, medical management also helps to prevent future ergonomic problems. In an effective program, health care providers should be part of the ergonomics team, interacting and exchanging information routinely in order to prevent and properly treat CTDs and other ergonomic problems. Ideally, a physician or occupational health nurse with training in the prevention and treatment of ergonomic injuries and illnesses should supervise the program. In the absence of medical personnel, the hygienist may be asked to oversee this aspect of the ergonomics program. In such instances, the hygienist would be wise to at least seek consultation with appropriately trained medical professionals. As a minimum, the medical management program should address issues such as injury and illness record keeping, early recognition and reporting of ergonomic problems, systematic evaluation and referral, conservative treatment and return to work, follow-up monitoring, and the adequacy of staffing and facilities.

4. *Training and education.* The purpose of training and education is to ensure that employees are sufficiently informed about ergonomic hazards to which they may be exposed and thus are able to participate actively in their own protection. Training allows managers, supervisors, and employees to understand ergonomic and other hazards associated with a job or production process, their prevention and control, and their medical consequence. The hygienist can play an important role in this training process, effectively contributing to the overall reduction of risk posed by exposure to such hazards. Ergonomics training can be *general* in nature covering the nature and variety of CTDs that could be encountered in the workplace, including general methods of hazard recognition. Training can also be *job-specific*, provided to those workers where known ergonomic hazards exist

in a given job, task, or function. Whatever the case, the training provided should be adequate to ensure proper understanding of ergonomic health hazards.   .

SUMMARY

Ergonomics is basically the study of how human performance is affected by both the dynamic and static conditions that exist in the work environment. The focus of ergonomics is to fit the person to the job and vice versa. In recent years, the study of ergonomics has become a major element in the overall workplace safety and health process. More often than not, the responsibility for assuring the control of ergonomic hazards in has fallen to the industrial hygienist. However, ergonomics is an extremely broad and complex issue that reaches far beyond the scope of basic industrial hygiene. It has application in engineering and design. It considers the influence of forces such as inertia, gravity, and power on the movement of muscles and joints, and it is concerned with the collection and scientific evaluation of *anthropometric data* to evaluate the biological and mechanical relationships that exist between man and machine *(biomechanics)*. From the industrial hygiene perspective, these concepts are certainly interesting but hardly useful in the practical application of the *basic* industrial hygiene principles of recognition, evaluation, and control. Therefore, this chapter provided a brief introduction to the study of ergonomics and how it evolved over the years to become a major concern throughout industry today. Since recognition of ergonomic hazards is so essential to their control and prevention, this chapter also focused on the most frequently encountered ergonomic health hazards. These include *repetitive motion disorders* and, more specifically *cumulative trauma disorders* (CTDs) such as *tendonitis, Raynaud's disease,* and the most common CTD, *carpal tunnel syndrome* (CTS). Suggested methods to prevent and control exposures that can lead to such problems were also provided.

Finally, the basic elements of a recommended ergonomics program were presented and discussed. The initial step toward the establishment of such a program is to conduct a *work site analysis* or assessment to identify existing ergonomic hazards and conditions. Such an assessment is a major aspect of the basic industrial hygiene function as well. Although the problems presented by ergonomic hazards are indeed everyone's concern, the hygienist can be of particular value in helping to recognize, evaluate, and control occupational exposure to such hazards.

10

Industrial Hygiene Sampling and Monitoring

INTRODUCTION

Throughout this text, the subject of industrial hygiene *sampling and monitoring* has been briefly discussed, with specific reference to a particular chapter topic or subject matter. For example, in the discussions related to noise exposure or exposure to radiation, information on sampling and monitoring specific to those types of exposures was provided. In fact, during any given day, the industrial hygienist may be required to monitor and sample for radiation, flammable or explosive atmospheres, oxygen deficiency, and/or toxic substances in the workplace.

This text has also provided brief discussion on some of the common types of air sampling equipment and techniques. However, since sampling and monitoring in general, and *air sampling* in particular, form the basis of the industrial hygiene hazard *evaluation* process, this chapter is provided as a specific reference for the industrial hygiene professional to supplement that which has already been detailed in the previous chapters of this Part.

Sampling and monitoring of the workplace are critical to determine if the workplace is safe. These must be accomplished in a careful manner to ensure that reliable information is obtained. This information can then be used to provide statistically reliable analyses for proper decision making.

MONITORING AND SAMPLING OBJECTIVES

Workplace air is monitored and sampled for a variety of reasons. The objectives will vary with the situation and the actual, potential, or anticipated hazards. For example, the requirements for air monitoring at a factory or manufacturing facility with a fixed or limited number of known hazardous commodities will be very different than those at an uncharacterized hazardous waste site or during/following an uncontrolled released of hazardous materials.

Monitoring includes the collection of samples that are representative of actual employee exposure. In essence, monitoring provides a "snapshot" of the exposure during a specific work time that represents the overall daily, weekly, or monthly exposure. The data are then compared to existing standards or regulations to determine if the employer is in compliance.

Monitoring of hazardous agents in the workplace includes a number of acceptable techniques and procedures. These techniques, established and tested within industry and government, are developed into procedures. These procedures detail the specific protocol for performing the sampling or monitoring.

While there are, indeed, a variety of conditions that may require air monitoring, the basic impetus for monitoring remains the same: to provide information regarding the type and relative quantity of hazards present (Maslansky and Maslansky, 1993). Once the information obtained through monitoring has been evaluated and properly analyzed, it can be used to assess the risks posed to workers, the general public, and the environment.

Monitoring information can also be used to assist and even facilitate the selection of personal protective equipment (PPE). In fact, when monitoring data show the risks of exposure are specific to given areas in the facility, these data can be used to justify the delineation of those areas where PPE will be required. Aside from the determination and evaluation of hazard exposure risk, other reasons for monitoring and sampling are to ensure that a facility or procedure is in compliance with applicable regulations, to evaluate the effectiveness of mitigation techniques, to assess the efficiency of emissions control, and to determine the need for specific medical monitoring of exposed personnel. Air monitoring can be used to confirm suspicions regarding the type of hazards present or to rule out other potential hazards that may also exist.

It should also be acknowledged that valid air monitoring data can be used to eliminate questions or allegations of workers regarding exposure, as well as to provide documentation of adherence to occupational and environmental regulatory standards. When such data are used for regulatory or legal considerations, however, it becomes even more critical that air sampling and monitoring be conducted by trained personnel in a manner that will withstand legal challenge.

AIR SAMPLING INSTRUMENTATION AND EQUIPMENT

Selecting equipment and sampling media is the first step in implementing effective sampling strategies. To decide on the proper equipment, the hygienist must first understand the specific sampling needs. For example, the prudent hygienist will not rely solely on verbal information regarding the nature and degree of chemicals present, unless the source is highly reliable. It is preferable to review labels and material safety data sheets (MSDS) or request the name and phone number of the supplier. If specific chemicals have been identified, the hygienist should attempt to determine if any of them will interfere with one another in the sampling methods that have been selected. Some companies will purchase a certain amount of equipment based on anticipated sampling needs. However, due to the high cost of some instruments or the rarity of use of others, a necessary instrument may have to be rented or specially purchased. Also, instruments that have been stored for a period of time or used by others should be checked out to ensure that they are in good working condition, that their batteries are fully charged (if applicable), and that calibrations are accurate. This is especially important whenever equipment has been rented.

Instrument Calibration

Air monitoring instruments are calibrated by the manufacturer to respond accurately to one particular vapor or gas. Calibration is a process of adjusting the instrument response until the reading corresponds to the actual concentration present. Initial calibration is performed at the factory, using multiple concentrations of calibrant gas. After factory calibration, the instrument will respond accurately only to the calibrant gas within its detection range. Subsequently, the meter will not respond accurately to other gases or vapors. Rather, it will provide a *relative response* reading that may be higher or lower than the actual concentration present. This, however, does not mean that the meter can only be used for situations where the calibrant gas or vapor is present. It simply means that the meter responds to other gases or vapors using the calibrant as a reference. When detecting gases other than the calibrant gas, the concentration read is often expressed as calibration gas equivalents, response equivalents, equivalent units, or simply units. Depending on the type of instrument, some manufacturers may provide *relative response curves* or *conversion factors* for gases and vapors other than the calibrant gas. When instruments are to be purchased, the prudent hygienist should inquire as to the availability of such information from the manufacturer.

Once the instrument has been obtained, *calibration checks* prior to each use are necessary to ensure that the meter remains in calibration. This is especially true in the event that a meter has remained idle for relatively long periods of time (weeks or even months). The calibration check is typically performed using a single concentration of calibrant gas. For example, a carbon monoxide detector may be factory calibrated to accurately respond to carbon monoxide concentrations between 1 ppm and 200 ppm. A calibration check can be performed by the operator using a single, known concentration of calibrant gas, such as 25 ppm carbon monoxide. If the instrument responds correctly to 25 ppm, then the user can be reasonably assured it will accurately respond to other concentrations within the detection range of 1–200 ppm. This check is the only means of demonstrating that the instrument is still working properly. If the meter does not respond as expected to the calibrant gas, appropriate adjustment must be made until an accurate reading is obtained. Also, since actual or anticipated environmental and field or site conditions can affect the operation of many instruments, it is often recommended that the calibration check be performed on-site under the temperature conditions of ambient use. This is often referred to as a *field calibration check.*

Airborne Contaminant Monitoring

Airborne contaminants can exist in a wide variety of forms. Normally, contaminants are found in the form of vapors, mists, fumes, particulates, or fibers (see Chapter 5). To effectively evaluate the nature of an exposure, the industrial hygienist must monitor the *breathing zone* area of the worker. This will provide the most accurate "picture" of the exposure. It should be noted that there is no way to measure such exposures with 100 percent accuracy. Even through the utilization of such measures as biological sampling (i.e., sampling the body itself), accuracy will still fall short of total reliability. Therefore, the hygienist must often utilize procedures that have been thoroughly tested and have been proven to provide reliable results within certain statistical parameters.

Hence, sampling is typically accomplished using procedures that have been proven to consistently provide the most reliable results. There are basically two types of techniques for measuring airborne contaminants: *direct reading instrumentation* and *constant flow sampling pumps.* Direct reading instrumentation provides instant or nearly instant readings for a particular airborne contaminant. The results are usually expressed in *parts per million* (ppm) or *milligrams per cubic meter* (mg/m^3). Constant flow sample pumps draw air through a sample media to capture the contaminant. The samples are then analyzed by a laboratory that can provide results in a minimum of 24 hours (see Chapter 11 for additional information).

Direct Reading Instruments

By far, the simplest method of determining the airborne concentration of most chemical contaminants is to use instruments that instantly provide a readout of the concentration in ppm or mg/m^3. Direct reading instrumentation includes highly sophisticated, expensive equipment of relatively low technology, and is the least costly type of air monitoring equipment. This section will provide brief discussion of some of the most common types of equipment and their individual uses, advantages, and disadvantages. The user must understand each type of equipment's limitations and be careful to use each in accordance with the manufacturer's requirements.

Direct reading instrumentation is tested by various different organizations. Tests for accuracy, reliability, interferences, and safety are conducted. Originally, such tests were performed exclusively by governmental agencies or professional societies. However, direct reading instrumentation tests can also be accomplished by independent testing laboratories. The user should carefully review such testing data prior to purchase of any equipment.

Colorimetric indicator tubes (CITs) are widely used in industrial hygiene and industrial settings. CITs operate by drawing a calibrated amount of air through a tube filled with an absorbent material (Figure 10-1). The air can be drawn through the tube using a variety of different types of hand-operated pumps or, in some cases, using automatic air pumps. The absorbent material inside the tube is treated to be sensitive to the specific contaminant being sampled and will change colors as air is drawn through the tube. The tube is marked to indicate various concentrations. As air passes through, the tube changes color to match up to a line and corresponding ppm level. Once the total amount of required air is drawn, the tube will show the concentration. There are different tubes for each contaminant. Once the tube is used, it is disposed of.

CITs have been in use for many years and have been highly successful in giving early warnings for chemicals such as carbon monoxide, hydrogen sulfide, and others. In the past few years, these tubes have become very specialized to detect a wide variety of chemicals.

There are advantages to using CITs. Primarily, CITs are capable of quick response times. There is no waiting for laboratory analysis or other interpretations. CITs are simple to use. They are excellent for use by non-professionals to measure for possible problems.

Normally, CITs for various chemicals will use the same sample pump. Hence, many types of testing can be accomplished using the same pump. This makes them cost effective as well. There are approximately four manufacturers of CITs from which the user can select. It is recommended that the user select a single manufacturer so that there is no cross-utilization of sample pumps. CITs must be used in accordance with the manufacturer's specifications to ensure that results are valid.

CITs also have some disadvantages that should be noted. Calibration of the tubes is accomplished by the manufacturer, resulting in a risk of inaccuracy. The accuracy of the tubes is a major concern. The accuracy for each individual tube should be known prior to use. This information is normally provided by the manufacturer. Typical accuracy is usually estimated at ±25 percent.

CITs may react in different ways with other contaminants. For instance, a tube that detects hydrazine (N_2H_4) will react faster to ammonia in low concentrations. Also, the user must be very careful when monitoring with CITs around other personnel. Workers in the area may see the tube change and overreact. Users should

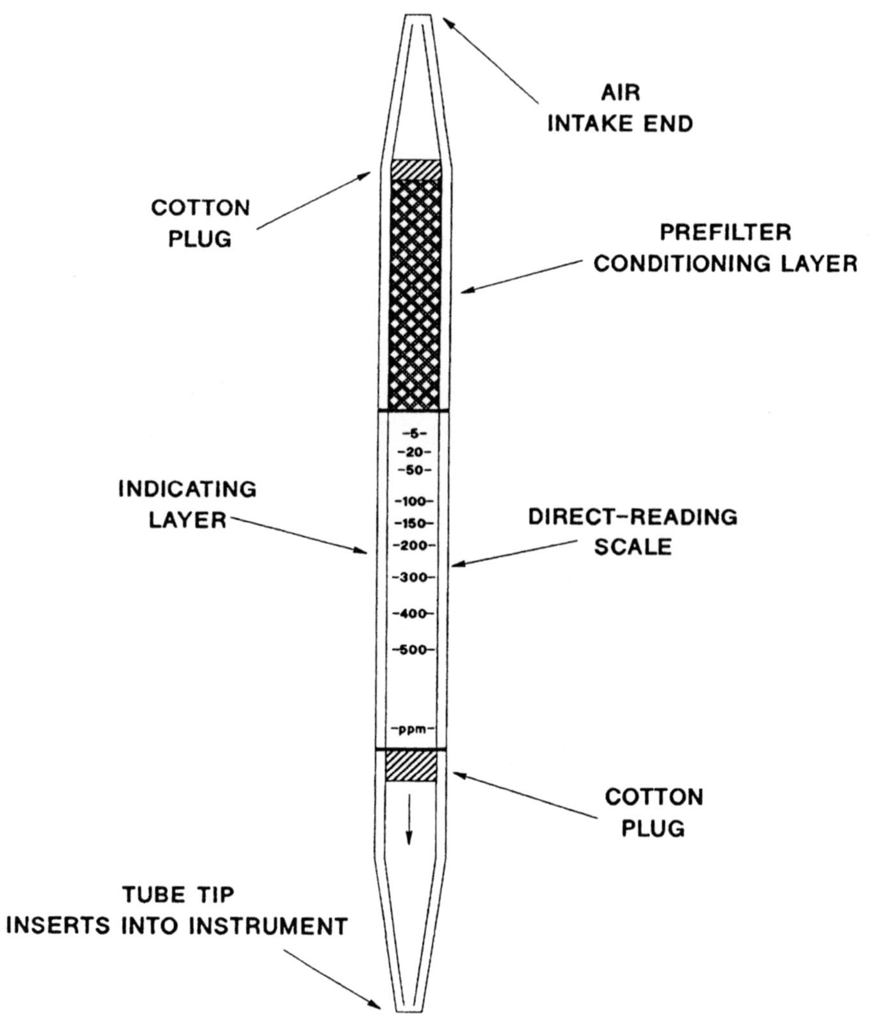

FIGURE 10-1 A colorimetric indicator tube (CIT)

also become familiar with any possible interferences that may render CIT data useless or inaccurate.

Another disadvantage is that CITs are designed for a quick, grab sample. It is difficult to determine the overall exposure in accordance with the OSHA PELs since the standards are established for an eight-hour work shift. Therefore, when attempting to determine overall exposure, CITs should be used concurrently with more definitive sampling techniques. Legal issues for accuracy and shift exposure should guide the user to perform more extensive sampling, rather than relying solely on CITs.

Detection accuracy is a definite concern when using CITs. Overall, manufacturers are very much attuned to the need for accuracy and initiate in-depth quality control programs. However, this does not mean that the user should rely solely on the manufacturer to perform *quality control* (QC). It is highly recommended to implement a user-specific QC program to test the reactivity of the tubes. This can be accomplished by purchasing calibration gases to perform tests. In small, one-time instances, this may not be feasible. However, companies that utilize a high number of tubes should test their accuracy. If a specific problem is found, the entire shipment of tubes should be returned to the manufacturer for exchange.

Since CITs are disposable, proper care must be taken to ensure proper disposition. In many cases, discarded CITs are considered hazardous waste and must be disposed of as such. The hygienist should confer with the manufacturer to verify the proper disposal techniques.

Electrochemical sensors are another form of direct reading instrumentation. There are numerous types of monitors, meters, analyzers, etc., on the market for detection of airborne chemicals. One of the technologies utilized today is based on electrochemical response to a specific chemical. The basic detection method is for the chemical to be drawn across a specific sensor material that will generate an electrical reaction. This electrical response corresponds to a ppm value on the meter. These instruments provide very quick response; even faster than CITs. Instruments of this type are very useful in early warnings for leaks or releases. The meters can be connected to alarms to provide audible and/or visual notification when a specific level is met. Many meters are equipped with their own alarms.

Air is drawn to the sensor by a pump or is transmitted by diffusion. These instruments are now available as personnel monitors that can be clipped onto a belt to warn the worker of the presence of toxic contaminants. This is a very useful technique for confined space entry where carbon monoxide, hydrogen sulfide, or other contaminants may be present.

Many meters are equipped with multiple monitoring capability where three or four contaminants can be monitored at the same time. This adds to their diversity and flexibility.

As is the case with CITs, there are also disadvantages to using this type of equipment. The user must know the potential interferences, responsiveness, and operating limitations of each type of unit. Many meters are sensitive to radio frequency interference. This is very important when hand-held radios are used, because they could set off a false reading very quickly.

Calibration is another concern with such instruments. The user must calibrate electro-chemical instruments in accordance with the manufacturer's recommendations. This calibration should be checked routinely, and instruments should be placed on a cycle of testing and calibration. Test gasses should be maintained to provide quick checks.

Organic vapor detectors are another general type of monitor. Although these have been placed under the category of direct reading instrumentation for ease of discussion, there are different types of sensors utilized for detecting volatile organic compounds (VOCs), that may or may not provide direct and instantaneous response. These technologies include photo-ionization, flame ionization, gas chromatography, and others. VOC detection is a major concern as a result of environmental testing requirements for hazardous waste remediation and detection. Because of their wide range of uses, organic vapor detectors are extremely versatile. Their use, however, must be carefully planned, especially when working in hazardous waste remediation or investigative studies. Organic vapor detectors will react to numerous VOCs. It is therefore critical to understand the limitations the manufacturer places on detection. Sampling must be accomplished in a manner that is consistent with the manufacturer's literature.

Constant Flow Air Sampling Pumps

Constant flow sample pumps are utilized to perform detailed sampling of a given environment. Sampling is performed by drawing a known quantity of air through a sample medium. The medium is analyzed by an accredited laboratory, which provides results. This technique is very accurate and provides results to determine OSHA compliance. OSHA compliance officers often utilize this method of air sampling to determine if the PELs are exceeded in accordance with OSHA-developed procedures.

SAMPLING STRATEGIES

Generally speaking, an acceptable strategy for air sampling will be one that will *prioritize* the specific needs of the situation. This ensures the maximum possible benefit from the sampling session(s) while optimizing the use of the resources required to perform the sampling (e.g., personnel, equipment, time, etc.). The best strategies are those that are most easily implemented and, of course, are most cost effective.

Historically, sampling that is performed for limited periods (such as one or two days) has concentrated on worst-case exposures. In such instances, decisions to conduct sampling are based on the periods when production is highest, the areas where the most concentrations of the materials to be sampled are used, or the jobs involved are those using the most toxic compounds. If the primary objective of a particular sampling session is to prevent worker overexposure, a sampling strategy should be developed that takes into consideration the nature of the contaminant. A common example would be a chemical's biological half-life. The biological half-life is one method of determining the likelihood of a compound remaining in the body. The longer the half-life, the more likely it is that the compound will accumulate in the body. In fact, it has been suggested that to have air sampling results more closely reflect the potential to exceed body burden, the minimum sampling time should be equal to 0.3 times the biological half-life (Roach, 1977; Ness, 1991). In this regard, there are three common types of substances that may require such sampling strategy:

1. Chemicals or substances with a long half-life, such as coal dust or asbestos, that will result in cumulative biological effects over a lifetime of exposure
2. Chemicals or substances that produce long-term effects because they have exposure thresholds above which there is a dose rate dependency
3. Compounds that have short half-lives, such as chlorine or ammonia, but cause irritation and acute effects in high levels

The type of *toxic effect* can also affect the sampling strategy. For instance, grab sampling using a series of short samples can provide useful information as to whether high levels of compounds that are irritants are present at the work site. Sampling strategies will also differ depending on whether the sampling is for chronic or acute exposures. *Integrated sampling* techniques are usually the preferred approach for *chronic exposures* to cumulative poisons such as lead or mercury. Integrated sampling is an approach that does not rely on air sampling alone to determine chronic exposures. Wipe samples, bulk samples, and biological sampling are combined with air sampling techniques to ensure a complete and effective sampling strategy for chronic exposures. Situations that can pose a potential for *acute exposure* to fast-acting chemicals at relatively low levels should be sampled over shorter periods. Sampling for chlorine and hydrogen sulfide in a confined space, for example, would require continuous real-time monitoring. The two common strategies for sampling such acute exposures are to:

1. Collect samples during periods of highest exposure (if known)
2. Collect short-duration samples at regular intervals to minimize bias and identify unpredictable peaks

Once the purpose and corresponding strategy have been decided, the sampling period must be selected. This will depend on three main factors:

1. The length of the workday, the shift, or the actual task to be performed
2. The expected concentration of the commodity to be sampled

3. The standard for comparing the resulting data (e.g., ACGIH-TLV, OSHA-PEL, NIOSH, etc.)

Wherever possible, it is recommended that exposures on all shifts be characterized through sampling since variations can (and often do) occur. The day shift may be the busiest, whereas the evening shift and swing shifts often have fewer employees. However, in some cases, the lack of supervision on evening shifts may lead to periodic lapses in good work practices and exposures may subsequently increase. Also, the work shift is generally defined as eight hours. When collecting samples to compare to an eight-hour time weighted average (TWA) such as OSHA's, sample periods must reflect this.

The best way to determine personnel exposures is to perform air sampling and analysis using air sampling pumps. The following steps are provided to discuss the basic procedures for performing personnel exposure monitoring.

Pre-survey

A number of tasks need to be completed prior to performing air sampling. The area should be thoroughly evaluated to identify employees with the highest and lowest potential for exposure. This evaluation should include a walk-through investigation by an industrial hygienist or other safety and health professional. The list of all hazardous chemicals should be reviewed. Each material safety data sheet (MSDS) for the hazardous materials should be evaluated to identify the specific chemical agents. Particular attention should be given to the toxicity of the chemicals and their properties, such as vapor pressure, the form the chemical takes (solid, liquid, gas), flash point, etc. The walk-through should include sketches of work locations, ventilation sources, location of chemical materials and overall work patterns. The work process should be thoroughly documented.

Direct reading instruments should be used whenever possible while performing the walk-through to identify specific processes that are emitting contaminants at high or suspect levels.

Discussions or interviews with supervisors and/or employees should be conducted to determine any complaint areas. As a general rule, employees should always be interviewed in the presence (or with the concurrence) of their supervisors. Employees who work the same process in other areas or on other shifts may provide insight into any perceived or real problems.

Each process should be evaluated to determine which employees are exposed to chemicals. Quantities used and the types of chemicals should be noted. It is also useful to note the chemicals by category because they may be sampled either by employee or by area. For instance, toluene and xylene can be sampled on the same sample media and analyzed at the same time. Heavy metals can also be analyzed on one sample filter.

Developing A Sample Plan

The next major step in performing air sampling is to develop the sample plan. This will form the basis of the sampling strategies to be used. As indicated above, sampling priority should, in most cases, concentrate on employee exposures. This means that the air sampling equipment will be placed on the worker, with the sampling media located near his or her breathing zone. Remember, the objective is to obtain an air sample that most closely resembles the concentration that the worker is receiving.

The industrial hygienist must determine who to place the sampling equipment on. Normally, it is advisable to sample more than one person. By sampling more than one person, the results will provide more conclusive data, and if a sample is damaged, there will be a backup. It is also advisable to sample with an area sample; that is, sampling in the general area, preferably at a normal breathing zone level. This is useful for determining the overall extent of chemical contamination in the work environment.

The next important step is to determine the specific commodities that require sampling. Information from the walk-through should be carefully reviewed to determine these chemicals or commodities. Once this has been decided, the sampling methods must be determined. NIOSH has developed procedures for sampling and analysis that are available through the U.S. Government Printing Office. OSHA also has a technical manual, used by OSHA industrial hygienists for sampling, that is available through the government.

The industrial hygienist should contact the laboratory well before performing sampling to discuss intended strategy, potential interferences, sample media, shipping requirements, required sample volume, and any other concerns. Information on using an industrial hygiene laboratory is provided in Chapter 11. It is very important to establish a good working relationship with the laboratory. This simple step will also help to ensure that the samples provided are sufficient to accurately determine personnel exposures.

Prior to performing sampling, the industrial hygienist should verify that the equipment is operational, has properly charged batteries, and that all tubing, holders, etc., are available. The supervisor or contact person in the work area to be sampled should be notified to ensure that there are no schedule problems. It is very disconcerting to show up to perform sampling and find out that there is no one performing the operation that you need to sample.

Performing the Sampling

All sampling should be performed in a manner that provides the least disruption to the work. Early, before the sampling, the sample pumps should be calibrated using a primary standard and a representative sample media. Do not use the sample media used for calibration for sampling, because the sample integrity will be invalid. Primary standards are available from a number of manufacturers that facilitates the calibration process. The user simply adjusts the flow rate to that required by the procedure.

The sample pump and its associated tubing must be affixed to the employee. The sample media should be positioned in the breathing zone area. If, for instance, the media is a tube, it can be clipped to the worker's collar, with the tubing positioned up the back. Tape, such as duct tape, should always be available. It can be used to keep sampling tubing and media in place.

It is also recommended that the sampling process be explained to the employee. This should always be accomplished with concurrence of the manager or supervisor of the area.

All sampling activities should be carefully documented. Pre-surveys, contacts with the laboratory, and sampling activities should be recorded. If the sampling should be disclosed in court, all information as to how the sampling was performed will be crucial.

When sampling, the calibrated flow rate must be recorded. Sample start time and sample stop time must also be documented. During the sampling process, the sample pump must be checked frequently to ensure that it is operating properly. Most sample pumps have rotameters or some other type of indicator that allows uninterrupted verification of its proper operation. All observations made during the sampling process should also be recorded, including the description of the work going on at the time of the sampling. Once the sample process is complete, the sample is sealed and prepared for shipment. Immediately after sampling, each sample pump must have the flow rate checked with the same media used for the calibration. If the sample flow rate has changed more that 15 percent, the sample results may not be accurate and a resample is recommended. The flow rate should be averaged with the beginning and ending flow rate and multiplied by the run time to determine the sample volume. The sample volume for each sample must be provided along with the sample to the laboratory so that the results can be calculated in parts per million or milligrams per cubic meter. Chapter 11, on Industrial Hygiene Laboratory Services, provides additional information on shipping samples to laboratories.

Overnight shipments provide less opportunity for accidents in shipping and also increase the response time for results.

Reporting the Results

Once the laboratory provides the results to the industrial hygienist, it is important to check the calculations to ensure that the results make sense. If results seem extremely high or low based on what was observed during the sampling process, sample again without providing a full report. It is better to be confident in the results prior to making firm recommendations.

If the sample results indicate an extremely dangerous condition, the hygienist should call or visit the contact person (the supervisor or manager) and describe the problem. Also, the hygienist should document any verbal interactions with supervisors. A short interim report should be provided by the hygienist, detailing the dangerous situation and what actions should be taken.

The final report should be as simple and direct as possible. The following is a recommended outline:

1. *Executive summary.* Provide a brief, high-level, low-detail summary of the sampling event, the nature of any problems, and the means of correcting them.
2. *Background.* This makes up the body of the report, since it will provide a detailed narrative of the entire sampling event, the reasons for it, and any historical data pertinent to the situation or work condition.
3. *Methodology.* This portion of the report describes the methods and techniques used to conduct the sampling event. Here is where the instrumentation and equipment used will be noted. Any standard procedures that were followed (NIOSH, OSHA, etc.) should also be listed here. For legal considerations, this portion of the report can be most important because it establishes the methods, techniques, strategies, and procedures used during the sampling event.
4. *Findings/conclusions.* This is where the hygienist will provide a listing of all the findings, determinations, discoveries, and other pertinent information regarding the actual results of the sampling event. Laboratory data can be helpful in completing this portion of the report. It is sometimes helpful to include actual copies of laboratory reports as an appendix. The hygienist can also draw conclusions based on available data and provide them in this section of the report.
5. *Recommendations.* No report of operational problems is complete without the provision of workable recommendations or solutions. The hygienist should evaluate all data available and provide recommended ways to correct the problems noted in number 4. If possible, it is always best to provide management with more than one alternative solution to a given problem.

It is also recommended that a draft be developed first and shared with the ultimate recipients of the final report. Many times, especially in corporate organizations, certain issues will have a stronger impact if they are reported in a certain manner. This may include reviews by legal representatives or members of upper management. Also, by letting management review your report in the draft stage, there will be no surprises when the final report is submitted. On that note, however, it is morally and ethically important that the industrial hygienist not vary from the facts and findings. How it is reported can be altered to meet management's wishes, but content should never be changed.

SUMMARY

This chapter provided a brief discussion on the various air monitoring and sampling techniques that are commonly used in the industrial hygiene profession. The importance of developing sampling strategies when approaching air monitoring situations cannot be over emphasized. The types of equipment that can be used vary almost as much the situations that arise requiring their use. However, in general terms, two types of instrumentation are typically utilized. These are the *direct reading instrumentation* and the *constant flow air sampling pumps*. The most commonly encountered direct reading instrument is the *colorimetric indicator tube* (CIT), which provides good accuracy for an on-site assessment of exposure conditions. Constant flow sampling pumps can be used to obtain highly accurate measurements of workplace exposure conditions.

The sample report was also briefly discussed in this chapter. A recommended format that can be modified to accommodate each individual user's requirements was provided.

III

Supplementing the Industrial Hygiene Program

INTRODUCTION TO PART III

Thus far, the *Basic Guide to Industrial Hygiene* has provided the information considered to be essential to the practice of industrial hygiene on the most fundamental of levels. In Part I, Understanding Industrial Hygiene, the reader received an overview of the profession, how it developed, the basic purpose and objectives of industrial hygiene, the regulatory agencies and their involvement, and an overview of human anatomy and how it can be affected by occupational health hazards. Part II built on this knowledge, as it introduced and discussed the various types of hazards that are commonly encountered in today's working environment. Also provided was information on how to *recognize* when these hazardous conditions are present and how they can be *evaluated* on-site by the hygienist. Methods and techniques to *eliminate, reduce,* or *control* these exposures were also briefly discussed.

In the practice of industrial hygiene, it frequently becomes necessary to utilize the services of external agencies and professional organizations to ensure the overall adequacy and efficiency of the industrial hygiene program. While there are many aspects of the program that can be handled internally in a quite effective and proficient manner, there are also those times when the hygienist must seek external assistance. Knowing when it is necessary to obtain such help is an important part of the industrial hygienist's profession. Therefore, in Part III, two extremely important areas that commonly (and sometimes necessarily) are the focus of external assistance are discussed.

First, Chapter 11 provides basic information concerning the selection and use of *industrial hygiene laboratories.* No matter how proficient and effective an internal program may be, the time will come when samples (air, water, etc.) must be processed by an *accredited* industrial hygiene laboratory. Information on the accreditation process, operational protocol, chain of custody requirements and quality control, sampling techniques and methods, and other requisite criteria will be provided.

Chapter 12 concludes this Part with discussion on the selection and use of additional expertise in the form of *professional consulting services.* The decision to use a professional industrial hygiene consultant is strictly internal to the company, based on specific needs that may extend beyond the internal capabilities of existing staff personnel. Therefore, Chapter 12 should not be considered an endorsement for the use of such services. The only intent is to provide the reader with basic information on the issues and concerns that should be considered once a decision has been made to utilize an external consultant. These include legal considerations, credential verification, and the development of selection criteria based on individual needs and requirements.

On completion of Part III, readers should have obtained sufficient information to confidently seek external services that may be required to supplement their industrial hygiene programs. Coupled with the material presented in the previous two Parts of this text and the information in the appendices, readers will be able to move forward in their industrial hygiene endeavors with a degree of competence, confidence, and understanding of basic industrial hygiene.

11

Industrial Hygiene Laboratory Services

INTRODUCTION

Even the most basic industrial hygiene program will eventually require the laboratory analysis of some type of samples. In fact, proper health hazard *evaluation* (a key element of the overall industrial hygiene process) may not be possible without the use of such services. Air samples taken to determine the specific concentration of airborne contaminants are, by far, the most common type of samples to be evaluated by industrial hygiene laboratories. However, in some instances there may also be needs to evaluate drinking water, soil contaminants, and/or the presence of harmful agents on the surfaces of structures (usually gathered using an absorbent *wipe* material).

The results of a laboratory analysis can often make the difference between a successful industrial hygiene program and one that fails to accurately determine the extent of health hazards in the workplace. The latter can lead to further and more serious endangerment of employee health, violations of regulatory requirements, and litigation arising as a result of these shortcomings. Hence, strict *quality control* of all samples must be enforced at the laboratory, as well as by the industrial hygienist responsible for obtaining, packaging, and mailing (if applicable) the samples to be analyzed. Also, to ensure a standard approach to the analysis of the wide variety of samples a laboratory may receive, certain specific methods and procedures have been prescribed by such agencies as NIOSH, the ACGIH, the AIHA, and many others, depending on the nature of a given sample. Also, the hygienist should be cautious if using a laboratory other than those accredited by the American Industrial Hygiene Association.

This Chapter will provide information on the methods and techniques associated with laboratory analyses, with specific focus on operational protocols (such as sample *chain of custody* procedures) and the factors that should be considered when selecting an industrial hygiene laboratory for sample analysis. It is noted that the information provided here is by no means comprehensive, and the hygienist should be prepared to address concerns specific to the particular operational requirements. As a minimum, however, the guidelines established in this chapter should help form an understanding of industrial hygiene laboratory services that are critical to the success of most industrial hygiene programs.

SELECTING THE LABORATORY

The support of a quality industrial hygiene laboratory is critical to assuring the reliability of a health and safety program. Industrial hygiene sampling must be in accordance with accepted procedures and protocols. Laboratory analysis is critical to measuring personnel exposures to chemicals in the workplace. Analysis of these samples must be provided by a laboratory that is proficient in the state-of-the-art analysis techniques capable of surviving the scrutiny of potential legal proceedings.

In general, samples are collected based on methodology developed by NIOSH and OSHA. Some procedures are developed by laboratories directly and validated utilizing industry-accepted protocols. These procedures are normally used when a specific NIOSH or OSHA procedure does not exist.

Selecting a laboratory is a process that must be taken very seriously. It is a good idea to *prequalify* more than one laboratory to provide backup and to allow for quality control testing (discussed later in this chapter). When selecting a laboratory, two issues must be considered. First, the safety and health professional must know exactly what will be sampled. This can be determined early by analyzing each *material safety data sheet* (MSDS), extracting the name of the hazardous chemical(s), and compiling a list of these chemicals. This list can be further categorized by the types of chemicals, such as *volatile organic compounds* (VOCs), for example.

Second, the safety and health professional *must* utilize a laboratory that is accredited by the *American Industrial Hygiene Association* (AIHA) for the specific chemical type(s) to be analyzed. The listing of accredited laboratories can be obtained by writing to the AIHA (see Appendix A). Accreditation is important because it assures the user of the validity of the sample results. This assurance is a result of the laboratory's participation in a testing program where samples are submitted by the AIHA to the laboratory for testing. These samples are submitted without informing the laboratory of the known concentrations. Sample results must consistently fall within a specific range for the laboratory to become accredited and to maintain accreditation. Another reason laboratory accreditation is important is that the results of sampling and analysis may have to be defended legally or professionally. In such instances, laboratory accreditation is recognized as a standard. For example, sample results may indicate that a process is generating airborne concentrations in excess of the OSHA *permissible exposure limit* (PEL). The reliability of the sample results is critical to provide management with substantiation for implementing engineering controls. Equally important is the concern that the sample results may indicate levels below the PEL when, in reality, employees are exposed to a much higher level that can result in serious health impacts.

Figure 11-1 depicts the following recommended selection flow process that should be utilized when selecting an industrial hygiene laboratory.

1. Compile a list of chemicals that may be sampled.
2. Contact the AIHA for a listing of accredited laboratories.
3. Compare the listing of the laboratories to the chemicals and compile a list of those laboratories that qualify.

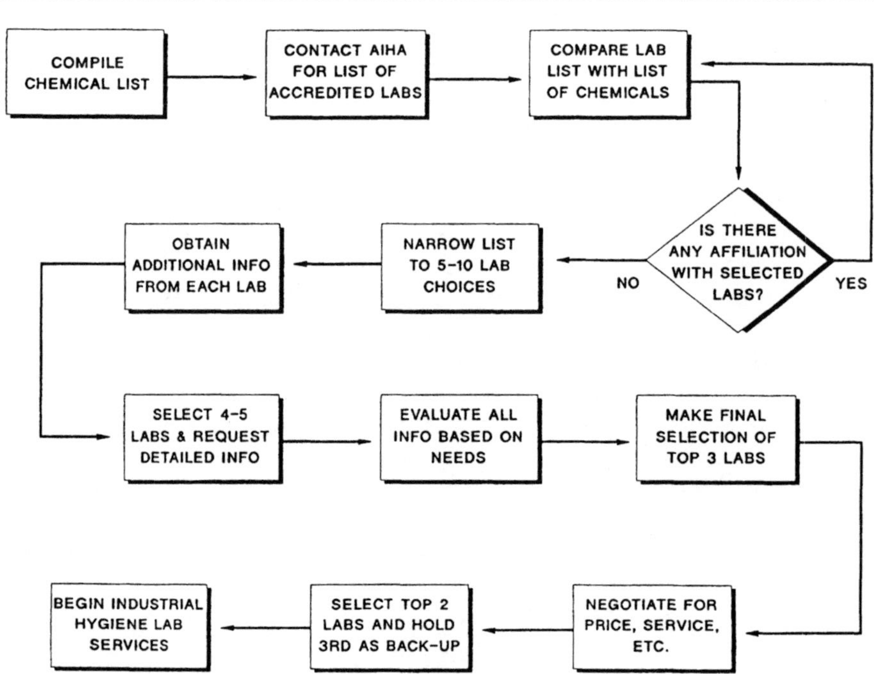

FIGURE 11-1   Flow chart of recommended process for selecting industrial hygiene laboratory services

4. Avoid selecting any laboratories that may be affiliated with your company or a company for which you are performing work (if the results are challenged, there could be an inference that there is a conflict of interest).

5. Narrow the list to approximately 5 to 10 potential laboratories. It may be preferred to select a local laboratory so that delivery of samples can be accomplished directly, establishing a more accountable *chain of custody* (see discussion below). However, with the reliability of overnight commercial mail service, this may not be a serious concern.

6. Request information directly from each laboratory under consideration. Review the laboratories and eliminate those whose capabilities do not match your requirements. There may also be a wide variety of rates between the laboratories. However, the evaluator should be careful not to eliminate a potential laboratory due to cost at this point in the process. Prices can be negotiated if necessary, based on proposed quantities or other considerations.

7. Select a final four to five laboratories for serious consideration and develop a *requirements list* for them to provide additional information. This should include a copy of the laboratory's quality control plan, a list of clients with names and telephone numbers of contacts, and a breakdown of unit analysis costs to detail such things as media costs, shipping, and forms. Many labs will include all shipping costs, the costs of the sample collection media (tubes, solutions, filters, etc.), analysis forms, and can provide results by facsimile.

8. Once the information is received, the reviewer should contact the laboratory's clients for information concerning turnaround time, quality of results, and responsiveness. In addition, all information should be reviewed and each laboratory should be ranked in overall ability and performance experience. If a laboratory refuses to provide the above information, it should be dropped from consideration.

9. Once the information is reviewed and the laboratories are ranked, the top three should be selected for negotiating the lowest unit price based on the services provided. Contact all laboratories and inform them that they are under consideration for the project and you would like their best price. Emphasize that more samples will be forthcoming if the service is deemed satisfactory.

10. Once the information is provided, the top two should be selected and notified. The third candidate may still be used if either of the other two do not meet their obligations.

Another general consideration is to use a laboratory that has performed analysis for NIOSH, OSHA, or other government agencies. This provides additional credibility if a case should go to legal proceedings.

SAMPLING PROTOCOLS

OSHA compliance safety officers perform personnel exposure monitoring in accordance with guidelines provided in the OSHA *Field Operations Manual.* Generally, laboratory analysis is also performed in accordance with OSHA guidelines or those provided by NIOSH. These guidelines and procedures include details as to the method of analysis, the method for sampling, the precision of the results, the required sample volume to ensure accurate results, the required sample flow rates, the sampling media, information as to interferences, and packaging/shipping information.

Prior to sampling, it is beneficial to contact the laboratory and explain the type of sampling that will be performed and to brief them on what is involved, including information on any other chemicals that may be present. One or two weeks advance notice is a recommended lead time, allowing for any problems such as scheduling conflicts, equipment calibration allowances, etc. However, lead time is not a requirement, especially in emergencies or quick response cases. If the laboratory provides the sampling media, it should be ordered early enough to ensure delivery. It is strongly recommended that you order the sampling media from the laboratory, as the quality of the media is routinely tested by the laboratory as part of its quality control program.

Once sampling is performed and completed, careful consideration of the storage requirements must be made to ensure that the samples are not contaminated or damaged. Storage times should be minimized. When ready for shipment, the sample forms should be completed, the samples properly marked, and the shipment made by overnight mail. As a general rule, the longer the time between sampling and analysis the higher the risk that re-volitazation or other loss of sample accuracy may occur. Sample periods should be planned to allow for shipping of samples to the laboratory as soon as possible.

On receipt of the laboratory's analysis, the person who performed the sampling should review the data for calculation errors. For instance, the sample results may be provided in total micrograms per sample and in milligrams/cubic meter. The results should be checked against the sample volume to verify accurate results.

SAMPLE HANDLING AND CHAIN OF CUSTODY

Sample integrity should be a major concern of the hygienist, not only in those cases where litigation is even remotely possible, but also from a purely occupational health assurance perspective. Simply stated, the hygienist will be unable to meet the objectives of the industrial hygiene program if there is low confidence in the *quality control* and precautionary measures to ensure sample integrity. The combination of collection media, sampling pump, tubing, and a flow rate measuring device is called the *sampling train*. To prevent mix-ups with samples, the process of quality control must begin with the industrial hygienist long before the samples are sent to the laboratory. Sample holders such as tubes, cassettes, and/or impingers must always be labeled with a unique number before placing them into the sampling train. The last step is to prepare the samples for shipment to an industrial hygiene laboratory. Sample seals are often necessary. OSHA compliance officers seal their samples prior to shipment to ensure sample integrity throughout transport. Samples collected for litigation purposes may also require seals. If properly constructed and applied, seals can provide proof that the samples were not tampered with during transit or storage (the seal should use a glue that cannot be removed without detection). Another use of the sample seal is to ensure that the caps and plugs on sorbent tubes remain in place. Seals minimize the opportunities for contamination of samples and help prevent leakage of the sampling media during shipment. The industrial hygienist can take other actions to guarantee sample integrity prior to shipment. For instance, samples should never be left in a work area. They should be stored in a safe and secure location and should not be easily accessible. Samples should also be kept away from elevated temperatures, such as in a glove compartment or car trunk, to prevent any unwanted alterations of their true nature.

When samples are submitted to a laboratory, there should always be an accompanying *sampling data sheet*. If some samples taken at the same time happen to be incompatible, such as air samples and bulk samples, they may have to be shipped separately. In such cases, separate data sheets should accompany each shipment. The sampling data sheet is often accompanied by or integrated into a *chain of custody* record. Generally, the information shown on the example chain of custody form in Figure 11-2 represents the minimum suggested data that should be incorporated into any analytical request.

Quality Control During Sampling and Analysis

The importance of assuring the quality and integrity of any sample cannot be overstated. Because of this critical aspect of the industrial hygiene program, quality control actions must be implemented from the time the sample is taken, through its analysis at the laboratory, and on to its final disposition.

Quality control during the entire sampling and analysis process is necessary to ensure that the exposure concentration measurements represent the real employee exposure. Quality must be designed into the sampling process by ensuring that sufficient samples are collected and that they are clearly representative of the employee exposure. This must be accomplished by careful preparation and review of the employees' work tasks. Consideration must be made to evaluate any and all chemicals because of the potential for interferences.

INDUSTRIAL HYGIENE SAMPLE ANALYSIS REQUEST
AND
CHAIN-OF-CUSTODY VERIFICATION FORM

REQUESTING COMPANY
(Name/Address/Phone)

POINT-OF-CONTACT
(Name/Phone)

SAMPLE COLLECTED BY
(Person's Name)

SAMPLE DATE | SAMPLE REQUEST NUMBER | SAMPLE LOCATION | SAMPLE TYPE *(Media)*

DESCRIPTION OF SAMPLING METHOD USED:

POSSIBLE INTERFERENCES:

AIR VOLUME *(if applicable)* | FLOW RATE *(if applicable)* | SAMPLING EQUIPMENT *(model number and type)*

CALIBRATION DATE:

SAMPLE LIST:

NUMERICAL ORDER	CLIENT'S NUMBER	SAMPLE NAME	PROPERLY LABLED?	SHIPPING INTEGRITY VERIFIED?	AUTHORIZING SIGNATURES
1.			YES NO	YES NO	_____ Shipper / _____ Receiver
2.			YES NO	YES NO	_____ Shipper / _____ Receiver
3.			YES NO	YES NO	_____ Shipper / _____ Receiver
4.			YES NO	YES NO	_____ Shipper / _____ Receiver
5.			YES NO	YES NO	_____ Shipper / _____ Receiver

INFORMATION ON ADDITIONAL SAMPLES TO BE SHIPPED SEPARATELY:

OTHER INFORMATION, COMMENTS, OR SPECIAL INSTRUCTIONS:

DISPOSITION OF SAMPLE(S) AFTER ANALYSIS
Check Only One: ☐ RETURNED TO CLIENT ☐ HELD FOR STORAGE ☐ DESTROYED BY ANALYSIS ☐ OTHER: _____

DATE SHIPPED TO LAB | DATE ANALYSIS PERFORMED | SHIPPED TO *(laboratory name, address, and phone)*

DATE RECEIVED AT LAB | ANALYSIS PERFORMED BY *(Name)*

FIGURE 11.2 A sample chain-of-custody form that can be used to ensure proper sample integrity and quality control

The quality of the sampling process can also be enhanced by sampling for numerous contaminants on the same sample media. For example, charcoal tubes can sample for a wide variety of organic compounds that can be analyzed at the same time. By using the approach of *multiple sampling,* quality is enhanced due to less handling, less storage time, and less overall sampling time.

The industrial hygienist should develop procedures that outline how quality will be maintained throughout the sampling and analysis process. These procedures should detail sampling equipment, calibration techniques, and processes that routinely need sampling. They could include a listing (by process) of each chemical. Along with these procedures, a copy of the laboratory's quality control plan should be kept on file for review. Any discrepancies noted in quality attributable to the laboratory should be reported to the laboratory immediately. It is the user's right to require that the laboratory provide status of discrepancies.

The validity of results provided by the laboratory must continually be verified by the user. One common method used to identify sampling and analytical errors is the use of *sampling blanks* whenever the industrial hygiene effort requires laboratory analysis. There are basically two types of blanks: *field blanks* and *media blanks.* Field blanks, by far the most common, are clean sample media taken to the sampling site and handled in the same fashion as the actual sample media, except that no air is drawn through them. Media blanks are unopened samples sent to the laboratory with the true samples. During analyses, blanks are used to measure the signal contribution created by the reagents used by the laboratory when preparing the samples for analysis. They are also used to measure any contamination that may have occurred during handling, shipping, and/or storage before analysis. The specific method being used should be consulted concerning the actual number and type of blanks required. A general rule is at least one blank for each day or shift during which sampling is conducted (Ness, 1991). The NIOSH recommended practice for the number of field blanks is two blanks for each ten samples, with a maximum of ten field blanks for each sample set.

Another technique used to determine the stability of a sample or to provide information on the presence of any interferences, is the *spiked sample.* Here again, there are basically two types of spiked samples: *laboratory spikes* and *field-exposed spikes.* Spiked samples are generally prepared by adding a specific (known) concentration of a chemical to media similar to that expected in the field. A microliter syringe is used to pipette a sample of a compound, such as an organic solvent, onto the collection medium, in this case a charcoal tube, while a pump pulls air through the tube. The most common media to be spiked are passive sorbent badges, sorbent tubes, and impinger solutions. The laboratory spike (i.e., one that has been prepared at the laboratory) is taken to the site, but not exposed, and then sent back to the laboratory with the actual samples for analysis. This allows the spiked sample to be handled, transported, stored, and analyzed in the same way as the actual samples, but with no exposure to field elements. Because the amount of contaminant on the spiked sample is known in advance, any changes detected on analysis will provide critical information on the stability of the sample set. Field-exposed spikes are similar to the laboratory spikes except the field spikes are prepared at the site and analyzed along with the regular samples. Information on sample stability or the presence of any interfering contaminants will show-up on the samples but not on the field spikes.

A third method of assuring the quality of laboratory analyses is the use of *blind samples*. These are bogus samples mixed in with the actual samples and sent to the laboratory for regular analysis, with no instructions provided as to the legitimacy of the samples. While the hygienist will know what the results of a blind sample should be, the laboratory will have no clue. The hygienist assigns bogus flow rates, tracking numbers, and other such information and requests the laboratory to analyze the blind samples for a specific contaminant. Results other than those expected may indicate a problem with the laboratory's analysis methods and render the actual sample analysis results as suspect, in which case sampling should be repeated. Such methods may appear deceptive but in reality are very effective in demonstrating the accuracy of a particular laboratory's analytical capabilities. The use of blind sample analysis is actually a quite common practice in quality control assurance. In fact, many laboratories cannot obtain or maintain certain certifications, such as that granted by the American Industrial Hygiene Association (AIHA) or the National Institute for Drug Abuse (NIDA), without passing blind sample tests on a periodic basis.

Finally, *parallel sampling* can be accomplished by sending side-by-side samples to different laboratories. While the results should not be exactly the same, they should be reasonably close in accuracy. Extreme differences (in excess of 50 percent) may indicate a major problem. Sampling should be repeated.

THE LABORATORY REPORT

The specific content of the laboratory's analytical report will depend on the nature and function of the laboratory. As a minimum, however, the report should include a description or reference to the method(s) of analysis used on the particular samples, an explanation of any deviations or special circumstances encountered during the analysis of the sample set, estimates of the limits of detection and quantitation associated with the analytical procedures used, the sampling methods and media employed and how the limits were derived, and the actual results of the analysis.

The laboratory report should always be dated and signed by the analyst and anyone else responsible for its approval. The name of the analyst should be printed as well for easy identification. The name of the client is usually at the top of the report. In some cases, a company will have all of its laboratory results delivered to one person for consistency. Regardless of who sends in the samples, the results will always go to the designated person. This practice helps ensure proper tracking of sample information. It also establishes a single point of contact (POC) between the laboratory and the client. Having a single POC is especially important when discrepancies or inaccuracies are detected or when simple clarifications concerning the results of sample analyses are required.

If the client provides a reference number (i.e., a project number), it should also be referenced on the report with the client's individual sample numbers and any corresponding laboratory numbers. The laboratory number is a unique number assigned by the laboratory when it first receives the samples for analysis to prevent them from being mixed up at the laboratory. This information will assist the hygienist in determining which results correspond to which samples without having to retrieve and cross reference the sample submission sheet.

When samples are submitted to the laboratory, the hygienist (i.e., the client) should provide a description of each sample along with the unique sample identification number. The information can include the name of the employee who was sampled, a description of the work area that was sampled, or the type of bulk material that is being sent for analysis. If this information is also included on the laboratory report, it assists the client in readily identifying the samples. Use of the client's own sample number on the report greatly facilitates the entire process.

The type of analysis should be listed according to the type of instrumentation that was used. The number and source of the analytical method should be listed, for example, *NIOSH 7400* is used for asbestos. Any modifications to these methods that may have been implemented by the laboratory should also be explained. For instance, there are currently two different rules for counting asbestos fibers: "a rules" and "b rules." For occupational health requirements, OSHA requires the use of "a rules." If the sample/analysis is not related to an occupational exposure, modifications to the method may be in order. The modification of a particular method does not necessarily mean that results are in doubt. Laboratories often do this because they have learned from experience that a certain modification will provide more accurate analysis under a given set of circumstances.

Any particular detection limits of a given analysis should be listed. The limit of *detection* is defined as the amount of the substance being analyzed that can be distinguished from background. The limit of *quantitation* is the amount of the substance under analysis above which the precision of the reported results is better than that of a specified level. Since there are numerous methods of determining these quantities, it is advisable to let the laboratory decide on the method to be used for a given situation. For example, some laboratories will calculate a new detection limit for each sample based on volume only. If very large sample volumes are collected, this procedure can result in very small detection limits, thus providing a sensitivity not normally achieved by the method of analysis being used. It is important to note that the reported results should not exceed the allowable range for the method used unless the method has been validated for lower levels prior to sampling. The results of blank analyses should also be reported along with the actual samples. Depending on the method by which the samples are analyzed, any applicable regulatory standard(s) may also be listed in the report. For example, in the case of industrial hygiene laboratories, the OSHA permissible exposure limit (PEL) is usually used.

In some cases, an individual report is generated for each sample. More common, however, is to find separate reports generated for each type of analysis, with all samples from each category listed on each report. Whatever the format, report pages should be numbered so that the total number of pages appears on each page of the report with the individual page number (e.g., "page 1 of 10,"). In this way, the client will immediately notice any missing pages.

Depending on how thorough the laboratory is, additional supporting documentation (such as a copy of the chromatogram) may be attached to the report. Any problems, oddities, or other unusual conditions that were observed during analysis, such as unexpected discolorations, peaks (in the case of gas chromatography), or contaminated blanks, should also be reported.

The industrial hygienist should be alert for any interferences that may have existed at the sampling site and should provide this information to the laboratory prior to analysis. Such information can sometimes account for the oddities that may result during analysis and alert the laboratory to the possibility that complications may be present. Also, when defects or problems associated with the sampling media itself are observed, the hygienist should also advise the laboratory. For example, *breakthrough* (the presence of 25 percent or more of a contaminant in the rear portion of a sorbent tube) should be noted. It is an indication that the sample has passed through the sampling tube and, subsequently, may not provide accurate results during analysis.

SUMMARY

Regardless of the basic nature of any industrial hygiene program, the use of an industrial hygiene laboratory to analyze exposure samples will eventually be necessary. This chapter offered key information on the selection and use of industrial hygiene laboratory services. A ten-step process for laboratory selection was suggested, with recommended criteria for evaluating the available services that may be available from a variety of laboratories. The importance of using a laboratory that has achieved *accreditation* from the American Industrial Hygiene Association (AIHA) cannot be overstated. This becomes especially important should any legal challenges develop as a result of the analysis itself or in support of personnel exposure proceedings.

The critical issue of *quality control* and the importance of preserving *sample integrity* throughout the sampling and analysis process was explored, with particular emphasis on assuring proper accountability throughout the sample *chain of custody.* Several methods of assuring proper quality were explained in this chapter. These include the use of *blind samples* (purposely submitted false samples to test the accuracy of the analysis), *spiked samples* (purposely loaded sample media to determine the accuracy of the analysis), *blank samples* (empty sample media that has been handled the same way as real ones to evaluate chain of custody handling integrity), and *parallel samples* (those taken at the same time but sent to different laboratories for analysis). *Multiple sampling* (collecting more than one contaminant at the same time, on the same media) can also be used to save time and enhance quality. To ensure that a proper level of information is being reported by the selected laboratory, this chapter also contained information on the *laboratory report,* what it should include (as a minimum), and what the hygienist should expect.

Without exception, the services of an accredited industrial hygiene laboratory is an essential element of any successful industrial hygiene program. In fact, the overall success or failure of the entire occupational health effort can rest on the credible analysis of personnel exposure sampling.

12

Obtaining Additional Assistance

INTRODUCTION

Regardless of how well planned or comprehensive a particular industrial hygiene program may be, there may come a time when external assistance will be required. Such assistance may be necessary to ensure the optimum reduction of hazard exposure risk for a particular situation. Because of their *technical expertise,* outside experts can help demonstrate the overall effectiveness and adequacy of the organization's industrial hygiene program through audits or surveys. Moreover, as *expert witnesses,* they can help strengthen a company's position in the face of impending legal proceedings. This is not to suggest that the use of such assistance is essential for program success in *all* cases. However, there are certain benefits to using independent expertise in the form of industrial hygiene consulting services. There may also be disadvantages, depending on the nature and extent of the services to be provided.

In most instances, consultants will be brought in to assist in the evaluation and control of an already recognized hazard. There may, of course, be times when consultative services become necessary to help in the recognition process as well, but this is not as common.

In an effort to facilitate the decision-making and selection process, this chapter will explore the use of independent private industrial hygiene consulting services.

WHEN TO USE A CONSULTANT

The use of consulting services in the United States has increased substantially in recent years. This is primarily the result of an increasingly volatile economy that has forced businesses into adopting new strategies aimed at the reduction of operating costs. In the industrial hygiene arena, as in the safety and health field in general, consultants have become necessary for a number of reasons, the most common of which are as follows:

1. *Technical expertise.* In the practice of modern industrial hygiene, there are a wide variety of areas of hazard concern that may often require specialized experience and training to properly evaluate. Simply stated, the need to find a consultant with such specialized knowledge is based on the fact that no matter how well-educated and diverse a particular company's staff industrial hygiene professional may be, he or she cannot be *all-knowing* in *all fields* at *all times.* Examples include health physics (ionizing and non-ionizing radiation), noise abatement, design engineering, ventilation, and asbestos abatement and control, to name only a few. Given enough time and operational resources, it is probable that many industrial hygiene professionals could work through the issues associated with these specific areas of concern. However, each professional must also realize that using a consultant who has specific experience in the appropriate field of study can save a substantial amount of time and money. Also, it likely that experienced consultants have already had the opportunity to learn from previous mistakes throughout their careers. They bring this valuable experience to each new job, eliminating the learning curve and allowing their clients to realize the objective of hazard exposure reduction much more quickly and effectively. Consultants often have access to special equipment and/or facilities that may be needed to assess a particular problem. Because such equipment can be quite expensive, it may be easier to justify the use of a consultant than the purchase of the equipment to do the job.

    In short, because of their *technical expertise,* industrial hygiene consultants can assess situations to assist in the hazard recognition process. They can recommend control measures when required, and suggest alternative compliance measures. Finally, they can also design and even supervise the installation of these measures and evaluate their effectiveness.

2. *Independent third-party assessment.* Consultants can often provide an independent (unbiased) third-party assessment of compliance requirements, especially when new requirements will require large expenditures of capital resources. In most organizations, funding of unforeseen (unplanned) safety and health requirements are often associated with the company's financial overhead. In other words, work that must be done over and above pre-established budgetary constraints is added to the bottom line (is deducted from profits). Even when this unplanned (i.e., unforecast) work is necessary to ensure compliance with safety and health requirements, it is, nevertheless, unplanned. In many cases, securing the financial resources necessary to accomplish this work can be a major problem for

the staff industrial hygienist. For example, if a multi-million dollar modification to bring a manufacturing facility in compliance with new specifications will also cause a production shut-down, it will probably be extremely difficult to win management support for the project. An independent private consultant can often provide the knowledgeable and unbiased assessment of the situation that may be required to convince management that such modifications are, indeed, necessary. Also, because of their specialized experience, the consultant may be able to help find the most effective and least costly method of assuring compliance.

In summary, industrial hygiene consultants can be used to provide an unbiased, *third-party assessment* of compliance requirements. They can also keep their client's management informed and aware of both current and proposed federal regulations and how these will effect the operation of their client's business enterprise.

3. *Expert witness.* Litigation associated with worker's compensation, product liability, and/or alleged negligence generally requires the services of a consultant who is accepted as an *expert* in the field. The dispute resolution process in the United States permits certain individuals to render opinions based on data rather than merely reciting information. These opinions are sought to explain past, present, and even future events. Federal rules of evidence ordinarily do not permit witnesses to testify as to opinions or conclusions. Expert witnesses are exceptions to this rule. An expert can be defined as a person who, by virtue of relevant training and experience in some art, science, profession, or calling, is capable of doing things other people cannot do. In a court of law, expert witnesses may state an opinion as to relevant and material matter in which they profess to be expert and may also state their reasons for the opinion. Juries are instructed to accept the opinions of the expert witness as fact (Feder, 1991). Hence, the use of an expert in such situations can be extremely beneficial to one side and damaging to the other. Experts are not only proficient in what they do, they are also smooth and efficient in the actions they take. Experts are knowledgeable of specific methods and techniques that can be used to solve problems or tasks. They are extremely adept at sifting through irrelevant information in order to uncover the basic issues or actual problems. Experts can use their specialized experience to formulate solutions to seemingly complex problems (Johnson, et al, 1984). There are a number of sources that can be referenced when attempting to verify an expert's ability to assist in a litigious industrial hygiene situation (or any other for that matter). These include:
   a. Recommendations from other professionals in the field
   b. Published works (by or about the expert)
   c. Technical products (developed by an expert)
   d. Academic degrees held by the expert

e. Formal credentials (licenses, awards, or honors)
f. Teaching experience
g. Supervisory work (over others in the field)
h. Membership in specific organizations and societies

Using an expert industrial hygiene witness can be a tremendous advantage to the company that has become involved in litigation. When such situations arise, companies should seriously consider the use of experts to assist in the favorable resolution of disputes concerning industrial hygiene issues and concerns.

SELECTING A CONSULTANT

Selecting a consultant who is qualified to provide specific services requires careful evaluation of the circumstances that led to the need for such services. For example, if a specific problem is associated with asbestos abatement, the hygienist should evaluate the exact nature of the problem before selecting a consultant. If abatement is the only concern, a consultant who is experienced in this arena would suffice. However, if there is a question of alleged employee exposure or if there is an impending OSHA citation and/or penalty that must be resolved, then the talents of the selected consultant may have to be more specific than that of the general asbestos abatement consultant. In the first situation, the consultant need only be experienced with the regulations and requirements associated with asbestos abatement. In the latter case, the consultant must be well versed in the medical and legal aspects of asbestos exposure, in addition to the general characteristics of asbestos and the requirements for proper abatement. In other words, the situation will dictate the specific *need,* and this need will form the basis for consultant selection.

To identify a candidate to satisfy the consulting need, it is recommended that the hygienist first look to professional societies and organizations. Many have listings of experienced consultants who are trained in specific and general areas of expertise. For instance, the American Society of Safety Engineers (ASSE) has a Consultants Division. In this listing of professionals, the prospective client can identify consultants by expertise and area of specialization. The National Safety Council (NSC) can provide similar information about safety and health consultants. The American Industrial Hygiene Association (AIHA) can also provide information on experienced industrial hygiene consultants. Once again, depending on the specific need, other specialized organizations can be contacted. The Health Physics Society (HPS), for example, can help locate qualified consultants to assist with radiation problems or concerns. Similarly, the Illumination Engineering Society (IES) of North America should be contacted when issues concerning lighting require the use of a professional consultant. The appendices of this text provide contact information for numerous professional organizations that can help locate qualified consulting services in a wide variety of specialized areas.

Many insurance companies use consultants, including industrial hygiene professionals, for their loss prevention programs. A company should therefore look to its own insurer for lists of known consultants and their qualifications or areas of specialty. A comparison between these qualifications and the specific need for a consultant may help locate an expert who can provide the required services.

Finally, colleges and universities are another source of information regarding consulting services. Universities that are well known for their academic programs in a given field of interest may be able to help locate qualified consultants. Through their contacts with professional organizations, other universities, and their own adjunct faculty, universities can often provide a listing of individual experts in a given field such as asbestos, explosive safety, radiation, design, ventilation, and so on.

Regardless of the sources considered for candidates, selection of a consultant should always be based on one primary objective: the qualifications of the consultant specific to the project at hand. The size of a consulting firm and the length of time it has been in business are seldom indications of professional capability. In many respects, the actual selection process is similar to that described in Chapter 11 under Selecting the Laboratory. The process should be as follows:

1. Methodical (i.e., selection process based on pre-established criteria)
2. Specific to an identified need (i.e., selection based on qualifications of the individual in relation to the need for such services)
3. Considerate of time (availability of consultant to provide services on schedule), cost (selection of best value services), and ethics (fair selection of consulting services from those bidding)

As with the selection of a laboratory, it is advisable to develop a list of three to four potential consulting candidates before making a final selection. While specific selection criteria should be based on the particular need for the consultant, the following general questions/considerations should always be evaluated before the selection is made:

1. What is the experience (years, types of services, success rate, relevant teaching assignments, etc.) of the consultant and how is it related to the project or task at hand? It is suggested that a client list be obtained so previous customers of the consultant can be contacted for reference.
2. What is the consultant's educational background? What schools were attended and what degrees are held? Is this educational experience relevant to the task at hand? Has the consultant attended special conferences or seminars related to industrial hygiene?

3. What is the consultant's status? Is this an independent private, full-time consultant or someone associated with a larger firm? If part-time, does the consultant's primary employer know of and approve of these consulting activities?
4. In what professional societies and organizations is the consultant a member or, better still, an office holder? Does the consultant hold any professional certifications (such as the Certified Industrial Hygienist, the Certified Safety Professional, the Certified Health Physicist, etc.), and are these certifications relevant to the task at hand?
5. Has the consultant ever published (books, articles) in the field of industrial hygiene? Has the consultant ever provided presentations on industrial hygiene topics to peers at professional conferences or seminars?
6. What is the consultant's fee structure? Are charges to be hourly, weekly, monthly, or based on a standard estimate for the entire job?
7. What laboratory will the consultant use (if applicable), and how was this laboratory selected? Are the laboratory fees included in the overall consulting fee, or will these fees be extra?

While there are certainly many more questions to be asked, starting with these general inquiries should help narrow the selection pool to one candidate.

CONSULTING AGREEMENTS

Once a consultant has been located and selected, it is important that the arrangements for services be documented and agreed upon by both parties. The *consulting agreement* must include provisions regarding the protection of trade secret information that the consultant may be given access to during the performance of the consulting services. It should clarify whether the consultant carries sufficient insurance to cover errors and omissions, injuries or illnesses incurred as a result of their consulting activities, and any other significant coverage that may be required for this type of service. A statement to clarify the limitations of any liability for which the consultant will sustain as a direct result of their services to the client, as well as that which will be afforded the consultant by the client. A *hold-harmless* statement that can provide some level of indemnification for both the consultant and the client is also common. A provision that allows for cancellation of the agreement by the client based on specific and agreed-upon criteria can also be included. Finally, a firm and clear *statement of work* (SOW) should also be established. The SOW should define the specific actions that the consultant will be required to perform under the agreement and a description of the product expected from the consultant at the conclusion of the work (e.g., final report of findings, recommendations, solutions, etc.). A sample consulting agreement is provided as Figure 12-1.

CONSULTING SERVICES AGREEMENT

This exclusive CONSULTING SERVICES AGREEMENT is made as of this ___ day of ___, 199__ by and between ___________________, hereinafter referred to as the "CONSULTANT," and ___________________, hereinafter referred to as "COMPANY."

I. PERIOD OF AGREEMENT

This agreement shall begin on ___ day of ____, 199__ and shall terminate on the ___ day of ____, 199__. However, this agreement may, during the period of its duration, be terminated with or without cause in whole or in part at any time by either party by giving two (2) days written notice.

II. STATEMENT OF WORK

CONSULTANT shall, as an independent contractor and not as an employee of COMPANY, perform the work directed. CONSULTANT shall perform the services described in Attachment A of this agreement, to the best of his abilities and in an efficient manner within the time schedule established in Attachment B of this agreement.

III. PAYMENT

COMPANY shall, during the term of this agreement, pay CONSULTANT at the rate of __________ per ____ (hour/day/week/month/job). CONSULTANT shall maintain and keep a record of time applied to the work of COMPANY in accordance with the requirements of COMPANY. Time spent in travel hereunder shall not be deemed to be spent on work required hereunder. Payment to CONSULTANT shall be tendered by COMPANY within 30 days of receipt of signed time sheet and invoice from CONSULTANT.

IV. EXPENSES

Travel will be expended as required to perform the duties of this agreement. All travel and related expenses will be charged in accordance with the Joint Travel Regulations and the Federal Acquisition Regulations. CONSULTANT will invoice COMPANY with original receipts for allowable travel expenses.

V. PROPRIETARY INFORMATION

CONSULTANT will not, without written authorization from COMPANY, disclose to others either during the time of his consulting services or thereafter except as required by CONSULTANT's services to COMPANY, any confidential information, matter, or thing connected with the business or development of COMPANY that is not common knowledge. CONSULTANT will not take any COMPANY documentation or things without written consent of COMPANY.

VI. NON-COMPETITION WITH COMPANY

CONSULTANT agrees that he will not offer, provide similar services either directly or indirectly for any firm (other than COMPANY) participating in the activities detailed in Attachment A for six (6) months after the date this agreement terminates.

VII. AUTHENTICATION

The COMPANY and CONSULTANT, as parties to this AGREEMENT, hereby agree to the conditions and provisions contained herein, as witnessed by signature this ___ day of ____, 199__.

______________________________   ______________________________
Signature for COMPANY   Signature for CONSULTANT

FIGURE 12-1 Sample cover page, without stated Attachments, for a consulting services agreement. Particulars can be specific for a company or situation. Attachments to the agreement should spell out all details.

POTENTIAL DISADVANTAGES

Although working with a consultant can be extremely beneficial in terms of hazard evaluation and control, there are some aspects that, if not properly monitored or managed, could severely downgrade the overall efficiency and effectiveness of the industrial hygiene effort. Such situations, however, are rare and, if all parties involved take care to honor the provisions of the consulting agreement, such unfortunate circumstances are completely avoidable.

1. *Expense.* Simply speaking, good consulting services are seldom cheap. The selling feature of using such services should be a comparison of the cost verses the benefit of using a consultant. The hygienist may have to convince management that the benefits far outweigh the initial costs and, in the long run, using a consultant may actually prove the least expensive route to assuring a healthy workplace. Also, by establishing and agreeing on an acceptable fee structure and payment schedule before services begin (i.e., in the consulting agreement), problems associated with add-on fees and unbudgeted costs after the job begins can be avoided.

2. *Conflicting commitments.* It is advisable that the hiring company determine from the onset that the consultant has and anticipates no schedule conflicts during the contracted period of time. This simple action will avoid delays in the task due to other demands on the consultant's time and resources. Without a doubt, consultants are in the business to make money and at times may take on more assignments than they are capable of handling. Once again, such issues should be resolved *before* work begins to ensure the uninterrupted provision of services from the consultant.

3. *Conflicts of interest.* Another potential drawback is to find a consultant who, after work has begun, has a particular conflict with the work to be performed. For example, an expert witness may be called to help resolve a dispute over an alleged worker's compensation claim. After accepting the case and signing the agreement, the expert may find that he or she has done work for the opposing counsel on a previous case. Or, even worse, the opposing counsel may discover that the hired consultant has voiced an opinion on a similar case that contradicts the current statement of position. The potential for such conflicts can be minimized with a thorough and complete review of a consultant's qualifications *before* services begin.

There may indeed be other less-than-obvious disadvantages to using a consultant that are not covered here. However, achieving the maximum benefit from a consultant and avoiding problems is largely a matter of assuring full understanding of the requirements (by both parties to the agreement), by defining clear responsibilities and establishing and maintaining clear lines of communication throughout the performance of the work.

SUMMARY

This chapter has provided some basic information concerning the use of consulting services to augment the overall industrial hygiene effort. Consultants can provide *technical expertise* when such is lacking for a specific program or problem area (e.g., asbestos, radiation, noise, etc.). Their specialized training and expertise can often help resolve such problems in less time and for less cost. Also, when an *unbiased assessment* is required to convince frugal management of the need to perform a specific action or implement a specific process, a consultant can provide an evaluation of the facts and render an opinion. Consultants can also provide alternative solutions to many problems. Perhaps the most common use of the consultant is in the legal arena. As an *expert witness,* the consultant is able (and permitted by law) to enter an opinion as fact during legal proceedings. Using an expert to assist in the establishment of a legitimate worker's compensation claim, or defend against a bogus one, is quite common in today's litigious society.

Selecting a consultant first requires a thorough understanding of the requirements for such services. Consultants should be selected based on the need at hand. Once the need has been established, prospective clients can look to professional and trade organizations, their own insurance company, or colleges and universities to help locate qualified professional consultants. Once selected, the terms and conditions of the work to be performed should be specifically detailed in a *consulting agreement.* The agreement should be signed by both the client and the consultant before services begin. To avoid problems that can arise due to conflicts in interests, work scheduling, or any other reason, the client should carefully review the consultant's qualifications to perform the work at hand.

When the performance of the industrial hygiene function at any given facility or organization needs help, external assistance from a professional consultant may be the answer. While using a consultant in *all* cases for *all* circumstances is hardly necessary, it is important to know when to bring in a consultant. As with the use of the industrial hygiene laboratory (covered in Chapter 11), use of consultant services may not always be avoidable. Hence, this chapter provided some basic guidelines that should assist the reader in making the decision to hire a consultant and how to proceed from there.

Appendix A

Sources of Additional Education/Training

The following is a compilation of sources where the interested reader may obtain additional information and/or training in the areas of industrial hygiene and occupational health management. There are, of course, many more excellent organizations and societies. However, those listed here should provide the reader of this *Basic Guide to Industrial Hygiene* additional information presented at the next technical level.

PROFESSIONAL ORGANIZATIONS

There are numerous national and international professional organizations that either specialize in the practice of industrial hygiene or have established sections, divisions, or special memberships designed to serve those who are interested in the study of industrial hygiene. A common practice of most membership-based organizations and societies is the provision of educational materials, publications, and other media to their members at discounted prices. Included among these are the following:

American Industrial Hygiene Association (AIHA)
2700 Prosperity Avenue, Suite 250
Fairfax, VA 22031

A non-governmental organization founded in 1939 composed of industrial hygienists employed in private industry, government, and academia. Industries are also eligible for associate membership. The AIHA is dedicated to the prevention of workplace-related illnesses and injuries affecting the health and well-being of workers or the community. The Association establishes workplace environmental exposure limits and operates a national laboratory accreditation program. Their *American Industrial Hygiene Association Journal* is published monthly and contains manuscripts written by member hygienists and usually presents the latest developments in industrial hygiene practices and techniques. The AIHA Annual Industrial Hygiene Conference is, perhaps, the premier national educational conference designed specifically for the enhancement and professional development of the industrial hygienist. It is co-sponsored by the American Conference of Governmental Industrial Hygienists.

American Board of Industrial Hygiene (ABIH) and the
American Academy of Industrial Hygiene (AAIH)
4600 West Saginaw, Suite 101
Lansing, MI 48917-2737

The ABIH is an organization responsible for certifying industrial hygienists. The organization offers comprehensive and specialized tests to qualified applicants and is responsible for determining whether individuals are qualified to practice industrial hygiene. Other objectives are to encourage the study, improve the practice, and elevate the standards of the industrial hygiene professional. The AAIH resides within the ABIH as an organization composed of board-certified industrial hygienists.

American Association of Occupational Health Nurses (AAOHN)
3500 Piedmont Road, Northeast
Atlanta, GA 30305

An organization dedicated to promoting the field of occupational health nursing, formerly named the American Association of Industrial Nurses. It has numerous chapters at the state and local levels throughout the United States. Their annual educational conference often features topics of interest to the industrial hygienist.

American Conference of Governmental Industrial Hygienists
6500 Glenway Avenue, Bldg. D-5
Cincinnati, OH 45211

A professional, non-governmental organization founded in 1938, composed of industrial hygienists employed in the government and academia. The ACGIH establishes threshold limit values (TLV) for certain chemicals, and co-sponsors (with the American Industrial Hygiene Association) the annual American Industrial Hygiene Conference. Their primary function is to encourage the exchange of experience among governmental industrial hygienists, and to collect and make available information of value to their members. They are an excellent source of information of interest to the industrial hygiene and occupational health profession.

American Industrial Hygiene Foundation
c/o American Industrial Hygiene Association
2700 Prosperity Avenue, Suite 250
Fairfax, VA 22031

An organization that operates under the auspices of the American Industrial Hygiene Association that provides fellowships to graduate industrial hygiene students, promotes the development of graduate schools of industrial hygiene, and encourages qualified individuals to enter the field.

American Public Health Association
1015 18th Street, Northwest
Washington, DC 20036

An organization founded in 1872 dedicated to protecting and promoting personal and environmental health by exercising leadership in the development and dissemination of health policy. The organization represents all disciplines and specialties in public health, including industrial hygiene, and is the largest public health association in the world (1993 membership exceeding 50,000).

American Society of Safety Engineers (ASSE)
1800 East Oakton Street
Des Plaines, IL 60018

An international, not-for-profit, multi-disciplinary, professional organization with more than 30,000 (1994) members consisting of primarily individual safety professionals dedicated to the advancement of occupational safety, health and environmental professions. Organized in 1911 as the United Association of Casualty Inspectors and incorporated in 1915, ASSE is one of the oldest sustaining professional membership societies in the United States. Their monthly publication, *Professional Safety,* often includes articles of interest to the industrial hygiene or occupational health professional. Their growing list of technical publications also contains numerous titles dealing with the subject of industrial hygiene. On the national level, the ASSE holds an annual Professional Development Conference (PDC), and quite often offers specialized training and post-conference seminars on industrial hygiene and related topics (ergonomics, cumulative trauma, chemical safety, etc.). Chapters engage in a number of activities ranging from monthly meetings with focused presentations to their own specialized seminars and training sessions on a variety of topics. A list of chapters grouped by state and listed by name can be obtained from their national office at the address above.

Human Factors Society (HFS)
P.O. Box 1369
Santa Monica, CA 90406

The HFS is a multi-disciplinary society composed of psychologists, engineers, physiologists, and other related scientists and professionals concerned with the use of human factors in the development of systems and devices of all kinds. *Human Factors,* their official journal, is published bimonthly.

Illuminating Engineering Society of North America (IES)
345 East 47th Street
New York, NY 10017

Although a non-governmental and, therefore, non-regulatory organization, this society is the scientific and engineering stimulus in the field of lighting. Its guidelines for proper light illumination, as published in their *IES Lighting Handbook,* has become the accepted reference for standard lighting requirements. In addition to the *Handbook,* the IES publishes a monthly magazine entitled *Lighting Design and Application;* a quarterly *Journal of Illumination Engineering;* and more than fifty other publications on specific lighting concerns in areas such as mining, roadway, office, and emergency lighting. Through the society and its many publications, the industrial hygienist can obtain reference material on all aspects of lighting conditions.

National Safety Council (NSC)
1121 Spring Lake Drive
Itasca, IL 60143

The NSC is dedicated to assuring the advancement of safety and health of persons, both on and off the job. Their many excellent seminars, programs, and publications offer the profession quality resources that are unmatched on a national level. In addition to a wide variety of activities and resources in the safety and health profession, the NSC offers assistance, training, and printed materials to aid the industrial hygiene professional. Their renowned National Safety Congress, which occurs each fall, features a host of effective training sessions and meetings, many of which are of particular interest to the industrial hygiene profession. Their textbook, *Fundamentals of Industrial Hygiene,* is highly recommended for the reader wishing to pursue a level of industrial hygiene knowledge beyond that presented in this *Basic Guide.*

Health Physics Society
1340 Old Chain Bridge Road
McLean, VA 22101

A professional society composed of persons active in the field of health physics, which is the profession concerned with the protection of man and his environment from radiation hazards. The HFS also sponsors the American Board of Health Physics, which is engaged in the voluntary certification (by examination) of health physics practitioners.

Compressed Gas Association (CGA)
1235 Jefferson Davis Highway
Arlington, VA 22202

Composed of firms that are engaged in the production and distribution of compressed, liquefied, and cryogenic gases, and those who manufacture related equipment. The CGA submits recommendations to government agencies aimed at the improvement of safety standards and methods of handling, transporting, and storing gases. It also acts as advisor to regulatory agencies and other organizations concerned with compressed gas safety.

American National Standards Institute (ANSI)
1430 Broadway
New York, NY 10018

An organization responsible for developing consensus standards on a wide variety of matters, including occupational health and safety. ANSI has established standards for respiratory protection, exposure limits to chemicals and physical agents, and standards for personal protective equipment (including fall protection). Many of these standards have been incorporated into those established as law by OSHA.

Industrial Health Foundation, Inc.
34 Penn Circle, West
Pittsburgh, PA 15206

A nonprofit organization dedicated to the advancement of healthful working conditions in industry. Its members include industrial firms or organizations, including trade organizations. They provide engineering, occupational medicine, toxicological, and technical information services upon request. Also make available technical information on a wide variety of industrial health issues and concerns.

PUBLICATIONS AND BOOKS

In addition to the many excellent publications available through the various technical and professional organizations listed above (and those noted in the bibliography of this text), reference books on the subject of industrial hygiene include the following recommended titles:

1. Title: *Industrial Hygiene*
   Author/Year: R. W. Allen, M. D. Ells, and A. W. Hart, 1976
   Publisher: Prentice-Hall, Inc., Englewood Cliffs, NJ

2. Title: *Basic Industrial Hygiene Manual*
    Author/Year: R. S. Brief, 1975
    Publisher: American Industrial Hygiene Association, Akron, OH

3. Title: *Recognition of Health Hazards in Industry*
    Author/Year: W. A. Burgess, 1981
    Publisher: John Wiley & Sons, Inc., New York, NY

4. Title: *Occupational Health Hazards*
    Author/Year: W. G. Dubenspeck, 1974
    Publisher: Exposition Press, Hicksville, NY

5. Title: *Chemical Exposures*
    Author/Year: N. A. Ashford and C. S. Miller, 1991
    Publisher: Van Nostrand Reinhold, New York, NY.

6. Title: *Noise and Hearing Conservation Manual*
    Author/Year: R. L. Berger, 1990
    Publisher: American Industrial Hygiene Association, Akron, OH

7. Title: *Safety and Health for Engineers*
    Author/Year: R. L. Brauer, 1990
    Publisher: Van Nostrand Reinhold, New York, NY

8. Title: *Lead-Based Paint Hazards*
    Author: V. M. Coluccio, 1994
    Publisher: Van Nostrand Reinhold, New York, NY

9. Title: *Common Sense Toxics in the Workplace*
    Author/Year: I. R. Danse, 1991
    Publisher: Van Nostrand Reinhold, New York, NY

10. Title: *Ergonomic Design For People at Work*
    Author/Year: Eastman Kodak Company, 1991
    Publisher: Van Nostrand Reinhold, New York, NY

11. Title: *Quick Selection Guide to Chemical Protective Clothing*
    Author/Year: K. Forsberg and S. Z. Mansdorf, 1993.
    Publisher: Van Nostrand Reinhold, New York, NY

12. Title: *Pesticides in Drinking Water*
    Author/Year: D. I. Gustafson, 1993
    Publisher: Van Nostrand Reinhold, New York, NY

13. Title: *A Practical Guide to Chemical Spill Response*
    Author/Year: J. W. Hosty and P. Foster, 1990
    Publisher: Van Nostrand Reinhold, New York, NY

14. Title: *1,001 Chemicals in Everyday Products*
    Author/Year: G. R. Lewis, 1994
    Publisher: Van Nostrand Reinhold, New York, NY

15. Title: *Hawley's Condensed Chemical Dictionary*
    Author/Year: R. J. Lewis, 1993 (12th Edition)
    Publisher: Van Nostrand Reinhold, New York, NY

16. Title: *Hazardous Chemicals Desk Reference*
    Author/Year: R. J. Lewis, 1993 (3rd Edition)
    Publisher: Van Nostrand Reinhold, New York, NY

17. Title: *Rapid Guide to Hazardous Chemicals in the Workplace*
    Author/Year: R. J. Lewis, 1994 (3rd Edition)
    Publisher: Van Nostrand Reinhold, New York, NY

18. Title: *Sax's Dangerous Properties of Industrial Materials*
    Author/Year: R. J. Lewis, 1992 (8th Edition)
    Publisher: Van Nostrand Reinhold, New York, NY

19. Title: *The VNR Dictionary of Environmental Health and Safety*
    Author/Year: F. S. Lisella, 1994
    Publisher: Van Nostrand Reinhold, New York, NY

20. Title: *The Ergonomics Edge*
    Author/Year: D. MacLeod, 1994
    Publisher: Van Nostrand Reinhold, New York, NY

21. Title: *Air Monitoring Instrumentation*
    Author/Year: C. J. Maslansky and S. P. Maslansky, 1993
    Publisher: Van Nostrand Reinhold, New York, NY

22. Title: *Surface and Dermal Monitoring for Toxic Exposures*
    Author/Year: S. A. Ness, 1994
    Publisher: Van Nostrand Reinhold, New York, NY

23. Title: *Air Monitoring for Toxic Exposures*
    Author/Year: S. A. Ness, 1991
    Publisher: Van Nostrand Reinhold, New York, NY

24. Title: *A Comprehensive Guide to the Hazardous Properties of Chemical Substances*
    Author/Year: P. Patnaik, 1992
    Publisher: Van Nostrand Reinhold, New York, NY

25. Title: *Fundamentals of Industrial Hygiene*
    Author/Year: B. A. Plog (Editor), 1988 (3rd Edition)
    Publisher: National Safety Council, Chicago, IL

26. Title: *Chemical Exposure and Disease*
    Author/Year: J. D. Sherman, 1988
    Publisher: Van Nostrand Reinhold, New York, NY

27. Title: *Noise Control, A Guide for Workers and Employers*
    Author/Year: R. L. Stepkin and R. E. Mosely, 1984
    Publisher: American Society of Safety Engineers, Des Plaines, IL

28. Title: *Physical and Biological Hazards in the Workplace*
    Author/Year: P. H. Wald and G. M Stave, 1994
    Publisher: Van Nostrand Reinhold, New York, NY

29. Title: *The Law of Occupational Safety & Health*
    Author/Year: G. Z. Nothstein, 1981
    Publisher: Macmillan Publishing Company, New York, NY

30. Title: *Chemistry of Hazardous Materials*
    Author/Year: Eugene Meyer, 1990
    Publisher: Prentice Hall, Englewood Cliffs, NJ

PUBLICATIONS AND PERIODICALS

The following is a partial listing of the many magazines and periodicals that are either dedicated to the industrial hygiene profession or often feature up-to-date information of interest to the industrial hygiene practitioner. Some are available at no charge to safety and health professionals, while others are offered for a modest subscription fee. Interested readers should inquire the respective publishers listed below for further information. Of course, many more sources could be listed here. Many trade journals for disciplines such as mechanical and electrical engineering often feature articles on safety and health.

1. Title: *Industrial Safety & Hygiene News*
   Publisher: Chilton Company, Chilton Way, Radnor, PA 19089
   Frequency: Monthly

2. Title: *American Industrial Hygiene Journal*
   Publisher: American Industrial Hygiene Association, 475 Wolf
                Ledges Parkway, Akron, OH 44311
   Frequency: Monthly

3. Title: *Human Factors*
   Publisher: Human Factors Society (HFS), P.O. Box 1369, Santa
                Monica, CA 90406
   Frequency: B-monthly

4. Title: *Industrial Hygiene News Report*
   Publisher: Flournoy & Associates, 1845 West Morse Avenue,
                Chicago, IL 60626
   Frequency: Monthly

5. Title: *Occupational Safety and Health Reporter*
   Publisher: Bureau of National Affairs (BNA), 1231 25th Street, NW,
                Washington, DC 20037
   Frequency: Monthly

6. Title: *American Journal of Public Health*
   Publisher: American Public Health Association, 1015 Fifteenth
                Street, Washington, DC 20005
   Frequency: Monthly

7. Title: *Applied Industrial Hygiene*
   Publisher: American Conference of Governmental Industrial
                Hygienists (ACGIH), 6500 Gleenway Avenue, Bldg. D-7,
                Cincinnati, OH 45211
   Frequency: Monthly

8. Title: *Occupational Hazards*
   Publisher: Penton-IPC, 1100 Superior Avenue, Cleveland, OH 44114
   Frequency: Monthly

9. Title: *Occupational Health Nursing*
   Publisher: American Association of Occupational Health Nurses, 79
                Madison Avenue, New York, NY 10016
   Frequency: Monthly

10. Title: *Occupational Health & Safety*
    Publisher: Stevens Publishing Corp., 5002 Lakeland Circle, P.O. Box 7573, Waco, TX 76710
    Frequency: Monthly

11. Title: *Professional Safety*
    Publisher: American Society of Safety Engineers, 1800 East Oakton Street, Des Plaines, IL 60018-2187

12. Title: *Safety & Health*
    Publisher: National Safety Council, 1121 Spring Lake Drive, Itasca, IL 60143-3201
    Frequency: Monthly

13. Title: *Journal of Toxicology: Clinical Toxicology*
    Publisher: Marcel Dekker, Inc., 305 East 45th Street, New York, NY 10017
    Frequency: Bi-weekly.

14. Title: *OSHA Up-to-Date*
    Publisher: National Safety Council, 1121 Spring Lake Drive, Itasca, IL 60143-3201
    Frequency: Monthly

TRAINING SEMINARS AND ORGANIZATIONS

There are a number of professional training and consulting organizations that offer periodic and/or regularly scheduled training sessions and seminars on the subject of industrial hygiene. Many of those listed below will provide industrial hygiene training at the beginner, novice, intermediate, and expert levels.

    Sponsor: American Industrial Hygiene Association (AIHA)
    Address: 475 Wolf Ledges Parkway, Akron, OH 44311

As discussed early in this Appendix, the AIHA is, perhaps, the premier authority on the industrial hygiene profession. They offer numerous training seminars throughout the year on a wide variety of general and specific topics of concern to the industrial hygiene professional. Their yearly American Industrial Hygiene Conference (co-sponsored by the ACGIH) is offered each spring. It provides the highest level of educational opportunities in a single location; more than any other professional conference.

Sponsor: American Society of Safety Engineers
Address: 1800 East Oakton Street, Des Plaines, IL 60018-2187

The American Society of Safety Engineers is an international organization of more than 30,000 safety and health professionals. The society sponsors numerous training seminars on a variety of topics, including industrial hygiene and occupational health (ergonomics, radiation, etc.) throughout the year at various locations around the United States. Their yearly Professional Development Conference (PDC), held each June in a selected U.S. city, also offers short training sessions on a number of topics of interest to the industrial hygiene practitioner. For example, their 1994 PDC in Las Vegas was followed by the first annual International Ergonomics Symposium. In addition to these many training sessions and seminar opportunities, the ASSE also makes available to its membership (for a modest charge) numerous pre-packaged educational materials (slides/cassette combinations, workbooks, etc.) that have been developed to address specific areas of concern, including occupational health and industrial hygiene.

Sponsor: National Safety Council (NSC)
Address: 1121 Spring Lake Drive, Itasca, IL 60143-3201

The NSC is most known for the high-quality training and professional development courses it offers throughout the year, and for its many excellent publications. Through its well-established, nationwide network of local and regional chapters and offices, the excellent training opportunities offered by the NSC are accessible by virtually every practicing safety and health professional in the United States. Its annual National Safety Congress held each fall draws the largest gathering of safety and health professionals to one single professional conference. Training sessions are provided on virtually every topic of current concern to the practicing safety and health professional.

Sponsor: COMCO, Inc. Environmental & Safety Services
Address: 17121 Clark Avenue, Bellflower, CA 90706-5730

COMCO is a full-service safety and environmental consulting firm offering specialized training in most aspects of the safety and environmental professions. COMCO is capable of assisting its clients with all their safety training needs, including industrial hygiene and occupational health. Their highly competent staff can also assist in the development of specific industrial hygiene programs based upon individual client need.

Sponsor: Summit Training Source (STS), Inc.
Address: 6504 28th Street, S.E., Grand Rapids, MI 49506

STS develops high-quality training packages (films, workbooks, study materials, etc.) which can be purchased to complement any organization's specific training requirements. Their materials cover a broad spectrum of safety and health issues, including packages on industrial hygiene and related topics (i.e., ergonomics).

Sponsor: Tel-A-Train, Inc.
Address: P.O. Box 4752, 309 N. Market Street, Chattanooga, TN 37405

Like STS, Tel-A-Train is a highly reputable training firm that offers professional videotape training on a wide variety of subjects of interest to the safety and health professional. Films on chemical safety and hazard communication, for example, have been adopted by literally hundreds of companies and organizations as a major element of their right-to-know training programs.

Sponsor: Envirosafe Training and Consultants (ETC)
Address: 811 Camp Horne Road, Pittsburgh, PA 15237

ETC provides training and consulting in all areas of environmental and occupational health and related fields. ETC can assist clients with both training and consulting needs. ETC is approved and certified by a number of states in asbestos and lead abatement, worker, supervisor, inspector, management planner, project designer and related training, including NIOSH 582 equivalent. Consulting services are provided in all areas of industrial hygiene and environmental science.

Sponsor: Gannon University
Address: University Square, Erie, PA 16541

Gannon University offers a certificate program in Environmental and Occupational Science and Health. Through this program, undergraduate and graduate level courses are offered in the areas of industrial hygiene, radiation health, epidemiology, toxicology, environmental and occupational health, industrial hygiene sampling, environmental management and engineering, and environmental and occupational regulations. These courses and other programs are offered during regular semesters and summer programs. Graduate and undergraduate programs can be structured in relation to the programs offered.

Sponsor: Barfield & Associates, Inc.
Address: 435 Simms Way, Merritt Island, FL 32952

Barfield & Associates is a full service safety, industrial hygiene, and environmental training and consulting firm. They can provide customized training and expert consultation on a wide variety of safety and environmental topics, including industrial hygiene services. In addition, Barfield & Associates specializes in the development of government contract proposals, not only in areas of safety, health and the environment, but in all areas of business development.

Sponsor: Granberry & Associates
Address: 2431 Aloma Avenue, Suite 276, Winter Park, FL 32972

Granberry & Associates is a highly reputable source of professional training and consulting services in the areas of chemical safety, occupational safety, industrial hygiene, and environmental compliance. Customized courses can be developed to meet the specific needs of any client. Expertise in all facets of occupational safety and health is available through Granberry & Associates.

Appendix B

Acronyms and Abbreviations

In the industrial hygiene profession, as in most technical and scientific arenas, numerous acronyms and abbreviations are used. The following is a reference listing of some of the most frequently encountered, in this text and in the industrial hygiene/occupational health profession in general.

AAAS	American Association for the Advancement of Science
AAEE	American Academy of Environmental Engineers
AAIH	American Academy of Industrial Hygiene
AAOHN	American Association of Occupational Health Nurses
AAP	Asbestos Action Program
AAS	Atomic Absorption Spectroscopy
ABIH	American Board of Industrial Hygiene
ABOHN	American Board of Occupational Health Nursing
AC	Advisory Circular
ACBM	Asbestos-Containing Building Material
ACEC	American Consulting Engineers Council
ACGIH	American Conference of Governmental Industrial Hygienists
ACIP	Advisory Committee on Immunization Practices
ACM	Asbestos-Containing Material
ACP	Air Carcinogen Policy
ACS	American Chemical Society
ACSH	American Council on Science and Health
ADA	American's with Disabilities Act
	American Dental Association
ADI	Acceptable Daily Intake

AEA	Atomic Energy Act
AG	Attorney General
AGA	American Gas Association, Inc.
AHA	American Hospital Association
AHERA	Asbestos Hazard Emergency Response Act
AIChE	American Institute of Chemical Engineers
AIDS	Acquired Immune Deficiency Syndrome
AIHA	American Industrial Hygiene Association
AIHC	American Industrial Health Council
AIHF	American Industrial Hygiene Foundation
AIP	Auto Ignition Point
AIS	Acceptable Intake for Subchronic Exposures
ALA	American Lung Association
ALC	Apparent Lead Concentration (during XRF testing)
AL	Action Level
ALD	Average Lethal Dose
ALJ	Administrative Law Judge
AMA	American Medical Association
ANEC	American Nuclear Energy Council
ANPR	Advanced Notice of Proposed Rule Making
ANS	Autonomic Nervous System
ANSI	American National Standards Institute
AO	Administrative Order
APA	Administrative Procedures Act
APHA	American Public Health Association
ASCE	American Society of Civil Engineers
ASCII	American Standard Code for Information Interchange
ASHAA	Asbestos in Schools Hazard Abatement Act
ASHRAE	American Society of Heating, Refrigeration, and Air Conditioning Engineers
ASME	American Society of Mechanical Engineers
ASSE	American Society of Safety Engineers
ASTM	American Society for Testing Materials
ASV	Anodic Stripping Voltametry
ATERIS	Air Toxics Exposure and Risk Information System
ATSDR	Agency for Toxic Substances and Disease Registry
BCG	Bacille Calmette-Guerin
BCSP	Board of Certified Safety Professionals
BEI	Biological Exposure Index
BID	Buoyancy Induced Dispersion
BLS	Bureau of Labor Statistics

BMR	Basal Metabolic Rate
BOD	Biological Oxygen Demand
BOE	Bureau of Explosives
BOM	Bureau of Mines
BSC	Biological Safety Cabinet
BTU	British Thermal Unit
C	Celsius, Centigrade (degrees)
CAA	Clean Air Act
CAAA	Clean Air Act Amendments
CAG	Carcinogenic Assessment Group
CAP	College of American Pathologists
CAS	Chemical Abstract Service
CAT	Computerized Axial Tomography
CC	Cubic Centimeter
CDC	Centers for Disease Control
CDI	Chronic Daily Intake
CEL	Ceiling Exposure Limit
CERCLA	Comprehensive Environmental Response, Compensation, and Liabilities Act
CEU	Continuing Education Unit
CFC	Chlorofluorocarbons
CFM	Cubic Feet per Minute
CFR	Code of Federal Regulations
CFS	Cubic Feet per Second
CGA	Compressed Gas Association
CHEMTREC	Chemical Transportation Emergency Center
CHRIS	Chemical Hazard Response Information System
CIAQ	Council on Indoor Air Quality
CIH	Certified Industrial Hygienist
CIS	Chemical Information System
CIT	Colorimetric Indicator Tube
CLC	Corrected Lead Concentration (during XRF testing)
CHP	Certified Health Physicist
CMA	Chemical Manufacturers Association
CMB	Chemical Mass Balance
CMV	Cytomegalovirus
CNS	Central Nervous System
CO	Carbon Monoxide
CO_2	Carbon Dioxide
COD*	Chemical Oxygen Demand
COC	Chain-of-Custody

COH	Coefficient of Haze
COHN	Certified Occupational Health Nurse
COHST	Certified Occupational Safety and Health Technologist
CPC	Chemical Protective Clothing
CPR	Cardiopulmonary Resuscitation
CPSC	Consumer Product Safety Commission
CRAVE	Carcinogenic Risk Assessment Verification Exercise
CSP	Certified Safety Professional
CTD	Cumulative Trauma Disorder
CTS	Carpal Tunnel Syndrome
dBA	Decibel-A Scale
dB	Decibel
DBT	Dry Bulb Temperature
DCS	Decompression Sickness
DHHS	Department of Health and Human Services
DOC	Department of Commerce
DOL	Department of Labor
DOJ	Department of Justice
DOT	Department of Transportation
DNA	Deoxyribonucleic Acid
EAP	Employee Assistance Program
EBL	Elevated Blood Lead
ECD	Electron Capture Detector
ECP	Exposure Control Plan
ED	Effective Dose
EEC	European Economic Community
EIS	Environmental Impact Statement
EOP	Emergency Operations Plan
EPA	Environmental Protection Agency
EPID	Epidemiological Studies
ERT	Emergency Response Team
ERV	Expiratory Reserve Volume
F	Fahrenheit (degrees)
FACM	Friable Asbestos-Containing Materials
FAM	Friable Asbestos Material
FAR	Federal Acquisition Regulations
FCC	Fluid Catalytic Converter
	Federal Communications Commission
FDA	Food and Drug Administration (USA)

FEA	Federal Energy Administration
FEMA	Federal Emergency Management Agency
FIFRA	Federal Insecticide, Fungicide and Rodenticide Act
FLP	Flash Point
FM	Friable Material
FOI	Freedom of Information (Act)
FOM	Field Operations Manual
FR	Federal Register
FVC	Forced Vital Capacity
GAC	Granular Activated Carbon
GBT	Globe Temperature
GC/MS	Gas Chromatography/Mass Spectrograph
GFCI	Ground Fault Circuit Interrupter
GFF	Glass Fiber Filter
GM	Geiger-Mueller Counters
GSA	General Services Administration
GT	Globe Temperature
H_2O	Water
H_2O_2	Hydrogen Peroxide
H_2S	Hydrogen Sulfide
HAV	Hepatitis A Virus
HAZMAT	Hazardous Materials
HAZWOPER	Hazardous Waste Operations and Emergency Response
HBIG	Hepatitis B Immune Globulin
HBV	Hepatitis B Virus
HC	Hazardous Constituents
HCl	Hydrogen Chloride
HCP	Hazard Communication Program
	Hearing Conservation Program
HEPA	High Efficiency Particulate Air (filter)
HFS	Human Factors Society
HIV	Human Immunodeficiency Virus
HMIS	Hazardous Materials Information System
HPLC	High Performance Liquid Chromatography
HPS	Health Physics Society
HSV	Herpes Simplex Virus
Hz	Hertz
IAP	Indoor Air Pollutant
IARC	International Agency for Research on Cancer

IAQ	Indoor Air Quality
ICPES	Inductively Coupled Plasma Emission Spectroscopy
ICRA	Industrial Chemical Research Association
ICU	Intensive Care Unit
ICP	Inductively Coupled Plasma Spectrometer
IDLH	Immediately Dangerous to Life and Health
IES	Illumination Engineering Society (of North America)
IG	Inspector General
IP	Inhalable Particulates
IR	Infrared
IRV	Inspiratory Reserve Volume
keV	Kilo Electron Volts (radiation)
kHz	Kilohertz
kWH	Kilowatt Hour
KVp	Kilovoltage Peak
LASER	Light Amplification by Stimulated Emission of Radiation
LBP	Lead-Based Paint
LC	Lethal Concentration
LD	Lethal Dose
LEL	Lower Explosive Limit
LET	Linear Energy Transfer
LFL	Lower Flammability Limit
LOD	Level of Detection
LPG	Liquid Propane Gas
LPN	Licensed Practical Nurse
LRMS	Low Resolution Mass Spectroscopy
LVN	Licensed Vocational Nurse
MDA	4,4 Methylene Dianiline
MeV	Million Electron Volts
MED	Minimum Effective Dose
mg/m^3	Milligrams per Cubic Meter
MMWR	Morbidity and Mortality Weekly Report
M-M-R	Measles, Mumps, and Rubella Vaccine
MPC	Maximum Permissible Concentration
MPD	Maximum Permissible Dose
MPE	Maximum Permissible Exposure
MPO	Melting Point
MREM	Millirem
MSDS	Material Safety Data Sheet

MSHA	Mine Safety and Health Administration
MT	Metric Ton
mW	Milliwatt
MW	Microwave
NaOH	Sodium Hydroxide
NAR	National Asbestos Registry
NASA	National Aeronautics and Space Administration
NCAQ	National Commission on Air Quality
NCI	National Cancer Institute
NCRP	National Council on Radiation Protection and Measurement
NEC	National Electrical Code
NFPA	National Fire Protection Association
NIDA	National Institute for Drug Abuse
NIH	National Institutes of Health
NIOSH	National Institute for Occupational Safety and Health
NIST	National Institute of Standards and Technology
NLAC	National Lead Abatement Council
NMR	Nuclear Magnetic Resonance
NOHS	National Occupational Health Survey
NPL	National Priorities List
NRC	Nuclear Regulatory Commission
NRR	Noise Reduction Rating
NSC	National Safety Council
NTP	National Toxicology Program
NTSB	National Transportation Safety Board
O_2	Oxygen
O_3	Ozone
OA&R	Office of Air and Radiation (EPA)
ODC	Ozone Depleting Chemical
OHN	Occupational Health Nurse
OLDS	Ozone Level Depleting Substances
OSHA	Occupational Safety and Health Administration
OSHRC	Occupational Safety and Health Review Commission
PAPR	Powered Air Purifying Respirator
PbB	Blood Lead
PC	Potential Carcinogen
PCB	Polychlorinated Biphenyl
PCM	Phase Contrast Microscopy
PEL	Permissible Exposure Limit

PHA	Process Hazard Analysis
PHSA	Public Health Service Act
PL	Public Law
PM	Particulate Matter
PM_{10}	Particulate Matter, nominally 10 microns or less
PMR	Proportionate Mortality Ratio
PPB	Parts per Billion
PPE	Personal Protective Equipment
PPM	Parts Per Million
PSIA	Pounds Per Square Inch (Absolute)
PSI	Pounds Per Square Inch
PSIG	Pounds Per Square Inch (Gage)
PTS	Permanent Threshold Shift
QA	Quality Assurance
QC	Quality Control
QNFT	Quantitative Fit Testing
RAC	Radiation Advisory Committee
RAD	Radiation Absorbed Dose
RBE	Relative Biological Effectiveness
RCRA	Resource Conservation and Recovery Act
REC	Recognize, Evaluate, and Control
REL	Recommended Exposure Limit
REM	Roentgen Equivalent Man
RF	Radio Frequency
RFP	Request for Proposal
RH	Relative Humidity
RN	Registered Nurse
RNA	Ribonucleic Acid
RPG	Radiation Protection Guide
RTECS	Registry of Toxic Effects of Chemical Substances
RQ	Reportable Quantity
SARA	Superfund Amendments and Reauthorization Act
SCBA	Self-Contained Breathing Apparatus
SCFM	Standard Cubic Feet per Minute
SG	Surgeon General
SIC	Standard Industry Classification
SLM	Sound Level Meter
SO_2	Sulfur Dioxide
SOP	Standard Operating Procedure

SOW	Statement of Work
SSS	System Safety Society
STEL	Short-term Exposure Limit
STS	Standard Threshold Shift
TB	Tuberculosis
TD	Toxic Dose
TLD	Thermoluminescent Dosimeter (or "Detector")
TLV	Threshold Limit Value
TLV-C	Threshold Limit Value - Ceiling
TLV-skin	Threshold Limit Value - Skin Absorption
TLV-STEL	Threshold Limit Value - Short-term Exposure Limit
TSCA	Toxic Substances Control Act
TSP	Total Suspended Particulates
	Tri-sodium Phosphate (industrial detergent)
TTS	Temporary Threshold Shift
TU	Tuberculin Unit
TV	Television
TWA	Time-weighted Average
UEL	Upper Explosive Limit
UFL	Upper Flammability Limit
UL	Underwriter's Laboratory
UN	United Nations
USC	United States Code
UV	Ultra-violet
V	Volt
VC	Vital Capacity
VDT	Video Display Terminal
VOC	Volatile Organic Compound
VP	Vapor Pressure
VZV	Varicella Zoster Virus
WBGT	Wet Bulb Globe Temperature (or "Test")
WBT	Wet Bulb Temperature
WHO	World Health Organization
WSO	World Safety Organization
XRF	X-ray Fluorescence (lead paint analyzer)

Appendix C
Conversion Tables

In the practice of industrial hygiene it often becomes necessary to convert data from one form to another. Examples include temperature scales (Fahrenheit, Centigrade, Kelvin), metric system conversions (yards to meters), power (horsepower and BTU calculations), distance, force, energy and work, etc. This appendix has been developed to provide a quick reference to the most common of these conversion requirements. The tables presented here should facilitate the numerous conversions that may be necessary during the performance of industrial hygiene, as well as other engineering and scientific functions.

SURFACE AREA AND VOLUMES—GENERAL

R = Radius of a circle

d = Diameter of a circle

$\pi = 3.1416$

Circumference of a circle = $\pi d = 2\pi R$

Area of a circle = $\pi d/4 = \pi R$

Surface of a sphere = $\pi d = 4\pi R$

Volume of a sphere = $\pi d^3/6 = 4/3\pi R^3$

To convert mg/m^3 to parts per million (PPM) or vice versa:
$$mg/m^3 = (\text{Molecular Weight} \div 24.45) \times PPM$$
$$PPM = (24.45 \div \text{Molecular Weight}) \times mg/m^3$$

TEMPERATURE EQUIVALENTS AND CONVERSIONS

Scale	Symbol	Freezing Point of Water (° at 1 atm)	Boiling Point of Water (° at 1 atm)
Centigrade	C	0	100
Fahrenheit	F	32	212
Reaumur	R	0	80
Thermodynamic Kelvin Absolute Centigrade	K,A	273.18 ± 0.01	373.16 ±0.01
Approx. Absolute	AA	273	373
Rankin Absolute Fahrenheit	—	491.69	671.69

Conversion formulas:
C = (5/9) (F − 32) = (5/4)R − 32 = K − 273.16 = AA − 273
F = (9/5)C + 32 = (9/4)R + 32 = (4/5) (K − 273.16) + 32
R = (4/9) (F − 32) = (4/5)C = (4/5) (K − 273.16)
K = C + 273.16 = AA + 0.16 = (5/9) (F − 32) + 273.16
AA = C + 272 = K − 0.16 = (5/9) (F − 32) + 273
Rankin = F + 459.69

EQUIVALENTS AND CONVERSIONS

1 pound/minute = 7.58 grams/second (gm/sec) 453.6 grams/minute (gm/min) 27.215 kilograms/hour (kg/hr)	1 pound avoirdupois (lb) = 7000 grains (gr) 16 ounces (oz) 453.5923 grams (gm) 0.4535923 kilograms (kg)
1 pound/hour = 0.126 gm/sec 7.56 gm/min 0.4536 kg/hr 0.01667 lb/min	1 short ton = 2000 lb 0.892857 long ton 907.1846 kg 0.9071846 tonne (t)
1 gram (gm) = 15.4324 gr 0.003527.40 oz 0.002204623 lb	1 long ton = 2,240 lb 1.12 short tons 1016.047 kg 1.016047 g
1 kilogram (kg) = 1000 gm 35.2740 oz 2.204623 lb	1 metric ton, tonne (t) = 1000 kg 2204.623 lb 1.10231 short tons 0.9842107 long ton
1 grain (gr) = 0.0647989 gm 0.00228571 oz	1 ounce avoirdupois (oz) = 437.5 gr 28.3495 gm
1 cubic centimeter (cm^3) = 0.999972 milliliter (ml) 0.0610237 cubic inch 0.0338140 U.S. fluid oz	1 fluid ounce, U.S. = 1.80469 inches 29.5735 cm^3 29.5727 ml 1.0408 British fluid oz
1 cubic meter (m^3) = 999.972 liter (L) 35.3147 cubic feet 264.172 U.S. gallons 219.97 British gallons	1 fluid ounce, British = 1.7339 cubic inches 28.413 cm^3 28.412 ml 0.96076 U.S. fluid oz
1 milliliter (ml) = 1.000028 cm^3 0.0610255 cubic inches 0.33815 U.S. fluid oz 0.35196 British fluid oz	1 quart, liquid, U.S. = 57.75 cubic inches 32 U.S. fluid oz 946.353 cm^3 0.946326 L
1 cubic inch (in^3) = 0.554113 U.S. fluid oz 16.3871 cm^3 16.3866 ml	1 cubic foot (ft^3) = 1728 in^3 29.9221 U.S. quart 7.48052 U.S. gallons 28316.8 cm^3 28.316 L

EQUIVALENTS AND CONVERSIONS *(continued)*

1 milligram/cubic meter (mg/m^3) =
  1000 mg/liter
  35.314 mg/cubic foot
	1 milligram/cubic foot (mg/ft^3) =
	  28.32 mg/liter
	  0.02832 mg/m^3

1 liter (L) =
  10000.028 cm^3
  61.0255 cubic inches
  33.815 U.S. fluid oz
  1.05672 U.S. quart
  0.264179 U.S. gallon
  35.196 British fluid oz
	1 gallon, U.S. =
	  231 cubic inches
	  128 U.S. fluid oz
	  133.23 British fluid oz
	  0.8327 British gallons
	  3785.41 cm^3
	  3.78531 L

1 gallon, British =
  160 British fluid oz
  277.42 cubic inches
  1.2010 U.S. gallons
  153.72 U.S. fluid oz
  4546.1 cm^3
  4.5460 L
	1 liter/minute (L/min) =
	  0.06 m^3/hr
	  0.2640 gal/min
	  0.0353 ft^3/min
	  0.000589 ft^3/sec

1 m^3/hr =
  16.67 liter/min
  4.4 gal/min
  0.588 ft^3/min
  0.00989 ft^3/sec
	1 gallon/min =
	  3.78 liters/min
	  0.227 m^3/hr
	  0.1338 ft^3/min
	  0.00223 ft^3/sec

1 ft^3/min =
  28.32 L/sec
  1.699 m^3/hr
  7.50 gal/min
  0.01667 ft^3/sec
	1 ft^3/sec =
	  1690.0 L/min
	  102.0 m^3/hr
	  448.8 gal/min
	  60.0 ft^3/min

1 gram/sec =
  3.6 kg/hr
  0.13228 lb/min
  7.9367 lb/hr
	1 gram/min =
	  0.01667 gm/sec
	  0.060 kg/hr
	  0.0022 lb/min
	  0.1323 lb/hr

1 meter/second (m/sec, mps) =
  3.6 km/hr
  2.23694 mph
  3.28084 ft/sec
  196.850 ft/min
	1 mile per hour (mph) =
	  1.46667 ft/sec
	  0.44704 m/sec
	  1.609344 km/hr
	  88 ft/min

1 foot/min (fpm) =
  0.0113636 mph
  0.00508 m/sec
  0.018288 km/hr
	1 kilometer/hour (kph) =
	  0.277778 m/sec
	  0.621371 mph
	  0.911344 ft/sec

EQUIVALENTS AND CONVERSIONS *(continued)*

1 foot/second (ft/sec, fps) = 0.681818 mph 60 ft/min 0.3048 m/sec 1.09728 km/hr	1 Angstrom (Å) = 0.1 millimicron 0.0001 micron 0.0000001 millimeter 0.000000003937 inches
1 kilometer (km) = 0.621 miles	1 mile (mi), "statute" or "land" = 5280 feet 1.609 kilometers
1 meter (m) = 39.37 inches (exactly) 1.094 yards	1 mile (mil), "nautical" or "sea" = 6076 feet
1 centimeter (cm) = 0.3937 inches (exactly)	1 micron (μ) = 0.001 millimeter (exactly) 0.00003937 inches (exactly)
1 millimicron (m$^{\mu}$) = 0.001 micron (exactly) 0.00000003937 inches	1 foot (ft) = 0.305 meters
1 mil = 0.001 inch (exactly) 0.254 millimeter	1 nanometer (nm) = 10^{-9} meters
1 inch (in) = 2.540 centimeters (cm)	1 yard (yd) = 0.914 meters
Miles per hour (mph) to knots (kt) = mph ÷ 1.15	Knots (kt) to miles per hour (mph) = kn × 1.15
1 square centimeter (cm^2) = 0.155 square inches	1 square inch (sq in, in^2) = 6.452 cm^2
1 square meter (m^2) = 1.196 square yards 10.764 square feet	1 square kilometer (km^2) = 247.104 acres 0.386 square miles

POWER EQUIVALENTS AND CONVERSIONS

1 kilowatt (kW) =
  1.34 horse power (hp)
  737.54 ft-lb/sec
  0.948 Btu/sec
  0.2388 kcal/sec
	1 horse power (hp) =
	  0.746 kW
	  550 ft-lb/sec
	  0.707 Btu/sec
	  0.178 kcal/sec

POWER PER UNIT AREA

1 cal$_{15}$ cm^{-2} min^{-1} =
  1 langley min^{-1}
  0.069758 abs. watt cm^{-2}
  0.069745 int. watt cm^{-2}
  69.745 int. kW dekameter^{-2}
  3.6855 Btu ft^{-2} min^{-1}
  1440 cal$_{15}$ cm^{-2} day^{-1}
  5307.1 Btu ft^{-2} day^{-1}
	1 Btu ft min =
	  0.27133 cal$_{15}$ cm^{-2} min^{-1}
	  0.0189277 abs watt cm^{-2}

ILLUMINATION, BRIGHTNESS, ETC.

Total luminous flux from a source of unit spherical candlepower is 4π lumens.

Luminous efficiency: At wavelength of maximum luminosity, 0.555μ for photopic vision, the luminous efficiency is 680 lumens per watt, corresponding to a minimum "mechanical equivalent of light" of 0.00151 watts per lumen.

1 lux (lx) =
  1 lumen incident/m^2
  0.0001 phot (ph)
  0.09290 foot candle (ft-c)
	1 lambert (L) =
	  $1/\pi$ sb = 0.3183 sb
	  2.054 candle/square in (c/sq in)
	  929 ft-L

1 apostilb, in German (Hefner units) =
  0.09 mL
	1 apostilb, in international units =
	  $1/(\pi \times 10^4)$ sb or = 0.01 mL

1 phot (ph) =
  1×10^{-4} lumen incident/m^2
	1 millilambert (mL) =
	  10^{-3} L

1 foot candle (ft-c) =
  1 lumen incident/ft^2
  10.76 lx
	1 stilb (sb) =
	  1 int. c/cm^2
	  πL = 3.142 L
	  2919 ft-L

1 candle/in^2 (c/sq in) =
  0.1550 sb
  0.487 L
  452.4 ft-L
	1 footlambert (ft-L) =
	  0.0003426 sb
	  0.00107076 L
	  0.00221 c/sq in
	  3.142 c/sq ft

FORCE EQUIVALENTS

1 gram weight* =
    980.665 dynes
	1 pound weight* =
	    32.174 poundals
	    444822 dynes

1 kilogram weight =
    9.80665×10^5 dynes
	1 poundal =
	    13825.5 dynes

1 newton = 10^5 dynes	

*At standard gravity of 980.665 cm/sec^2 or 32.174 ft/sec^2

ENERGY AND WORK

1 erg =
    1 dyne-centimeter
    10^{-7} abs joule
    2.3892×10^{-8} cal$_{15}$
	1 absolute kilowatt-hour (abs kW-hr) =
	    3.6×10^6 abs joules

1 absolute joule (abs joule) =
    10^7 ergs
    0.23892 cal$_{15}$
	1 mean international kilowatt hour =
	    1.00019 abs kW-hr
	    3.60068×10^6 abs joules
	    3412.756 Btu

1 kilogram-meter (kg m) =
    9.80665 abs joules
	1 British thermal unit (Btu) =
	    252.08 cal$_{15}$
	    1055.07 abs joules
	    0.00029302 int kW-hr

1 15 gram calorie (cal$_{15}$) =
    4.1855 abs joules
	1 foot-pound (ft-lb) =
	    1.35582 abs joules

MULTIPLICATION MATRIX: LENGTH

	Meter	*Centimeter*	*Millimeter*	*Micron*	*Angstrom*	*Inch*	*Foot*
Meter	1	100	1000	10^6	10^{10}	39.37	3.28
Centimeter	0.01	1	10	10^4	10^8	0.0394	0.0328
Millimeter	0.001	0.1	1	10^3	10^7	0.394	0.00328
Micron	10^{-6}	10^{-4}	10^{-3}	1	10^4	3.94×10^{-6}	3.28×10^{-6}
Angstrom	10^{-10}	10^{-8}	10^{-7}	10^{-4}	1	3.94×10^{-9}	3.28×10^{-10}
Inch	0.0254	2.540	25.4	2.54×10^4	2.54×10^8	1	0.0833
Foot	0.305	30.48	304.8	304,800	3.05×10^9	12	1

MULTIPLICATION MATRIX: AREA

	Square meters	Square inches	Square feet	Square centimeters	Square millimeters
Square meters	1	1550	10.76	10000	10^6
Square inches	6.94×10^{-4}	1	6.94×10^{-3}	6.452	645.2
Square feet	0.0929	144	1	929	92903
Square centimeters	0.0001	0.155	0.001	1	100
Square millimeters	10^{-6}	0.00155	0.00001	0.01	1

MULTIPLICATION MATRIX: VOLUME

	Cubic feet	Gallons	Liters	Cubic centimeters	Cubic meters
Cubic feet	1	7.481	28.32	28320	0.0283
Gallons	0.1337	1	3.785	3785	3.79×10^{-3}
Liters	0.03531	0.2642	1	1000	1×10^{-3}
Cubic centimeters	3.53×10^{-5}	2.64×10^{-4}	0.001	1	10^{-6}
Cubic meters	35.31	264.2	1000	10^6	1

MULTIPLICATION MATRIX: VELOCITY

	Centimeters per second	Meters per second	Kilometers per hour	Feet per second	Feet per minute	Miles per hour
cm/sec	1	0.01	0.036	0.0328	1968	0.02237
m/sec	100	1	3.6	3.281	196.85	2.237
km/sec	27.78	0.2778	1	0.9113	54.68	0.6214
ft/sec	30.48	0.3048	18.29	1	60	0.6818
ft/min	0.5080	0.00508	0.0183	0.0166	1	0.0113610
mph	44.70	0.4470	1.609	1467	88	1

MULTIPLICATION MATRIX: FLOW RATES

	Liters per minute	Cubic meters per hour	Gallons per minute	Cubic feet per minute	Cubic feet per second
Liters per minute	1	0.06	0.2640	0.0353	5.89×10^{-4}
Cubic meters per hour	16.67	1	4.4	0.588	9.89×10^{-3}
Gallons per minute	3.78	0.227	1	0.1338	2.23×10^{-3}
Cubic feet per minute	28.32	1.699	7.50	1	0.01667
Cubic feet per second	1.69×10^3	1.02×10^2	448.8	60	1

MULTIPLICATION MATRIX: MASS

	Grams	Kilograms	Grains	Ounces	Pounds
Grams	1	0.001	15.432	0.03527	0.00220
Kilograms	1000	1	15432	35.37	2.205
Grains	0.0648	6.48×10^{-5}	1	2.29×10^{-3}	1.43×10^{-4}
Ounces	28.35	0.02835	437.5	1	0.0625
Pounds	453.59	0.4536	70003	16	1

MULTIPLICATION MATRIX: PRESSURE

	Pounds per square inch	Atmospheres	Inches (Hg) 0°C	Millimeters (Hg) 0°C	Feet (H_2O) 15°C	Inches (H_2O)	Pounds per square foot
lb/in^2	1	0.068	2.036	51.71	2.309	27.71	144
atm	14.696	1	29.92	760	33.93	407.2	2116
in (Hg)	0.4912	0.033	1	25.40	1.134	13.61	70.73
mm (Hg)	0.01934	0.0013	0.039	1	0.04464	0.5357	2.785
ft (H_2O)	0.4332	0.0294	0.8819	22.40	1	12	62.37
in (H_2O)	0.03609	0.0024	0.073	1.867	0.0833	1	51.197
lb/ft^2	0.0069	4.72×10^{-4}	0.014	0.359	0.016	0.193	1

MULTIPLICATION MATRIX: RADIANT ENERGY UNITS					
	erg	joule	Wsec	μWsec	gm-cal
erg	1	10^{-7}	10^{-7}	0.1	2.39×10^{-8}
joule	10^7	1	1	10^6	0.239
Wsec	10^7	1	1	10^6	0.239
μWsec	10	10^{-6}	10^{-6}	1	2.39×10^{-7}
gm-cal	419×10^{-7}	4.19	4.19	419×10^6	1

Glossary

GLOSSARY OF TERMS

The following are definitions for many of the terms that appear in this text or those which may be encountered during the normal practice of industrial hygiene. The majority of these terms were adapted from the latest edition of The VNR Dictionary of Environmental Health and Safety (Lisella) unless otherwise noted.

Abatement   The removal or elimination of a nuisance; the actions taken to effect same.

Abel Test   A colorimetric test that involves the use of moist potassium iodide paper which turns violet or blue in the presence of gasses evolved from nitroglycerin, nitrocellulose, and nitroglycol.

Abiotic   Indicating the absence of life; non biological.

Absolute Humidity   The mass of water vapor present in a unit volume of the atmosphere, measured either as grams per cubic meter or actual partial pressure.

Absolute Pressure   Pressure that has been measured relative to zero pressure.

Absorbed Dose   For any ionizing radiation, the energy imparted to matter by ionizing particles per unit mass of irradiated materials at the point of exposure. (*see also* **RAD**).

Absorption   (1) *Toxicological.* The ability of a substance to penetrate the body of another; the movement of a chemical from the site of exposure (oral, dermal, respiratory) across a biologic barrier and into the bloodstream or lymphatic system. (2) *Chemistry.* The process by which one material is pulled into and retains another to form a blended or homogeneous solution. (3) *Physiology.* The process by which porous tissues such as the skin and intestinal walls permit the passage of liquids and gases into the bloodstream. (4) *Radiation.* The process whereby the number of particles or quanta in a beam of radiation is reduced or degraded in energy as it passes through some medium. The absorbed radiation may be transformed into mass, other radiation, or energy by interaction with the electrons or nuclei of the atoms on which it impinges.

Acaricide   Chemical used to kill ticks and mites.

Accelerator   In radiation, a device for imparting a large amount of kinetic energy to charged particles, such as electrons, protons, deuterons, and helium ions.

Accident   An unplanned, unforeseen, and unwanted or undesired event that results in physical harm (illness, injury, or death) and/or property damage.

Accident Analysis   A concerted, organized, methodical, planned process of examination and evaluation of all evidence and records identified during investigation of accidents.

Accident Investigation   A detailed and methodical effort to collect and interpret facts related to an individual accident, conducted to identify the causes and develop control measures to prevent recurrence; a systematic look at the nature and extent of the accident, the risks taken, and loss(es) involved; an inquiry as to how and why the accident event occurred.

Acclimation   The process by which humans develop a tolerance to various environmental or occupational stressors. The process of becoming accustomed to new conditions.

Acid   A compound with pH between zero and seven. As pH decreases from seven to zero, acidity increases. Acids contain hydrogen as an essential constituent.

Acid Mantle   The lipid (oily) outside layer of the skin structure, composed of oil and sweat, easily removed by washing. The acid mantle normally has a pH less than seven.

Acid Rain   Precipitation contaminated with sulfur dioxide, nitrous oxide, and other chemicals from power plants and/or industrial sites.

Acoustics   (1) The science of sound. (2) The cause, nature, and phenomena of the vibrations of elastic bodies that affect the organ of hearing. (3) The properties determining audibility or fidelity of sound in an auditorium.

Acquired Immune Deficiency Syndrome (AIDS)   A severe (life threatening) disease that represents the late clinical stage of infection with the *human im-*

munodeficiency virus (HIV). The HIV most often results in progressive damage to the immune system and various organ systems, especially the central nervous system. Body fluid-to-body fluid contact with an infected HIV carrier is required for transmission. HIV has been recovered from body fluids other than blood, such as tears, saliva, urine, bronchial secretions, spinal fluid, feces, vomitus, and others.

Acroosteolysis   A condition reported in workers exposed to vinyl chloride and manifested by ulcerating lesions on the hands and feet.

Action Level   A exposure limit usually set at 50 percent of the permissible exposure limit (PEL), as specified by the applicable OSHA standard. Exposures exceeding the action level typically require implementation of certain actions, such as medical surveillance, training, and monitoring programs, but not necessarily further controls (e.g., engineering controls) aimed at reducing exposures.

Activated Carbon (charcoal)   A form of carbon with a high potential for absorption of gases, vapors, and colloidal particles.

Activation Analysis   A method of chemical analysis, especially for small traces of material, based on the detection of characteristic radionuclides.

Activity   The rate of decay of radioactive material expressed as the number of nuclear disintegrations per second (*see* **Curie**).

Acuity   Of or pertaining to the sensitivity of receptors used in hearing and vision.

Acute Exposure   (1) *Chemical.* A sudden, short, rapid association with a chemical compound. (2) *Radiation.* Exposure of short duration, generally taken to be the total dose absorbed within 24 hours. (3) *Biologic.* A brief encounter with a pathogenic or nonpathogenic microorganism.

Acute Toxicity   The ability of a substance to cause poisonous effects resulting in severe biological harm or death soon after a single exposure or dose. Also, any severe poisonous effect resulting from a single short-term exposure to a toxic substance.

Adaptation   A change in an organism's structure or habit that helps it to adjust to its surroundings.

Adductor   The muscle that draws towards the mesial line of the body.

Adenalgia   A glandular pain.

Adenitis   Inflammation of gland or lymph nodes.

Adenocarcinoma   A malignant tumor that appears in glandular epithelium.

Adenoid   An enlarged mass of lymphoid tissue in the upper pharynx that hinders breathing.

Adenoma   A benign tumor originating in a gland.

Adhesion   Molecular attraction that holds surfaces of two substances in contact.

Adiabatic   Refers to a reaction that occurs without a gain or loss of heat.

Administrative Control   A measure initiated to reduce worker exposure to various stresses in the work environment. An example is limiting the amount of time an employee can work around health hazards.

Adrenal Gland   In mammals, a gland adjacent to the kidney that produces the hormone adrenaline, or epinephrine. The hormone influences the heartbeat rate, dilates blood vessels, increases blood sugar, and plays a major role in other physiological activities.

Adsorption   (1) Condensation of gases, liquids, or dissolved substances on the surface of solids. (2) The attraction and retention of atoms, molecules, or ions on the surface of a solid.

Advection   Movement caused by the motion of heat, air, water, or another fluid. It specifically refers to the horizontal movement by wind currents of chemical pollutants or heat.

Aeroallergen   Airborne material, such as particulates, pollen, dusts, or dander that may precipitate an allergic response in susceptible persons.

Aerobic   Describes an environment with molecular oxygen present; organisms that live or grow in the presence of molecular oxygen; reactions that occur in the presence of molecular oxygen.

Aerosol   A dispersion of solid or liquid particles of microscopic size in a gaseous medium. Smokes, fogs, fibers, dusts, and mists are examples of common aerosols.

Agent   A biological, physical, or chemical entity capable of causing disease.

AIDS   *See* **Acquired Immune Deficiency Syndrome**.

Air Dose   In radiation, a dose of x-rays or gamma rays, expressed in roentgens, delivered at a point in free air. In radiological practice, it consists of the radiation of the primary beam and that scattered from surrounding air.

Airflow   The volumetric rate at which air flows through a space, usually measured in cubic feet per minute (CFM) or cubic meters per second (CMS). Also, the speed at which air moves through a space, usually measured in feet per minute or meters per second.

Air, Makeup   Air that replaces other air exhausted from a space. Insufficient makeup air is one possible cause of insufficient exhaust airflow.

Air Pollutant   Any substance in air that could, in high concentration, harm people, animals, or vegetation, or damage non-living material. Such pollutants may be from solid particles, liquid droplets, gases, or any combination of these.

Air Purifying Respirator   A device worn by an individual that filters the air to be breathed and returns it to an acceptable state. Also known as chemical cartridge respirators, this type of respirator relies on the person's own breathing force to draw air through the filter medium, or it may utilize a powered blower to provide breathing air (i.e., a powered air purifying respirator or PAPR).

Air Titration   A field analytical method involving the use of an impinger or bubble to draw air through a liquid reagent that changes color in direct proportion to the concentration of the contaminant in the air. Not as precise as laboratory methods.

Aliphatic Hydrocarbon   One of the major groups of organic compounds characterized by a straight- or branched-chain arrangement of carbon atoms. This group is composed of three subgroups   alkanes (paraffins) that are saturated and relatively unreactive; alkenes that contain double bonds and are reactive; and alkynes (acetylenes) that contain triple bonds and are highly reactive.

Alkali   A compound that has the ability to neutralize an acid and form a salt (*see* **Base**).

Allergen   Any of a wide variety of substances or environmental conditions that may provoke an allergic reaction. Also referred to as sensitizers.

Allergy   An unusual or exaggerated response to a particular substance in a person sensitive to that substance.

Alpha Emitter   A radioactive substance that gives off alpha particles during the decay process. Also referred to as alpha decay.

Alpha Particle   A specific particle, consisting of two protons and two neutrons (a helium nucleus), ejected spontaneously from the nuclei of some radioactive elements. It has low penetrating power and short range. Even the most energetic alpha particles will generally fail to penetrate unbroken skin. The danger arises when matter containing alpha-emitting isotopes is introduced into the lungs or intestinal tract.

Alpha Radiation   A stream of alpha particles.

Alpha Ray   A strongly ionizing and weakly penetrating radiation stream of fast-moving helium nuclei.

Aluminum Silicates   Compounds containing aluminum, silica, and oxygen as main constituents.

Alveoli   Numerous small, terminal air sacs in the lungs where pulmonary capillary blood is in close juxtaposition to the alveolar gas, permitting the rapid exchange of carbon dioxide and oxygen in the lungs. There are approximately 300 million alveoli situated at the ends of small air passageways in the lungs. Alveoli are the main deposition site of respirable dust particles (1–10 microns in diameter) or respirable fibers (e.g., asbestos) that can result in various respiratory diseases such as silicosis and asbestosis.

Amalgam   Any mixture or alloy of mercury combined with other metals, such as zinc, gold, silver, or alloys.

Ames Test   A test used to determine the carcinogenicity of chemicals. It is often referred to as the Salmonella test. In the test, mutant strains of Salmonella Typhimurium cultured on a medium deficient in histidine are exposed to a potential carcinogen and liver extracts. Mutagenic bacteria will back-mutate to contain a functional histidine gene, permitting bacterial growth. The level of mutagenicity can be determined by the number of colonies that develop.

Amino Acid   An organic acid that is one of the building blocks in the formation or proteins.

Amplification   As related to radiation detecting instruments, the process (either gas, electronic, or both) by which ionization effects are magnified to a degree suitable for their measurement.

Anaerobes   Organisms unable to multiply in any environment that contains oxygen. Anaerobic microorganisms have oxygen-sensitive enzymes and cannot function in the presence of molecular oxygen. Some may be more air tolerant than others. Those severely affected by the presence of oxygen are called strict anaerobes or obligate anaerobes.

Anaerobic   Without oxygen. Also, refers to cells or organisms that can live without oxygen or processes that occur in the absence of oxygen.

Anatomic   (1) As used in this text, of or pertaining to human anatomy or any of its various components. (2) Relating to the science of the morphology or structure of organisms.

Anemia   A disorder of the blood as a whole; a deficiency in the number of red corpuscles or of hemoglobin.

Anemometer   An instrument used to measure the motion of wind or air that employs a pilot tube directed by a vane or rotor, or a pressure plate deflected against a spring or gravity.

Angstrom   A unit of length used chiefly in expressing short wavelengths. It is equal to 10^{-10} meter or 10^{-8} centimeter.

Annihilation Radiation   Photons produced when an electron and a positron unite and cease to exist. The annihilation of a positron/electron pair results in the production of two photons, each of which has at least 0.51 MeV energy.

Anoxia   The absence of, or a diminished amount of, oxygen in the blood, tissue, or a body of water. The deficiency of oxygen in organisms often results in an increased rate of breathing. Anoxia in humans is often accompanied by dizziness, rapid heartbeat, and headache. It can result in death.

Anthracosis   Also known as Collier's disease, Shaver's disease, miner's lung, or black lung; a pneumoconiosis resulting from the accumulation of carbon from inhaled smoke or coal dust in the lungs.

Anthrax   An acute bacterial disease usually affecting the skin. Also known as wool sorter's disease, rag picker's disease, or malignant edema, this disease is transmitted by contact with tissues of infected animals (cattle, sheep, goats, horses, and others) or contaminated hair, wool, or hides.

Anthropometry   The measurement of the human body, including body dimensions, range of motion of body members, and strength (including both static and dynamic measurements). The branch of anthropology that deals with the comparative measurements of the human body (*see* **Ergonomics**).

Antibiotic   A chemical substance produced by living organisms that inhibits the growth of or kills other organisms.

Antibody   A globulin found in tissue fluids and blood serum that is produced in response to the stimulus of a specific antigen, and is capable of combining with that antigen to neutralize or destroy it. Also referred to as immune substances.

Antigen   That portion or product of a biologic agent capable of stimulating the formation of specific antibodies.

Antiseptic   Chemical compounds that are capable of reducing the number of microorganisms on body surfaces. Used primarily on humans and animals, in contrast to disinfectants, that are used primarily on non-living surfaces, as a form of infection control (*see* **Disinfectant**).

Antitoxin   An antibody to the toxin of a microorganism, usually a bacterial exotoxin. Antitoxins combine with a specific toxin, in vivo or in vitro, with the consequent neutralization of toxicity.

Aphagia   The failure to eat that can result in illness or death.

Apocrine   Sweat glands that open into hair follicles. Apocrine sweat glands are limited to a few regions of the body, notably the underarm and genital areas.

Aromatic Hydrocarbon   A major group of unsaturated cyclic hydrocarbons containing one or more rings made up of six carbon atoms. This group, most notably benzene, is chiefly derived from petroleum and coal tar. The name is due to the strong and not always unpleasant odor characteristic of substances within this group.

Artery   A blood vessel that conveys blood from the heart to any part of the body.

Asbestos   A generic term used to describe a number of naturally occurring fibrous, hydrated mineral silicates differing in chemical composition. They are white, gray, green, or brown. Asbestos fibers are characterized by high tensile strength, flexibility, heat and chemical resistance, and good frictional properties. Chrysotile, crocidolite, amosite, anthophyllite, and actinolite are all forms of asbestos. Exposure to asbestos fibers is known to cause a variety of diseases, including asbestosis (a diffuse, interstitial non-malignant scarring of the lung tissues), bronchogenic carcinoma (a lung cancer), mesothelioma (a tumor of the lining of the chest cavity or lining of the abdomen), and cancer of the stomach, colon, and rectum.

Asbestos-Containing Materials (ACM)   Any materials containing asbestos that can be released on destruction or disturbance of the structural integrity of the material.

Asbestosis   Disease caused by prolonged exposure to asbestos fibers, that injures the lungs and diminishes their oxygen absorbing properties, thus reducing lung function (*see* **Asbestos**).

Aseptic Technique   Procedures designed to exclude infectious agents; laboratory or clinical techniques that do not result in the transfer of disease-producing microorganisms from one surface to another.

Asphyxia   The state of respiratory distress or suffocation due to the lack of respirable oxygen.

Asphyxiant   Any commodity capable of reducing or depleting the oxygen content of a space to the point of asphyxiation.

Asphyxiation   That point where oxygen content is no longer at a level capable of supporting life; a cause of death resulting from a lack of sufficient oxygen; suffocation.

Aspirate   To remove (by suction) a gas or body fluid from a body cavity, from an unusual accumulation, or from a container.

Associated Corpuscular Emission   The full complement of secondary charged particles (usually limited to electrons) associated with an x-ray or gamma-ray beam in its passage through air.

Asymptomatic   Without symptoms.

Athlete's Foot   *See* **Dermatophytoses.**

Atom   The smallest particle of an element that can not be divided or broken by chemical means. It consists of a central core called the nucleus, that contains protons and neutrons. Electrons move in orbital fashion in the region surrounding the nucleus.

Atomic Absorption Spectrophotometry   A method commonly used for the analysis of heavy metals.

Atomic Weight   Approximately the sum of the number of protons and the number of neutrons found in the nucleus of an atom, also known as the mass number.

Atrophy   Wasting away or diminution in the size of a cell, tissue, organ, or part, from defect, failure of nutrition, or lack of use.

Audible Sound   Sound containing frequency components between 20 and 20,000 Hz.

Audiogram   A graphic recording of hearing levels referenced to a normal sound pressure level as a function of frequency. Audiograms are used in the diagnosis and treatment of hearing loss (*see* **Audiometer**).

Audiometer   A frequency-controlled audio signal generator that produces pure tones at various frequencies and intensities and that is used to measure hearing sensitivity or acuity. Measurement of hearing threshold results in an audiogram, measured in decibels at selected frequencies (*see* **Audiogram**).

Audiologist   A professional, specializing in the study and rehabilitation of hearing, who is certified by the American Speech-Language-Hearing Association or licensed by a state board of examiners (OSHA).

Aural Insert Protectors   A form of hearing protector commonly known as earplugs. They are available in numerous configurations as foam, plastic, fine glass fiber, and wax-impregnated cotton. The three types are formable, custom-molded, and premolded.

Autoclave   A device that uses a combination of steam and pressure to bring about sterilization (i.e., the destruction of all microorganisms). Sterilization by autoclave occurs at 15 lb pressure, 120° C, for 15 minutes.

Avogadro's Number   One of the fundamental physical constants. It is expressed as $6.023 \times 1,023$ atoms per gram-atomic weight (or molecules per gram-atomic weight). It is an expression of the number of atoms in a gram-atomic weight of any element.

Bacillus   The shape of a bacteria cell, also commonly referred to as a rod. Bacilli (plural) are generally shaped like a cylinder. They may also be curved, spiral, or helical shaped. Bacilli that are curved are designated as vibrios; spiral rods are designated as spirilla; and helical rods are designated spirochetes.

Background Radiation   That which arises from radioactive material, other than that which is being measured. Background radiation due to cosmic rays and natural radioactivity is always present.

Background Noise   Noise that is coming from sources other than that which is being measured.

Bactericide   Any substance that kills bacteria.

Bacteriosis   Pertains to any disease or abnormal condition caused by a bacterium.

Bacteriostat   A substance that inhibits or retards the growth of bacteria but does not necessarily kill them.

Bagassosis   Similar to farmer's lung disease, except caused by exposure to bagasse, the dried stalks of sugar cane (*see* **Farmer's Lung Disease**).

Barrier Cream   A protective cream that may be used for preventing skin contact with harmful agents. Used a supplement (not a replacement) to personal protective equipment.

Basal Metabolic Rate (BMR)   Rate of heat production by the human body under neutral conditions.

Baseline Audiogram   The audiogram against which future audiograms are compared (OSHA).

Beam Axis   A line from the source through the centers of the x-ray fields.

Benign   Pertaining to the mild character of an illness or the non-malignant character of a neoplasm.

Beriberi   A disease caused by vitamin B (thiamin) deficiency, common among populations that survive on polished rice. It is characterized by loss of muscle power, emaciation, and exhaustion.

Berylliosis   Chronic poisoning caused by exposure to the dust or fume of beryllium metal, beryllium oxide, or soluble beryllium compounds. Symptoms include a loss of appetite and weight, weakness, cough, extreme difficulty in breathing, cyanosis, and cardiac failure. The disease may appear 5 to 20 years after exposure has ceased. It is usually progressive in severity, will cause fibrous growth in the lungs, can create kidney stones, and can be accompanied by enlargement of the heart, liver, and spleen.

Beta Decay   Radioactive change by emission of a beta particle. In beta decay, a neutron decays into a proton, with the emission of an electron (or beta particle); or, a proton transforms into a neutron and emits a positron. In both cases, the charge of the nucleus is changed without changing the number of nucleons.

Beta Particle   A small particle ejected spontaneously from the nucleus of a radioactive element. It has the mass of the electron, has a charge of either -1 or $+1$, and has a mass of $1/1,840$ that of a proton or neutron. It has low penetrating power and short range. The most energetic of beta particles can penetrate the skin (causing a skin burn effect) and other tissues.

Beta Ray   A stream of beta particles of nuclear origin more penetrating but less ionizing than alpha rays per unit length of travel; a stream of beta particles emitted in certain radioactive disintegrations.

Betatron   A circular electron accelerator providing a pulsed beam of high-energy electrons or x-rays by magnetic induction.

Bile   The yellowish-brown or green fluid secreted by the liver and discharged into the small intestine where it aids in the emulsification of fats, increases peristalsis, and retards putrefaction.

Binding Energy   The energy represented by the difference in mass between the sum of the component parts of a nucleus and the actual mass of the nucleus. It is the energy that holds the neutrons and protons together and, subsequently, the amount of energy required to separate the individual nucleons.

Bioaccumulative   Substances that increase in concentration in living organisms (that are very slowly metabolized or excreted) as they breathe contaminated air, drink contaminated water, or eat contaminated food.

Bioassay   Use of living organisms to measure the effect of a substance, factor, or condition by comparing pre- and post-exposure data.

Biochemical   Describes the event or action involving chemistry of living organisms and the chemical changes occurring therein.

Biohazard   Term applied to organisms or products of organisms that present a health risk to humans. Derived from a combination of the words biological and hazard.

Biohazard Area   Any area in which work has been or is being performed with biohazardous agents or materials.

Biological Exposure Index (BEI)   Set of reference values established by the American Conference of Government Industrial Hygienists as guidelines for the evaluation of potential health hazards in biological specimens collected from a healthy worker who has been exposed to chemicals to the same extent as a worker with inhalation exposure to the threshold limit value. The values apply to eight-hour exposures, five days per week.

Biological Half-Life   The time required for the body to eliminate one-half of an administered dose of any substance by regular process of elimination. This time is approximately the same for both stable and radioactive isotopes of a particular element.

Biological Hazard   *See* **Biohazard.**

Biomechanics   The study of the human body acting as a system under the laws of Newtonian mechanics and the biological laws of life.

Biopsy   The removal and examination of tissue from living mammals.

Biotic   Of or pertaining to life.

Biotoxicology   The scientific study of toxins produced by organisms, their effects, and the treatment of conditions they produce.

Bird Fancier's Disease   Extrinsic allergic alveolitis observed in some individuals who have been exposed to birds. The condition is accompanied by breathlessness or tightness of the chest, coughing, and wheezing. Extensive fibrosis can be seen in the chronic form.

Black Light   The region of the electromagnetic spectrum between 300 nm and 400 nm in the ultra-violet (UV) region. It is the region responsible for the added pigmentation of the skin (burning and tanning) following exposure to UV light.

Black Lung Disease   A disease contracted by coal miners, marked by varying degrees of pulmonary impairment, including x-ray abnormalities, cough, breathlessness, massive progressive fibrosis, formation of nodules and scar tissue in the lungs. Also known as coal miner's pneumoconiosis, Collier's disease, and Shaver's disease.

Blank Sample   An non-contaminated or otherwise clean sample media sent along to a testing laboratory along with actual sample results, used to help determine sample inaccuracies or other compromising conditions.

Blast Gate   A device that regulates airflow in duct work, similar to a damper, but usually operated by positioning a sliding metal plate across a duct.

Blind Sample   A sample media sent along to a testing laboratory that has been pre-conditioned at the sample site with known contaminant levels that are not reported to the laboratory and is used to determine the accuracy of a laboratory analysis.

Blind Spot   Normal defect of vision caused by the position of the optic nerve at the point where it enters the eye.

Bloodborne Pathogens   Pathogenic microorganisms that are present in human blood and can cause disease in humans. These pathogens include, but are not limited to, hepatitis B virus (HBV) and the human immune deficiency virus (HIV).

Blood Products   Any product derived from blood, including but not limited to blood plasma, platelets, red and white blood corpuscles, and other derived licensed products such as interferon.

Body Burden, Maximum Permissible   (1) *Radiation.* An amount of radioactive material in a critical organ such that the whole-body dose is 0.3 Rem per week or less; in case of alpha or beta emitter that is deposited in the bone, body burden is derived from the long-established maximum permissible body burden of radium (0.1 microcurie) adjusted for possible less uniform deposition. (2) *Biological.* The total amount of a substance stored in the body following exposure. The body burden of a particular substance is a function of its biological half-life and its biochemical uptake and elimination rate.

Body Fluids   Liquid emanating or derived from humans and include blood, dialysis, amniotic, cerebro-spinal, synovial, pleural, peritoneal, and pericardial fluids, as well as semen and vaginal secretions.

Bone   The skeletal tissue of vertebrates consisting of cells arranged in a matrix of collagen fibers and cells containing calcium and phosphate.

Bone Marrow   Soft material that fills the cavity in most bones. It manufactures most of the formed elements of the blood.

Bone Seeker   Any compound or ion in the body that migrates preferentially to the bone.

Botulism   A severe illness resulting from ingestion of the toxin from the strictly anaerobic bacillus Clostridium botulinium. The illness may cause blurred vision, sore throat, or other symptoms of a nervous system disorder. Since these toxins generally disrupt nerve impulse transmission, they are referred to as neurotoxins. Vomiting, diarrhea, or constipation may also occur. If death occurs, it is usually the result of respiratory paralysis.

Breakthrough   The presence of 25 percent or more of a contaminant in the rear portion of a sorbent tube (NIOSH, 1984).

Breathing Zone   Usually the air within a one-two foot radius surrounding a person's head.

Bremsstrahlung (radiation)   A German word meaning "braking radiation," it is the secondary x-radiation that is produced when a beta particle is slowed down or stopped by a high-density surface.

British Thermal Unit (BTU)   The quantity of heat needed to raise the temperature of one pound of water by one degree Fahrenheit.

Bronchial Tubes   Branches or subdivisions of the trachea (windpipe). A bronchiole is a branch of the bronchus that is a branch of the trachea.

Bronchitis   An inflammation of the bronchi or bronchial tubes.

Brucellosis   An illness caused by the bacterium of the Genus Brucella. Symptoms include fever, chills, headache, muscle aches, malaise, weakness, loss of appetite and subsequent loss of weight. Mortality is possible but rare. Contact with infectious materials such as animal blood is an important mode of infection for livestock growers, veterinarians, and processing-plant workers. Intact skin is an effective barrier, but cuts and abrasions provide a direct route of exposure. Inhalation and ingestion are also potential routes of infection.

Bubble Tube   A simple device used to calibrate air-sampling pumps.

Bulk Sample   As related to asbestos, a small portion of suspect building materials that is collected and sent to a laboratory for analysis by polarized light microscopy coupled with dispersion staining, or by electron microscopy for verification.

Bursitis   An inflammation of the joints of the body (*see also* **Cumulative Trauma Disorder**).

Byssinosis   A disease of the lungs caused by chronic exposure to cotton dust.

Calibration   Determination of variation from standard, or accuracy, of measuring instruments to ascertain necessary correction factors.

Calorie   The amount of heat required to raise the temperature of one gram of water by one degree Celsius.

Cancer   A malignant neoplasm.

Candlepower   A luminous intensity expressed in candelas; a candle one inch in diameter produces one candela in a horizontal direction.

Canopy Hood   An exhaust hood designed to capture contaminants or heated air rising from an open tank, placed some distance above the tank.

Capillary   A small, thin-walled blood vessel connecting an artery and a vein.

Capture Velocity   The air velocity at any point in front of a hood or at the hood opening necessary to overcome opposing or ambient air currents and to capture air contaminants at that point by causing them to flow into the hood.

Carbon Adsorber   An add-on device that uses activated carbon to adsorb volatile organic compound from a gas stream. These compounds can later be recovered from the carbon for analysis.

Carbon Dioxide (CO$_2$)   A minor component of air representing about 0.4 percent of the atmosphere, released by respiration and removed from the atmosphere by photosynthesis.

Carbon Monoxide (CO)   A colorless, odorless, tasteless gas formed as a product of the incomplete combustion of organic materials. This gas, which has an affinity for red blood cells approximately 220 times that of oxygen, causes a decrease in pulmonary and cardiac function on exposure.

Carboxyhemoglobin   Hemoglobin in which the iron is associated with carbon monoxide (CO).

Carcinogen   A substance known to cause cancer in humans and animals representing a broad range of organic and inorganic chemicals, hormones, immuno-suppressants, and solid-state materials.

Carcinogenic   Describes agents known to induce cancers.

Carcinoma   Malignant neoplasm composed of epithelial cells, regardless of the derivation.

Cardiac   Relating to the heart.

Cardiovascular   Of or pertaining to the heart and the blood vessels, or to their joint circulation characteristics.

Carpal Tunnel Syndrome (CTS)   A cumulative trauma disorder (CTD) often associated with activities involving flexing or extending the wrists, or repeated force on the base of the palm and wrist. The carpal tunnel is an opening in the wrist under the carpal ligament on the palmar side of the carpal bones in the wrist. The median nerve, the finger flexor tendons, and blood vessels all pass through this tunnel. Overuse of the tendons can cause them to become inflamed and swollen, creating pressure against the adjacent median nerve and resulting in CTS. Symptoms include tingling, pain, or numbness in the thumb and first three fingers.

Carpenter's Elbow   A type of cumulative trauma disorder (CTD) associated with repeatedly pushing the palm downward in such a way that a deviation of the ulnar nerve occurs. Symptoms include pain in the elbow, forearm, and hand.

Carrier Gas   Gases such as nitrogen, helium, argon, and hydrogen that are used in gas chromatography or other laboratory procedures to sweep (or "carry") another gas or vapor through a system.

Cascade Impactor   A device used to measure the size range of airborne particles based on the principle that a high velocity airstream striking a flat surface a a 90 angle will cause a sudden change in air direction and momentum. This will also cause the dust in the air to be deposited on a plate and to be separated from the airstream. A series of plates are used to capture different-sized particles, that can then be analyzed for total weight, particle count, and chemical composition.

Catalyst   A substance that increases the weight of a chemical reaction but is not permanently changed by the reaction.

Cataract   A clouding of the crystalline lens of the eye. The clouding obstructs the passage of light.

Caustic   Any substance that strongly irritates, burns, corrodes, or destroys living tissue.

Ceiling Exposure Limit (CEL)   The concentration of a chemical to which workers should not be exposed, even instantaneously, during any part of a working day.

Cell   The fundamental unit of structure and function in organisms.

Cellulitis   An inflammation of tissues that produces pain, edema, swelling, and functional difficulties. It may be caused by streptococcal, staphylococcal, or other organisms.

Centrifugal Collector   A mechanical system using centrifugal force to remove aerosols from a gas stream.

Centrifuge   A laboratory device used to subject substances in solution to centrifugal forces 20,00–25,000 times gravity.

Chain of Custody   In legal terms, regulatory agencies and employers must be able to verify the chain of possession and custody of any physical samples (air, water, soil, biological, etc.,) that may be used to support litigation. Procedures to assure this chain of custody include written records that can be used to trace possession and handling of the sample from its point of origin through analysis and its introduction as evidence. Without a continuous record of chain of custody, the validity of any sample or the results of any tests/analyses may be questioned.

Chain of Infection   A series of related factors or event that must occur before an infection will occur. These factors can be identified as host, agent, source, and transmission factors.

Chain Reaction   Any chemical or nuclear process in which some of the products of the process have an effect on additional particles of the reactants.

Chelate   A chemical compound in which a metallic ion is combined with a molecule with multiple chemical bonds.

Chelating Agent   A type of organic sequestering agent that reduces water hardness and inactivates certain metal ions in water. Sometimes used in detergent formulations to reduce the effects of metals in water.

Chelation   A treatment that removes harmful substances from the body. The chelating agent bonds to the contaminant that, due to the resulting poor absorption, is excreted from the body.

Chemical Asphyxiant   A substance that chemically interferes with the respiratory process. There may be sufficient atmospheric oxygen present, but the body is unable to utilize it because the physiological mechanism for use and transport of oxygen is blocked (such is the case with carbon monoxide).

Chemical Cartridge Respirator   An air-purifying respirator capable of filtering out chemical contaminants from the air that is breathed. It usually acts by the use of chemical sorbant pads encased in cartridges that are attached to the respirator facepiece.

Chemical Compound   A substance composed of two or more elements combined in a fixed and definite proportion by weight.

Chemical Dosimeter   A self-indicating device for determining total (or accumulated) radiation exposure dose based on color change accompanying chemical reactions induced by the radiation.

Chemical Emergency   An occurrence, such as a transportation accident, equipment failure, container rupture, or control equipment failure, that results in an uncontrollable release of a hazardous chemical into the environment or work place.

Chemical Protective Clothing   Clothes made from various materials that exhibit chemical-resistant properties.

Chemistry   The area of science that deals with the elements and atomic structure of matter and the compounds of the elements.

Chigger   The so-called red-bug or larva of the mite family Trombiculidae whose bite produces a welt with itching and dermatitis.

Chinese Restaurant Disease   Often called the Chinese restaurant illness or syndrome, this condition is due to the ingestion of large amounts of food containing monosodium glutamate (MSG), a flavoring additive. Symptoms include headaches, tightness in the face, and light-headedness.

Chloracne   A disfiguring skin condition noted among workers who have had significant contact with certain chemicals such as chlorinated diphenyls, chlorinated dioxins, and chlornaphthalenes.

Chlorofluorocarbons (CFCs)   Compounds composed of carbon, fluorine, chlorine, and hydrogen, such as trichlorofluoromethane. They are believed to have a deleterious effect on the stratospheric ozone layer.

Cholinesterase   An enzyme that hydrolyzes acetylcholine within the central nervous system. This enzyme can be depressed following exposure to organophosphate pesticide compounds. Workers using organophosphate pesticides should be routinely monitored for cholinesterase levels.

Chromatography   A practical analytical methodology involving the separation of complex mixtures and the detection of each component of the mixture.

Chromosome   One of the thread-like bodies (normally 46 in humans) of chromatin that are found in the nucleus and that are the bearers of genes.

Chronic Carrier   A person who continues to harbor an infectious agent without showing symptoms of the disease. Chronic carriers are possible in many illnesses. Salmonellosis is an example.

Chronic Exposure   (1) *Chemical.* Continual exposure to low levels of a chemical over a long period of time (usually three years or more), that can produce symptoms and disease. (2) *Radiation.* Exposure to radiation for long duration by fractionation or protraction. Generally any dosage absorbed over a 24-hour period or longer.

Chronic Toxicity   The capacity of a substance to cause long-term poisonous human health effects.

Chyme   The semi-liquid mass into which food is converted by the action of gastric secretions during the digestive process.

Cilia   Short, tiny, hairlike processes on the surface of protozoan or certain metazoan cells that, by their constant motion, accomplish locomotion or produce a water current. Examples are found in the bronchi and respiratory tract where they aid in the removal of dusts.

Circadian Rhythm   A biological activity that recurs in periods of 24 hours under natural environmental conditions. Sleep patterns in mammals and leaf movements in some plants are examples of circadian rhythms.

Circumaural Protector   A form of hearing protector commonly known as an earmuff, consisting of two cup-shaped devices that fit over the entire external ear and are sealed against the side of the head.

Clinical Laboratory   A workplace where diagnostic or other screening procedures are performed on blood or other potentially infectious materials.

Cloud Chamber   A device for observing the paths of ionizing particles, based on the principle that super-saturated vapor condenses more readily on ions than on neutral molecules,

Coal Tar   A black viscous liquid with a naphthalene-like odor that is obtained by the destructive distillation of bituminous coal and used as a raw material for dyes, solvents, and many other products. Coal tar is known to contain many carcinogens and, thus, its use has been extremely curtailed.

Cocarcinogen   A substance that works symbiotically with a carcinogen in the development of cancer.

Coccus   Spherical bacteria cells. Cocci (plural) may appear singly, in pairs called diplococci, in chains, or in grape-like clusters.

Collier's Disease   *See* **Anthracosis**; Black lung disease, Shaver's disease.

Collimation   The confining of a beam of particles or rays to a defined cross section.

Colorimetric Tube   *See* **Detector Tube.**

Compound   Chemical combination of two or more elements in a fixed and definite proportion by weight.

Compton Effect   The glancing collision of a gamma-ray with an electron wherein the gamma-ray gives up part of its energy to the electron.

Conduction   The transfer of heat by direct contact from one body to another.

Conductive Hearing Loss   A physical defect or condition of the outer or middle ear that interferes with the passage of sound. This can be the end result of physical obstruction within the outer or middle ear, a birth defect, the aging process, or disease, all of which can affect conversion of sound energy into mechanical energy. This type of hearing loss involves a reduction in the perception of loudness and not in clarity.

Confined Space   Any space not designed or intended for continuous occupancy that has a limited or restricted means of entry or exit, and that is subject to the accumulation of toxic or flammable contaminants or has an oxygen-deficient atmosphere. Confined spaces must be large enough for an employee to enter and perform assigned work. Where confined spaces are categorized, there are two levels of classification: permit-required confined space and low-hazard permit space.

Congenital   Refers to certain mental or physical traits, abnormalities, malformations, or diseases that may be either inherited or due to an influence that occurred between conception and birth.

Conjunctivitis   The inflammation of the conjunctiva (the mucous membrane lining of the inner surface of the eyelids and covering the front pat of the eyeball).

Contact Dermatitis   A delayed type of induced sensitivity of the skin resulting from cutaneous contact with a specific allergen.

Contact Irritants   Chemicals that produce visible signs of skin and eye irritation on contact. Rubber, plastics, resins, glues, cement, oil, and organic solvents are examples of contact irritants.

Contaminant   Any foreign material not normally found in a substance. Also, any physical, chemical, biological, or radiological substance or matter that has a diverse effect on air, soil, or water.

Contaminated Sharps   Any contaminated objects that can penetrate the skin including, but not limited to, needles, scalpels, broken glass, broken capillary tubes, and exposed ends of dental wire. In this context, the "contaminant" is normally considered to be blood, blood by-products, or other infectious materials.

Continuous Monitoring   Usually refers to air sampling or radiation monitoring conducted at locations where leaks may occur, or where hazardous materials are handled in high quantities.

Controlled Area (radiation)   A defined area in which the occupational exposure of personnel to radiation or radioactive material is under the supervision of an individual responsible for radiation protection.

Control Rod   A rod used to control the nuclear power of a nuclear reactor. The reactor functions through the fission of nuclear fuel by neutrons. The control rod absorbs neutrons that would normally produce fission in the atoms of the fuel. Pushing the rod into the reactor reduces the release of nuclear power and pulling the rod out increases the rate.

Convection   The transfer of heat from one place to another by moving fluid (a gas or a liquid). Natural convection results from differences in temperature.

Convection Heat Load   The amount of heat energy transferred between the skin and the air. Human skin is normally 95° F (35° C). Air in excess of that temperature will warm the body, whereas air below that temperature will cool the body.

Core   The heart of the nuclear reactor where the nuclei of the fuel undergo fission (spilt) and release energy. The core is usually surrounded by a reflecting material that bounces stray neutrons back to the fuel.

Corpuscle   A blood cell.

Corrosion   Physical damage, usually in the form of deterioration or destruction caused by the chemical or electrochemical action as contrasted with erosion caused by mechanical reaction.

Corrosive   Refers to a chemical agent that reacts with the surface of a material causing it to deteriorate or wear.

Corrosivity   The ability of a substance to produce corrosion.

Cosmic Radiation   Penetrating ionizing radiation, both particulate and electromagnetic, originating in outer space. Secondary cosmic rays, formed by interactions in the earth's atmosphere add to the general background radiation.

Count   The external indication of a device designed to enumerate ionizing events. It may refer to a single detected event or the total measured in a given time period.

Counter   A device for counting nuclear disintegrations used to measure radioactivity.

Criterion Sound Level   A sound level of 90 decibels (OSHA).

Critical Organ   The body organ receiving the radionuclide that results in the greatest overall damage to the body. Usually, but nor necessarily, it is the organ receiving the greatest concentration.

Cryogenic Gas   A liquefied gas that exists in its containers at temperatures far below normal atmospheric temperatures.

Cryogenic Liquid   A refrigerated liquid gas with a boiling point below −130° F (−90° C).

Cumulative Dose (radiation)   The total dose resulting from repeated exposures to radiation of the same region or of the whole body.

Cumulative Trauma Disorder   A collective term used to describe syndromes characterized by discomfort, impairment, disability, or persistent pain in the joints, muscles, tendons, and other soft tissues, with or without physical manifestations. It is often caused, precipitated, or aggravated by repetitive or forced motions that may occur in many differed occupational activities.

Curie (radiation)   A unit of measure formally defined as a quantity of any radioactive nuclide producing 3.7×10^{10} disintegrations per second. Now the curie (Ci) is officially a unit of activity rather than a quantity (i.e., a unit of radioactivity that is a measure of the rate at which a radioactive material emits particles). The new definition is 1 Ci $= 3.7 \times 10^{10}$ disintegrations $\times$ s^{-1}. The higher the rate of disintegration, the greater the hazard.

Cutaneous   Of or relating to the skin.

Cyanosis   Blueness or lividness of the skin, especially in the face and extremities, typically caused by imperfect oxygenation of the blood.

Cytology   Study of the function and structure of living cells.

Cytolosis   Disruption of cells, resulting in the destruction and breakdown of the cell membrane.

Cytotoxic   An agent that brings about destructive action on certain cells.

Damper   A device used to regulate airflow in ducts, often used to balance airflow in branch ducts.

Dander   Scales, dust, and dirt from the fur or feathers of animals that may cause allergic reactions in susceptible persons.

Daughter Products   Isotopes formed by the radioactive decay of some other isotope.

dBA   Refers to decibels measure on the A scale which is a frequency weighting network that approximates the response of the human ear.

Dead Time   In radiation, the time during which a Geiger-Mueller detector is insensitive to incoming radiation.

Decay Constant   The fraction of the number of radioisotope atoms that decay in a unit of time. The decay constant is 0.693/T, where T is the half-life.

Decay Product   A nuclide resulting from the radioactive disintegration of radionuclide or series of radionuclides. A decay product may be stable or radioactive.

Decay, Radioactive   The decrease in activity of any radioactive material with the passage of time, due to the spontaneous emission from the atomic nuclei of either alpha or beta particles, sometimes accompanied by gamma radiation.

Decibel   A means for expressing the logarithmic level of sound intensity, sound power, or sound pressure above an arbitrary reference value of 20 micropascals in air (*see* **dBA**).

Decomposition   The breakdown of dead organic material into smaller or simpler parts that are then recirculated. Bacteria, fungi, heterotrophic protists, and saprophagous insects are important in the process of decomposition.

Decompression Sickness   Illness or injury associated with exposure to high-pressure atmospheres followed by rapid exposure to normal pressure. Also known as the bends and caisson disease.

Decontaminate   To render safe or harmless by the removal or elimination of poisonous, noxious, or otherwise harmful agents.

Degradation   The process by which a chemical is reduced to a less complex form.

Degreaser   A chemical agent, usually a solvent, that is used to remove grease and oil from machinery. Because these chemicals will also remove the protective layer of oil on human skin, their use without protection can result in dermatitis.

Dehumidifier   A device for lowering the moisture content of air.

Delead   To remove lead from the tissues by use of a chelating agent that is then excreted in the urine. Also, the term applies to the removal of lead-based paint in dwelling units.

Deleterious   Refers to an agent (physical, chemical, or microbial) that is injurious or capable of causing harm.

Dendron   One of the branching processes of a nerve cell or neuron that conveys impulse.

Dengue   A viral disease carried by the Aedes mosquito. Also known a break-bone fever because of the intense joint pain associated with it.

Density   The ratio of the mass of a material to its volume expressed as g/cm^3, lb/ft^3, etc.

Deoxidation   The process by which oxygen is removed from a chemical compound.

Deoxyribonucleic Acid (DNA)   The type of nucleic acid that contains deoxyribose sugar and is found mainly in the chromosomes of animal and vegetable cells. DNA is considered to be the repository of hereditary characteristics and the auto-reproducing constituent of chromosomes and many viruses.

Depilatory   Chemical having the ability to remove or destroy hair.

Depolymerization   The breakdown of an organic compound into two or more less complex molecules.

DeQuervain's Disease   A type of tenosynovitis of the exterior tendons of the wrist or abductors of the thumb. The inflammation of the synovial lining is often pronounced under conditions of highly repetitive hand usage or is due to poor design of the workplace or tools (*see* **Cumulative Trauma Disorder**).

Dermal   Of or pertaining to the skin.

Dermatitis   Any inflammation of the skin surface from any cause. A skin abnormality resulting from an occupational exposure may also be referred to as industrial dermatitis, occupational contact dermatitis, professional eczema, cement dermatitis, chrome ulcers, oil acne, rubber itch, or tar warts.

Dermatophytoses   A group of diseases caused by fungi often found among farmers, animals handlers, pet and hide handlers, wool sorters, cattle ranchers, athletes. lifeguards, gymnasium employees, and animal laboratory workers. Ringworm of the hands and feet is the most common form and is usually prevented by recognition of the disease in animals, sterilization and proper laundering of towels, general cleanliness of showering facilities, and proper personal hygiene.

Dermatosis   Generic term for skin disorders, particularly those not involving inflammation.

Desiccant   Chemicals, such as silica gel, that absorb moisture and are typically used to promote or assure dryness.

Detector Tube   An air sampling device used to measure the concentration of various air contaminants. Consists of a glass tube filled with a solid chemical that changes color when it reacts with the air contaminant being sampled in combination with a hand-held pump device to draw air through the tube at a measured rate. Also known as colorimetric tube.

Deuteron   The nucleus of a deuterium atom.

Dew Point   The temperature at which air at a constant pressure and constant water-vapor content will be saturated. Cooling below the dew point usually results in frost or dew (when this temperature is below 0C, it is sometimes called the frost point.

Diabetes Mellitus   A disease characterized by excessive blood sugar levels due to a lack of the insulin hormone (*see* **Insulin**).

Diagnose   To isolate or recognize a disease or the cause of an illness.

Dialysis   The passage of solute molecules through a semipermeable membrane from higher concentration to a solution of lower concentration.

Diaphragm   In anatomy, the musculomembranous partition separating the abdominal and thoracic cavities that plays an important role in the respiration process.

Diarrhea   Fecal discharge of an abnormal frequency and liquidity. A condition often associated with food-borne illnesses and one of the most common symptoms of the gastroenteritis syndrome.

Differential Absorption Ratio   The ratio of concentration of an isotope in a given organ or tissue to the concentration that would be obtained if the same administered quantity of this isotope were uniformly distributed throughout the body.

Diffraction   The bending or breaking of a ray of light into its individual parts.

Dilution Ventilation System   A system of air-space ventilation that relies on the mixing of contaminated air with uncontaminated air for the purpose of controlling potential air borne health hazards.

Diphtheria   An acute, contagious disease of children that normally affects the membranes of the throat. The causative agent Corynebacterium diphtheriae is spread from the nose, mouth, and throat of infected persons.

Diplopia   An eyesight defect in which an object appears double.

Direct-Reading Instrument   An apparatus providing a direct readout of the contaminant level without further off-site laboratory analysis.

Disability   An impairment or defect of one or more organs or body members. For worker's compensation purposes, the following categories are generally used to determine the level of benefit to be awarded: (1) Permanent partial—A permanent physical impairment (loss of an eye, hand, etc.) that restricts the ability of the worker to perform certain jobs. Benefits are normally based on the percentage of disability incurred. (2) Permanent total—A disability that is so extensive it prevent the worker from obtaining or competing for a job. (3) Temporary partial—A condition that leaves the employee capable of performing some work and will probably improve to pre-injury or illness status over time and with treatment. (4) Temporary total—A disability that renders the worker incapable of working, but from which he or she is expected to recover fully.

Disease   A deviation from normal health status associated with a characteristic sequence of signs and symptoms and caused by a specific etiologic agent.

Disequilibrium   A loss of balance accompanied by swaying of the body and tremors sometimes experienced by workers exposed to whole body vibrations above 2 Hz.

Disinfectant   Chemicals used to reduce or kill microorganisms present on inanimate objects or surfaces.

Disintegration   In radiation, the process of spontaneous breakdown of a nucleus of an atom resulting in the emission of a particle and/or photon.

Distribution   The movement of a chemical substance or foreign material from entry site and throughout the body.

Dose   (1) *Radiation.* A quantity (total or accumulated) of ionizing (or nuclear) radiation. Exposure dose, expressed in roentgens, is a measure of the total amount of ionization that a quantity of radiation could produce in air. Absorbed dose, expressed in reps or rads, represents the energy absorbed from the radiation per gram of body tissue. Biological dose, expressed in Rems, is a measure of the biological effectiveness of the radiation exposure. (2) *Toxicology.* The total amount of a toxicant, drug, or other chemical administered to the organism.

Dose Rate   In general, the amount of ionizing radiation to which an individual would be exposed to that he or she would receive per unit of time.

Dosimeter   An instrument or device used to determine the amount of exposure to an agent or toxic chemical, usually over a period of time. Typically used to measure exposures to noise, radiation, and chemicals.

Dry-Bulb Temperature (DBT)   The temperature derived from a thermal sensor or a thermometer that is shielded from direct radiant energy. It is used for estimating comfort conditions and is also one of three ambient indices used for heat-stress analysis.

Dust   Airborne or settled particles usually formed by abrasion or arising from soil, bedding, or from surfaces such as floors and walls.

Dynamic Measurement   An aspect of anthropometry involving the correct location of controls, tools, and other items requiring worker manipulation.

Dyne   The unit of force that, when acting on a mass of 1 gram, will produce an acceleration of 1 centimeter per second per second.

Dysphagia   Difficulty in swallowing.

Dyspnea   Shortness of breath, or the sensation of it, due to labored breathing.

Eardrum   The tympanic membrane that separates the outer ear from the middle ear.

Earmuffs   Devices worn to protect against hearing loss in high-noise environments or to protect against exposure to cold (*see* **Circumaural Protectors**).

Earplugs   *See* **Aural Insert Protectors**).

Ear Protection   Any device worn by an individual to guard against hearing loss in high-noise environments. Some devices may also be used to protect against cold exposures and to prevent the entry of water into the ears.

Ecology   The relationship living things to one another and their environment, or the study of such relationships.

Eczema   Generalized term for an inflammatory process involving the epidermis and marked by itching, weeping, and crusting.

Edema   A condition in which body tissues contain an excessive amount of fluid.

Effective Dose (ED)   The amount of a toxicant (or drug) required to bring about a given functional change in an intact organism, at a biochemical site, or in an isolated tissue. Expressed in a proportion to the population affected (ED50, for example).

Electrochemistry   The science that deals with the use of electrical energy to bring about a chemical reaction or with the generation of electrical energy by means of chemical action.

Electrolyte   A chemical substance that breaks down into electrically charged particles (ions) when dissolved or melted.

Electromagnetic Radiation   A traveling wave motion resulting from changing electric or magnetic fields. The length of these waves can be relatively short (x-rays and gamma rays) or relatively long (ultra-violet, visible, and infra-red, through to radar and radio waves). All electromagnetic radiations travel with the speed of light in a vacuum. Generally speaking, the shorter the wave length, the more penetrating the radiation.

Electron   A negatively charged particle that is a fundamental constituent of all atoms. A unit of negatively charged electricity found in orbitals around the nucleus of the atom.

Electron Capture   A mode of radioactive decay in which an orbital electron merges with a proton in the nucleus. The process is followed by emission of an electron or photon.

Electron Volt   A unit of energy equivalent to that gained by an electron in passing through a potential difference of 1 volt. Often expressed in large units such as keV (thousand electron volts), MeV (million electron volts), BeV (billion electron volts.

Element   A pure substance that can not be broken down into a simpler substance by chemical change but whose atoms will disintegrate in simpler particles through physical decomposition when exposed to drastic bombardment with high-energy particles.

Elimination   The removal of a chemical substance from the body by metabolism or excretion. Also, as used in this text, the removal of health or physical hazard risk through control, substitution, or some other means.

Elutriator   An air-sampling device that uses gravitational force to remove non-respirable dust from the air sample.

Embryo   An organism in an early stage of development.

Emetic   An agent that induces or causes vomiting.

Emulsifier   A surface-active agent that promotes the dispersion of one liquid in another, such as small fat globules in water.

Endemic   Refers to diseases or infectious agents in the human population within a given geographic area that are constantly present or usually prevalent.

Endospore   A thick-walled structure formed within the cells of certain bacteria that allows the organism to withstand adverse environmental conditions, such as drying.

Endothermic Refers to a reaction in which the products contain more energy than the reacting materials, causing the absorption of energy as heat.

Engineering Controls Measures taken to prevent or minimize hazard exposure through the application of controls such as improved ventilation, noise reduction techniques, chemical substitution, equipment and facility modifications, etc.

Enteric Pertaining to the intestines.

Enteritis An inflammation of some portion of the intestines.

Entrant A person who has been authorized by their employer to enter a permit-required confined space.

Environmental Health The body of knowledge concerned with the prevention of disease through the control of biological, chemical, or physical agents in air, water, and food. Also concerned with the control of environmental factors that may have an impact on the well-being of people.

Enzyme An organic compound, frequently a protein, that accelerates (catalyzes) specific transformations of material, as in the digestion of foods.

Epidemic The occurrence of cases that are of similar nature in human populations in a particular geographic area and that are clearly in excess of the usual incidence.

Epidemiologist A person who applies epidemiological principles and methods to the prevention and control of diseases.

Epidemiology The study of the distribution and determinants of disease causation in human populations.

Epidermis The outer, non-vascular, non-sensitive layer of the skin that covers the true skin.

Epithelium Refers to cells that line all canals and surfaces that have contact with external air, and also cells that are specialized for secretion in certain organs, such as the liver and kidneys.

Erg A unit of work equal to the force of one dyne acting through a distance of one centimeter.

Ergonomics A multi-disciplinary activity that concentrates on the interactions between the human and their total working environment with consideration for the stressers that may be present in that environment, such as atmospheric heat, illumination, and sound, as well as all the tools and equipment used in the workplace. Also referred to as human factors and human factors engineering.

Erysipelas An inflammation of the skin marked by red patches with sharp border lines.

Erythema A abnormal redness of the skin, due to distention of the capillaries with the blood. It can be caused by a various agents such as heat, certain drugs, ultra-violet rays, and ionizing radiation.   .

Erythrocyte A red blood cell that contains hemoglobin and transports oxygen to body tissues.

Etiologic Agents Infectious microorganisms, viruses, or parasitic agents capable of producing infection and/or disease in a susceptible host.

Etiology The study or theory of the causation of disease; the sum of knowledge regarding disease causes.

Evaporation Rate The rate at which a material will vaporize (evaporate) as compared to the known rate of a standard material (such as normal-butyl acetate).

Excretion The removal of a substance or its metabolites from the body in urine, feces, or expired air.

Exothermic When applied to reactions, describes those that produce substances that have less energy than the reaction materials, resulting in a release of energy as heat.

Exotoxin A microbial toxin.

Experience Rating A method of adjusting worker's compensation rates using a three-year history of the employer's claim experience.

Expert Witness A witness qualified as a subject expert based on their knowledge, skill, experience, training, or education. Unlike other witnesses, an expert's testimony may be in the form of an opinion.

Expiration Exhaling of the lungs caused by the relaxation of the diaphragm and rib muscles that causes decreased chest cavity space, thus forcing air out through the trachea.

Expiratory Reserve Volume (ERV) The maximum amount of air that can be forcibly expired after a normal expiration.

Exposure (1) The amount of radiation or pollutant that represents a potential health risk. (2) The opportunity of a susceptible host to acquire an infection by either direct or indirect mode of transmission. (3) The amount of biological, physical, or chemical agent that reaches a target population. (4) The route by which an organism comes in contact with a toxicant (inhalation, ingestion, dermal absorption, injection).

Exposure Dose A measure of the x-radiation or gamma radiation at a certain place, based on the ability of the radiation to produce ionization. The unit of measure is the roentgen (R).

Exposure Dose Rate   The radiation exposure dose per unit time expressed as R/unit time.

External Radiation   Ionizing radiation in which the source is located outside the body and the radiation penetrates into deeper tissues.

Face Velocity   The average air velocity in the plane of an opening into an enclosure through which air moves, usually expressed in feet per minute or meters per second.

Faculative Anaerobe   Microorganisms that can multiply either in the presence or in the absence of oxygen. They can obtain energy either by respiration or by fermentation and do not require oxygen for biosynthesis.

Farmer's Lung Disease   A syndrome that consists initially of chills and fever, followed by impairment of lung function. It is normally caused by chronic exposure to moldy hay or other moldy organic material.

Federal Register   The official daily publication of the United States government that provide a uniform system for publishing Presidential and federal agency documents.

Fermentation   The breakdown of organic substance by microorganisms with a resulting release of energy.

Ferruginous Bodies   Bodies formed by fibers that have entered the lungs. These bodies can be formed by any kind of durable fiber including asbestos, fiberglass, and vegetable fibers of siliceous origin.

Fiberglass   A commercial, nonflammable fiber that is made from spun glass primarily used for insulation. Fibers of this material can penetrate the skin causing dermatitis in some people and, when airborne, can affect the lungs of some people. Fiberglass is resistant to most chemicals and solvent.

Film Badge   A pack of photographic film used for approximate measurement of radiation exposure for personnel monitoring purposes. Also called a film dosimeter.

Fission   The process whereby the nucleus of a particular heavy element splits into (generally) two nuclei of lighter elements with the release of substantial amounts of energy.

Fixative   A chemical, such as alcohol or formaldehyde, used for the preservation of biological materials.

Flanged Hood   A barrier placed around the periphery of a chemical hood to reduce air turbulence and hood entry pressure loss keeping the hood from drawing air from behind the hood face.

Flash Burn   An inflammation of the lens of the eye due to excessive exposure to ultra-violet radiation, usually from a welding arc.

Flammable Liquid   Any liquid having a flash point below 100° F (37.8° C).

Flammability Range   *See* **Upper Flammability Limit (UFL).**

Flash Point   The minimum temperature at which a flammable liquid gives off sufficient vapors in the air immediately above the surface of that liquid to ignite if an ignition source is introduced into that area. The lower the flash point, the more flammable the liquid.

Fluorescence   Phenomenon involving the absorption of radiant energy by a substance (usually a crystal) and its re-emission as visible or near-visible light.

Fluorescent Screen   A sheet of material coated with a substance (usually calcium tungstate or zinc sulfide) that will emit light when irradiated with ionizing radiation.

Fluoroscope   A screen mounted in front of an x-ray tube used for indirect visualization of internal body organs or internal structures of inanimate objects.

Flux   The number of visible-light photons, gamma-ray photons, neutrons, particles, or energy crossing a unit surface area per unit time. The units of flux are the number or particles (or energy, etc.) per square centimeter per second.

Fog   A term loosely applied to visible aerosols, less than 40 microns in diameter, that are liquid; formation by condensation is sometimes implied.

Foodborne Disease   Any disease that is transmitted through food contaminated with bacteria, viruses, fungi, parasites, or even some toxic chemicals.

Food Poisoning   A broad term including foodborne illnesses caused by the ingestion of foods containing microbial toxins or chemical poisons.

Foot Candle   The illumination resulting from the uniform distribution of a flux of one lumen (lm) on a surface area of one square foot. Hence, one foot candle equals on lumen per square foot.

Foot Lambert   The unit of photometric brightness equal to the uniform brightness of a perfectly diffusing surface emitting or reflecting one lumen per square foot.

Force   The push or pull that tends to impart motion to a body at rest, or to increase or diminish speed, or to change the direction of a body already in motion.

Fractionation   Any of several processes, apart from radioactive decay, that results in change in the composition of radioactive debris.

Frequency   The number of cycles, revolutions, or vibrations completed per unit of time. In sound, for example, the frequency describes the rate at which complete cycles of high- and low-pressure peaks are produced. The unit of measurement is cycles per second, or hertz (Hz). The normal human ear has a frequency range of 20 to 20,000 Hz at moderate sound pressure levels.

Friable   Refers to materials that have a tendency to crumble easily. Most often used to describe the condition that exists when asbestos fibers can potentially be released and become airborne presenting a respiratory hazard.

Frostbite   The destruction of tissue resulting from exposure to extreme cold or contact with extremely cold objects.

Frost Point   *See* **Dew Point**.

Fume   Small solid particles generated following the volatilization of a metal or plastic when their gaseous state condenses quickly on contact with cooler air. Welding, for example, causes the volatilization of metals into a gas, followed by condensation on contact with cooler air. This creates welding fumes typically on the order of 0.1–1 micrometer in diameter. In popular usage, the term fume is often incorrectly used to describe any type of air contaminant.

Fungus   A general term used to describe the diverse morphological forms of yeast, rust, mildew, and mold. Fungi (plural) are heterotrophs and obtain nourishment by absorption usually from dead or decaying organic matter. Some fungi are beneficial in foods and pharmaceutical development while other can cause pulmonary diseases. Fungi are found in soil, water, and air.

Fusion   A nuclear reaction characterized by the joining together of light nuclei to form heavier nuclei.

Gamma Ray   Electromagnetic radiations of high energy originating in atomic nuclei and accompanying many nuclear reactions (fission, radioactivity, and neutron capture). Physically, gamma rays are identical with x-rays of high energy. However, x-rays do not originate from atomic nuclei (*see* **X-ray**).

Ganglion   A mass of human or animal tissue containing nerve cells (neurons).

Gangrene   An infection caused by an anaerobic bacteria resulting in the destruction of body tissue.

Gas   A thin fluid, like air, capable of indefinite expansion but convertible by compression and cooling into a liquid and eventually a solid. Gases may be either elements (such as argon) or compounds (such as carbon dioxide).

Gas Amplification   As applied to gas-ionization radiation-detection instruments, the ration of the charge collected to the charge produced by the initial ionizing event.

Gas Chromatography   An analytical procedure involving a physical process for separating the components of complex mixtures.

Gas-Forming Bacteria   Organisms that ferment lactose in foods or other carbohydrates producing both acid and gas, that may render a food product unacceptable.

Gas Masks   A full-face respirator equipped with an air-purifying cartridge or canister that removes contaminants and renders air breathable to the user. (Not for use in oxygen deficient atmospheres).

Gas Pressure   The force, generally designated in pounds per square inch (psi), that is exerted by a gas on its surroundings.

Gastric   Pertaining to the stomach.

Gastroenteritis   Inflammation of the mucous membrane of the stomach and intestines that may be caused by various bacteria or viruses. Symptoms include diarrhea, abdominal cramps, nausea, vomiting, fever, malaise, muscle ache, and fatigue.

Geiger Counter   An electrical device that detects the presence of certain types of radioactivity. It consists of a needle-like electrode inside a hollow metallic cylinder filled with gas that, when ionized, sets up a current in an electrical field.

Geiger-Mueller Counter   A refined version of the Geiger counter that has an amplifying system and is used for detecting and measuring radioactivity.

Gene   A functional unit of heredity that occupies a specific location on a chromosome, capable of producing itself exactly at each cell division, and of directing the formation of an enzyme or other protein.

General Duty Clause   That part of the Occupational Safety and Health Act of 1970 that requires employers to provide each employee a place of employment free of recognized hazards that are likely to cause death or serious physical harm. It is often cited by OSHA to cover hazards for which a specific Standard does not yet exist. It is located at Section 5(a)(1) of the Act.

Genetic Engineering   A process of inserting new genetic information into existing cells in order to modify one of the characteristics of an organism.

Genome   A total set of chromosomes derived from one parent.

Genotoxic   Refers to the ability of a chemical to adversely affect the genome of living cells, such that on duplication, a mutagenic or carcinogenic event is expressed due to the alteration of the genome molecular structure.

Genotoxin   Chemical or radioactive substance known to cause or suspected of causing damage to the DNA in individual cells, thus causing mutations or cancer.

Germicide   Any substance that kills microbes, or an agent that destroys pathogenic microorganisms.

Gingivitis   Inflammation of the gums of the mouth.

Glare   The sensation produced by luminance within the visual field that is sufficiently greater than the luminance to which the eyes are adapted. This causes annoyance, discomfort, or loss of visual performance and acuity, a concern especially for individuals using video display terminals for extended periods.

Glass Blower's Cataract   An opacity of the rear surface of the lens in the eye caused by excessive exposure of the eyes to luminous radiation, primarily visible and infrared. Found in those occupationally exposed to furnaces or other hot devices for extended periods of time.

Globe Thermometer   A dry-bulb thermometer suspended in the center of a sphere that has been painted flat black and is used to measure radiant heat.

Globulin   Any of a group of proteins found in animal and vegetable tissues that can be precipitated from serum or plasma.

Goggle   A tight-fitting device worn over the eyes to provide splash and/or impact protection.

Goose Pimples   Also known as "goose flesh," a skin conditions marked by numerous small elevations around the hair follicles caused by the action of the arrectores pilorium ("raisers of hair") muscles.

Grab Sample   In industrial hygiene application, a type of air sample in which the air is admitted into a bag, vessel, or instrument instantaneously for subsequent analysis.

Gram   The basic unit of mass in the metric system. One gram is equal to 15.432 grains.

Gram-Atomic Weight   A mass in grams numerically equal to the atomic weight of the element.

Gram-Molecular Weight   Mass in grams numerically equal to the molecular weight of a substance.

Greenhouse Effect   A warming of the earth's atmosphere caused by a buildup of carbon dioxide or other trace gases. It is believe that this condition allows light from the sun's rays to heat the earth but prevent a counterbalancing loss of heat.

Granz Rays   X-rays produced at voltages of 5 to 20 kilovoltage peak (KVP).

Grit   Course nuisance dust particles that are larger than 75 microns in diameter.

Gross Alpha Particle Activity   Total activity, commonly measured in picocuries, due to emission of alpha particles. Generally used as a screening measurement for naturally occurring radionuclides.

Gross Beta Particle Activity   Total activity, commonly measured in picocuries, due to emission of beta particles. Used as a screening measurement for human-made radionuclides.

Guard   A device or shield placed over or in association with a piece of machinery, such as a drill, grinder, or saw, to protect the operator from physical harm.

Half-Life, Biological   The time required for the body to eliminate, by natural biological means, half of the material taken into it.

Half-Life, Effective   The time in which the quantity of a radioactive isotope in the body will decrease to half as a result of both radioactive decay and biological elimination.

Half-Life, Radioactive   The time for the activity of a given radioactive isotope to decrease to half of its initial value, due to radioactive decay. The half-life is a characteristic property of each radioactive isotope and is independent of its amount or condition.

Halide Meter   An instrument used for the direct measurement of halogenated hydrocarbons.

Halogen   Collective term for any of the five active non-metallic chemical elements fluorine, chlorine, bromine, iodine, and astatine.

Halogenation   The process whereby halogens are used for disinfecting purposes.

Halons   Bromine-containing compounds, normally used in firefighting methodologies, with long atmospheric lifetimes whose breakdown in the stratosphere is thought to cause ozone depletion.

Hazard   As used in this text, a risky, perilous, or dangerous condition or situation that could result in the exposure of individuals to unnecessary physical or health risks. Hazards can be biological, chemical, physical, mechanical, human-made, or naturally occurring.

Hazardous Chemical   According to OSHA (29 CFR 1910.1200), any chemical that is a health or physical hazard and for which there is statistically significant evidence that acute or chronic health effects may occur in exposed individuals.

Health Physics   The branch of radiological science dealing with the protection of personnel from harmful effects of ionizing radiation.

Hearing Conservation   The prevention or minimizing of occupational noise-induced hearing defects through the combined use of hearing protectors, training, the use of engineering and administrative control measures, annual audiometric testing, and the establishment of a written program.

Hearing Impairment   Loss of the ability to hear, either partially or completely.

Heat   The energy associated with a mass because of random motions of its molecules.

Heat Cramps   A condition related to work and/or exercise in hot environments that causes painful muscle spasms due to heavy swelling and the consumption of large amounts of water without adequate salt intake and adequate exercise/rest balance.

Heat Disorder   Any condition resulting from exposure to heat or hot work environments that results in an adverse effect on the health of the exposed individual. Such disorders include heat cramps, heat exhaustion, heat stress, and heat stroke. Also may be referred to as heat stress.

Heat Exchanger   Device for transferring heat from one fluid or body to another for the purpose of heating or cooling.

Heat Exhaustion   A potentially dangerous condition caused by work and exertion in high-temperature environments marked by mild elevation in body temperature, weak pulse, pale complexion, dizziness, fainting, profuse sweating, headache, low blood pressure, and cool, moist skin.

Heat of Fusion   The heat released by a liquid freezing to a solid, or that gained by a solid melting to a liquid without a change in temperature.

Heat Stress   Thermal stress on the body from the surrounding environment, including heat stroke, heat cramps, and heat exhaustion, caused by the body's inability to rid itself of excessive heat (*see* **Heat Disorder**).

Heat Stroke   A serious, potentially life threatening condition marked by a rapid rise in body temperature, hot dry skin, mental confusion, loss of consciousness, convulsions, coma, and the absence of sweating. The condition is caused by excessive physical exertion in hot environments by acclimatized individuals and dehydration. Recent intake of alcohol may expedite the onset of the condition.

Heat Syncope   A condition marked by fainting while standing erect and immobile in hot environments caused by the pooling of blood in dilated vessels of the skin and lower part of the body.

Heavy Metal   Metals such as arsenic, barium, cadmium, chromium, lead, mercury, and silver that do not rapidly break down in the body or the environment and thus can exert toxic effects because of their cumulative or residual properties.

Heme   The non-protein, iron-containing part of the hemoglobin molecule that carries oxygen and accounts for the color of blood.

Hemoglobin   The red pigment protein matter in the red blood corpuscles that carries oxygen from the lungs to the tissues and carbon dioxide from the tissue to the lungs.

Hemotoxin   Any substance that causes destruction of red blood cells.

Hepatitis B Virus (HBV)   A virus that causes inflammation of the liver. Can also occasionally be caused by toxic agents other than viral.

Hepatotoxic   Refers to an agent that produces damage to the liver.

Hertz   *See* **Frequency**.

Heterotroph   An organism that must obtain food energy by ingesting other organic material.

High-Efficiency Particulate Air (HEPA) Filter   A filter capable of removing 99.97 percent of all particles with a mean aerodynamic diameter of 0.3 micron. Often used to filter air in air-purifying respirators, vacuum systems, exhaust systems, and fans.

Histolysis   Process whereby tissue is broken down.

Homogeneous Radiation   A beam or flux consisting of radiation of the same kind and energy.

Homogeneous Reactor   A nuclear reactor is which the fissionable material and the moderator (if used) are combined in a mixture such that an effectively homogeneous medium is presented to the neutron.

Hood   A shaped inlet designed to capture contaminated air and conduct it into an exhaust duct and/or exhaust fan.

Hookworm   A parasitic worm that infests people and causes debilitation. Major infestations can cause anemia and retardation of mental and physical development. Adult hookworms feed on blood an tissue from the wall of the intestine. Eggs pass out in the feces, undergo a period of development in soil, and the larvae enter a new host by burrowing through the skin.

Hormone   A chemical substance found in one organ or part of the body and carried in the blood to another part. Hormones can alter the function and sometimes the structure of one or more organs.

Host   An organism, simple or complex and including humans, that is capable of being infected by a specific agent.

Hot-Wire Anemometer   A device, also known as a thermal anemometer, used to measure air velocity by the cooling effect or moving air over a heated element.

Human Immune Deficiency Virus (HIV)   Also called the human immunodeficiency virus, HIV is the virus that causes acquired immune deficiency syndrome (AIDS).

Humidity   The amount of moistness or dampness in the air (*see* **Relative Humidity**).

Hydrargyria   Chronic mercury poisoning.

Hydration   The process of absorbing or combining with water; the chemical addition of water to a compound.

Hydrocarbon   A compound composed solely of the two elements hydrogen and carbon.

Hydrogenation   The addition of hydrogen to a gaseous substance by the use of gaseous hydrogen combined with a catalyst.

Hydrology   The study of the distribution and movement of water.

Hydrolysis   The formation of an acid and a base from a salt by the ionic dissociation of water.

Hydrostatic Pressure   Pressure created by water at rest equally at any point within a confined area.

Hydrothermal   The generic term that refers to any geologic process involving heated or superheated water.

Hygiene   Refers to the science of health and the preservation of well-being (named for the Greek God Hygeia).

Hygrometer   An instrument used for the detection of atmospheric moisture.

Hygroscopic   Refers to substances that absorb water from the atmosphere.

Hyperbarism   A condition resulting from exposure to atmospheric pressure that exceeds the pressure within the body.

Hyperthermia   A marked sustained increase in body temperature due to the inability of the body to dissipate excessive heat generated through metabolic activity that can result in severe cellular damage and death if not treated promptly.

Hypothalamus   The portion of the brain that controls body temperature and produces hormones that affect the pituitary gland.

Hypothermia   Loss of body heat and decreased temperature due to extensive exposure to cold.

Ignitable   Capable of burning or causing a fire.

Ignition   The introduction of some external spark, flame, or glowing object that initiates self-sustained combustion.

Illness   A condition or pronounced deviation from the normal health state; sickness. Illness can be the result of disease or injury.

Illumination   The density of light flux incident on a surface.

Immediately Dangerous to Life and Health (IDLH)   Describes the concentration of a hazardous agent that will produce irreversible health effects or death following a 30-minute exposure.

Imminent Danger   Refers to a work situation in which there is an immediate danger of death or serious physical harm.

Immunodeficient   Lacking in the ability to produce antibodies in response to an antigen.

Inapparent Infection   Infection without recognizable clinical signs or symptoms.

Incendiary   A material that is primarily used to start fires.

Incidence (or incident) Rate   For OSHA record-keeping purposes, the number of injuries, illnesses, or lost workdays related to a common exposure base of 100 full-time workers (working 40-hours per week, 50 weeks per year).

Incubation   The growth and development of microorganisms.

Incubation Period   The time interval between effective exposure or a susceptible host to an agent (infection) and onset of clinical signs and symptoms of disease in that host.

Indicating Thermometer   A non-recording thermometer that allows the user to measure the temperature, generally on the Fahrenheit scale.

Indoor Air   The breathing air inside a habitable structure or conveyance.

Indoor Air Pollution   The presence of chemical, physical, or biological contaminants in indoor air in concentrations that could have an adverse effect on human health.

Indoor Air Quality   General term that applies to the assurance or the evaluation and assessment of indoor air pollution to determine if contaminant levels exceed established standards for a particular pollutant or set of pollutants.

Induced Radioactivity   Radioactivity produced in certain materials as a result of nuclear reactions that involve the formation of unstable nuclei.

Inductively Coupled Plasma Emission Spectroscopy (ICPES)   A method typically used for the simultaneous analysis of many heavy metals.

Industrial Hygiene   The art and science of anticipating, recognizing, evaluating, and controlling occupational and environmental health hazards in the work place and the surrounding community.

Inert Atmosphere   The atmosphere of a confined space that has been made non-flammable, non-explosive, or otherwise chemically non-reactive and, therefore, also generally incapable of supporting or sustaining human life.

Inert Gas   Any non-reactive gas (such as krypton, argon, helium, neon).

Infected Person   A person who harbors an infectious agent, whether or not the infection is accompanied by disease.

Infection   The entry and multiplication of an infectious agent that occurs in the body tissues of a human or animal and that results in cellular injury. Synonymous with the term infectious disease.

Infectious Agent   An organism, usually a microorganism, that is capable of producing infection or infectious disease.

Infectious Disease   A disease of humans or animals resulting from the invasion of the body by pathogenic agents and the reaction of the tissue to these agents and/or the toxins they may produce (*see* **Infection**).

Infestation   The lodgment, development, and reproduction of anthropods such as mites, ticks, or fleas on the surface of the body, in clothing, or in dwellings.

Inflammation   Normal tissue response to cellular injury or foreign material invasion, characterized by dilation of small blood vessels (capillaries) and mobilization of defense cells.

Ingestant   A substance capable of entering the body through the mouth or digestive system.

Infrared   Electromagnetic radiations of wave length between the longest visible red (7000 Angstroms or 7×10^{-4} millimeter) and about 1 millimeter.

Infrared Gas Analyzer   A real-time air-sampling device that measures the absorbency of inorganic and organic gases and vapors.

Inhalable Dust   *See* **Respirable Dust**.

Injury   Physical harm or damage to a person.

Inorganic Chemicals   Chemical substances or mineral origin, not basically of carbon structure.

Inspiration   The process of drawing air into the lungs.

Inspiratory Reserve Volume (IRV)   The maximum volume of air that can be forcibly expired following a normal inspiration.

Insulin   A sulfur-containing hormone produced in the pancreas of vertebrates. This hormone stimulates the conversion of glucose to glycogen and fat. An insulin deficiency results in excess blood sugar and causes the condition diabetes mellitus.

Integral Absorbed Dose   The energy imparted to matter by ionizing particles. The unit of measure is the gram-rad and is equal to 100 ergs.

Intensity, Radiation   The energy of any radiation incident on (or flowing through) a unit area, perpendicular to the radiation beam, in a unit of time.

Interferon   Low molecular weight protein produced by cells infected with viruses. It will block viral infection of healthy cells and suppress viral multiplication in cells already infected.

Internal Radiation   Nuclear radiation (alpha and beta particles and gamma radiation) resulting from radioactive substances inside the body. Important sources are iodine-131 in the thyroid gland, and strontium-90 and plutonium-239 in bone.

Inversion   An atmospheric condition in which a layer of warm air prevents the rise of cooler air. The cool air subsequently remains trapped below the warm air layer. This condition prevent the rise and dispersion of pollutants and can lead to adverse respiratory and cardiovascular health effects on the exposed population.

In Vitro   Refers to an experiment or procedure that is observable with a test tube, other laboratory equipment, or an artificial environment.

In Vivo   Occurring within the living body of a plant or animal. Also, those laboratory tests performed on live animals or human volunteers.

Ion   An atom or chemical radical (group of chemically combine atoms) bearing a positive or negative electrical charge caused by a deficiency or excess of electrons.

Ionization   The separation of a normally electrically neutral atom or molecule into electrically charged components. The term may also be used to describe the degree or extent to which this separation occurs. Ionization is the removal of a negatively charged electron from the atom or molecule (either directly or indirectly) leaving a positively charged ion. The separated electron and ion are then referred to as an ion pair (*see* **Ionizing Radiation**).

Ionization Chamber   A device consisting of two oppositely charged plates used to measure radioactivity.

Ionization Density   Number of ion pairs per unit volume.

Ionizing Radiation   Electromagnetic radiation (x-ray or gamma ray photons or quanta) or corpuscular radiation (alpha or beta particles, electrons, positrons protons, neutrons, or heavy particles) capable of producing ions by direct or secondary processes as it passes through matter.

Ion Pair   Two particles of opposite charge (*see* **Ionization**).

Irradiation   Exposure to radiation.

Irritation   A reaction of tissues to an injury that results in an inflammation; the response or reaction by tissues to the application of a stimulus.

Ischemia   A condition in which there is an insufficient amount of blood to a part of the body, due to a functional constriction or blockage of a blood vessel, that can result in damage to the affected area.

Isokinetic Sampling   An air sampling technique used to measure particulates and other contaminants in exhaust stacks.

Isotope   Forms of the same element having nearly identical chemical properties but differing in their atomic masses (due to different number of neutrons in their respective nuclei) and in nuclear properties such as radioactivity or fission.

Jaundice   An abnormal physical condition caused by bile pigments in the blood and characterized by yellowing of the skin and sclera of the eye and by lassitude and loss of appetite.

Joint   Articulation between two bones that may permit motion in one or more planes. They may become sites of concern with certain cumulative trauma disorders.

Joule   The amount of energy provided by one watt flowing for one second.

Jurisdiction   Generally, the authority of the court to hear and decide a case; or, the authority of a governing or responsible body or agency to create, decide, interpret, and implement policies as in "the authority having jurisdiction."

Kaposi's Sarcoma   An opportunistic neoplasm associated with acquired immune deficiency syndrome (AIDS).

K-Electron Capture   The process wherein the electron in the K shell of an atom is captured by the nucleus during a nuclear reaction. In this process, a characteristic x-ray is emitted.

Ketone   A class of liquid organic compounds derived by the oxidation of secondary alcohols. They are used as solvents in paints and explosives.

keV   An abbreviation for kilo-electron volt and equal to 1000 electron volts.

Kilovolt   The unit of electrical potential equal to 1000 volts.

Kinesiology   The study of movement of the human muscular system.

Kinetic Energy   The energy that a body possess by virtue of its mass and velocity, or the energy of motion. The equation for kinetic energy (K.E>) is   K.E. = $1/2mv2$, where m is mass and v is velocity.

Kuru   Disease in humans caused by a virus that affects the central nervous system and can be transmitted to subhuman primates.

Kyphosis   A posture of the lumbar spine caused by bending forward, such as over a work bench or poorly positioned computer terminal, and involving reverse curvature of the spine.

Labeled Molecule   A molecule containing one or more atoms distinguished by non-natural isotopic composition (with radioactive or stable isotopes).

Latency Period   The time interval between exposure to toxic chemical agents and the onset of signs and symptoms of illness.

Latent Heat of Fusion   The amount of heat required to convert a unit mass of solid to liquid at the melting point.

Latent Heat of Vaporization   The amount of heat required to convert a unit mass of substance from a liquid to a gas at a certain temperature.

Lateral Exhaust Hood   A slot hood typically used to exhaust air contaminants from an open surface tank, and requiring full access to the top of the tank. The slots are narrow rectangular openings, usually located in a plenum at the rear of the tank opening. Also known simply as a slot hood.

LC$_{50}$ A standard measure of acute toxicity, normally applied to inhalation hazards, designating the median lethal concentration of a chemical that is estimated to kill 50 percent of the exposed organisms in a specific period of time and conditions.

LD$_{50}$ A standard measure of acute toxicity, normally applied to ingestion and/or absorption hazards, designating the median lethal dose of radiation or chemical applied directly to experimental organisms that will kill 50 percent of the exposed population within a specific period of time and conditions.

LD LO   The lowest concentration and dosage of a toxic substance that kills test organisms.

Lesion   An abnormal localized change in the structure of an organ or tissue resulting from disease or injury.

Leukemia   A disease of unknown specific cause characterized by an overproduction of leukocytes and their precursors, and enlargement of the spleen. The disease is variable, at time running a more chronic course in adults than in children. Exposure to low intensities of ionizing radiation is thought to be one possible cause (*see* **Leukocyte**).

Leukocyte   A white blood corpuscle in the blood, lymph, or tissues that plays a major role in the body's defense against disease.

Linear Energy Transfer (LET)   The linear rate of energy loss locally absorbed by an ionizing radiation particle passing through a material medium.

Lipids   A comprehensive term for fats and fat-derived materials that denotes substances extracted from animal or vegetable cells by non-polar or fat-soluble solvents. Lipids are among the chief structural components of living cells.

Liter   In the metric system, a unit of measurements equivalent to 1.0567 quarts.

Local Exhaust Ventilation System   Air-handling system designed to capture and remove process emissions before they can escape into the work place or the environment generally consisting of a hood, conveying ductwork, an air-handling device, a fan, and an exhaust stack.

Loudness   An observer's impression of a sound's amplitude; a purely subjective assessment.

Low-Hazard Permit Space   A confined space in which there is an extremely low likelihood that an immediately dangerous to life and health (IDLH) or engulfment hazard could be present and in which all other potentially serious hazards have been controlled.

Lower Flammable Limit (LFL)   Often referred to as the Lower Explosive Limit (LEL), it is the lowest concentration of gas or vapor in the air that will propagate a flame if a spark or heat source is present.

Lumen   Used to represent total light output; the unit of luminous flux emitted through a unit solid angle from a uniform point source of one candela.

Luminance   A measure of the light emitted by a luminous body or reflected from a surface.

Lung Function Test   A test, usually employing a spirometer, that measures an individual's breathing capacity and, indirectly, their ability to wear a respirator.

Lymph   A clear to yellowish fluid found in tissues throughout the body and is eventually added to the venous blood circulation. It contains varying numbers of white blood cells and plays a major role in the body's defense against disease.

Lymphocyte   A white blood cell found in lymphatic tissues (e.g., lymph nodes, spleen, thymus) that is immunologically important and that attacks invading pathogens.

Macrophage   A typically large and long-lived cell originating in the bone marrow and distributed throughout the body that play an important role in immunity by either engulfing invading microorganisms, through the production of antibodies, or by presenting antigens to lymphocytes for destruction.

Malaise   A general feeling of fatigue or exhaustion; a lack of energy and/or desire to do anything; can appear as a symptom of disease and illness.

Malignant Tumor   A tumor capable of metastasizing or spreading cancerous cells from one part of the body to another.

Manometer   Instrument for measuring the pressure of any fluid or the difference in the pressure between fluids, whether liquid or gas.

Mass   The fundamental measure of the quantity of matter. Mass is different from weight in that it does not depend on gravitation force.

Mass Number   *See* **Atomic Weight.**

Mass Spectrometer   An electronic instrument used for the separation of electrically charged particles by mass.

Maximum Voluntary Ventilation   The volume of air breathed with maximum voluntary effort by an individual for a given period of time, usually 10–15 seconds, corrected to one minute. Also known as maximum breathing capacity.

Maximum Permissible Concentration (MPC)   The amount of radioactive material that can be tolerated in the environment or in the body without producing a significant injury.

Maximum Permissible Dose (MPD)   In radiation, the dose of ionizing radiation that, in the light of current medical knowledge, is not expected to cause detectable bodily injury to a person at any time during their lifetime.

Median Nerve   A major nerve that controls the flexor muscles of the wrist and hand. Its location in the carpal tunnel of the wrist makes it susceptible to injury or trauma as a result of overuse of tendons that pass through the same area. When the tendons swell, the nerve may be pinched causing severe pain in an illness known as carpal tunnel syndrome (CTS).

Medical Pathology   A disorder or disease (OSHA).

Megacurie   One million curies.

Mesothelioma   A rare neoplasm that grows as a thick sheet in the pleura of the lungs and in the peritoneum. This condition has been demonstrated in workers who have had extensive exposures to asbestos (*see* **Asbestos**).

Metabolism   The set of biochemical transformations that a chemical undergoes in the body, through which energy is made available for use by the organism.

Metabolite   Any product (foodstuff, intermediate, waste product) of metabolism.

Metal Fume Fever   An acute condition, usually of short duration, caused by the inhalation of finely divided fumes of zinc, magnesium, copper, or their oxides, and possible others produced during hot work (e.g., welding). Symptoms can appear four to 12 hours after exposure and consist of fever and chills. Most cases are the result of inhalation of zinc oxide from the welding of galvanized steel.

Metalizing   An industrial process involving the coating of parts with molten metal, usually aluminum, by means of vacuum deposition. The process often presents occupational health hazards from metal fumes, dust, heat, and non-ionizing radiation.

Metastasis   The transfer of malignant neoplastic cells from the original or parent site to a more distant one, with the resultant appearance of a neoplasm.

Metric Ton (MT)   1000 kilograms; equal to 2204.6 pounds avoirdupois or 2679.23 pounds troy.

MeV   A unit of energy commonly used in nuclear physics equivalent to 1.6 $\times 10^{-6}$ ergs. Approximately 200 MeV are produced for every nucleus that undergoes fission.

Mica Pneumoconiosis   A disease of the lung caused by excessive inhalation of mica dust usually over a period of many years.

Microaerophilic   Refers to organisms that are aerobic but require reduced concentrations of oxygen (pressures lower than 0.2 atmosphere) and elevated levels of carbon dioxide to grow.

Microbes   Microscopic organisms, such as algae, viruses, rickettsia, bacteria, fungi, and protozoa, some of which cause disease in humans, plants, and/or animals.

Microbial Growth Inhibition   Control of growth of microorganisms by various methods (such as heat, dehydration, refrigeration, chemicals, etc.), that can also be used to destroy them in some cases.

Microbiology   The science concerned with the study of microscopic and ultra-microscopic organisms (protistology).

Microcurie   One millionth of a curie.

Micron   A one millionth part of a meter; (i.e., 10^{-6} meter or 10^{-4} centimeter). It is roughly four one-hundred thousandth (4×10^{-5}) of an inch.

Microorganism   Any microscopic of submicroscopic organism, especially any of the viruses, rickettsia, bacteria, or protozoa.

Microscopy   By use of a microscope; investigation by means of a microscope.

Microwave   Nonionizing radiation that has a wave length of three meters to three millimeters (a frequency of 100 to 100,000 hertz), often used in cooking, radar, communications, and dosimetry.

Millirem   One thousandth of a rem (*see* **Rem**).

Milliroentgen   One thousandth of a roentgen (*see* **Roentgen**).

Minimum Erythemal Dose   The amount of energy (usually ultra-violet) expressed in microwatt-seconds per square centimeter of skin to which skin can be safely exposed.

Mist   Liquid particles, measuring 40 to 50 microns, that are generated by condensation from the gaseous state to the liquid state, or by the break up of a liquid into a dispersed state (splashing, foaming, or atomizing). In contrast, fog particles are smaller than 40 microns.

Mold   *see* **Fungus**.

Molecular Weight   The relative weight of a molecule of any substance as compared to the weight of an atom of carbon-12 (12.00000).

Molecule   Ultimate unit of quantity of a chemical compound that can exist by itself and retain all the properties of the original substance.

Momentum   The product of the mass of a body and its velocity, expressed in units of g-cm/s.

Monitoring, Radiation   Periodic or continuous determination of the amount of ionizing radiation or radioactive contamination present in an occupied space.

Monochromatic Radiation   Electromagnetic radiation of a single wave length, or in which all the photons have the same energy (e.g., lasers).

Monoenergetic Radiation   Particulate radiation of a given type (alpha, beta, neutrons, etc.) in which all particles have the same energy.

Mutagen   A chemical substance that has the ability to produce a change (mutation) in the genetic composition of the DNA in a cell. The change is capable of being passed on to succeeding generations. Mutations can also be brought about by radiation exposures.

Mutant   An individual that has been altered as a result of mutation (i.e., from a change in the character of a gene that is perpetuated in the subsequent division of the cell in which it occurs).

Mutation   A change in the characteristics of an organism produced by an alteration of the DNA of living cells.

Mycotoxin   A toxin produced by a mold growing on a specific substrate, many of which are known to be potent carcinogens.

Myelin   The white, fatty substance that forms a sheath around certain nerve fibers.

Natural Gas   a mixture of naturally occurring gases of hydrocarbon and non-hydrocarbon components (the main component is methane) usually associated with petroleum deposits.

Near Miss   A situation in which personnel injury or illness, or equipment, property, or environmental damage is narrowly averted.

Necrosis   The death of one or more cells or a portion of a tissue or organ, usually resulting from irreversible damage to the affected area.

Negative Air Machine   A device consisting of a fan and ductwork and, often, a high-efficiency particulate air (HEPA) filter. It is used to move air from one area to another to maintain a state of negative air pressure inside a contaminated area, preventing leakage of contaminants into other areas.

Negative Air Pressure   Air pressure in a room or duct that is less than the pressure in adjacent areas.

Neoplastic   Of or pertaining to a neoplasm or a cancer.

Neoplasm   Term commonly used to describe a cancer but technically means any new growth of cells that is more or less unrestrained and not governed by the usual limitations of normal growth. The growth is benign if there is some degree of growth restraint and no spread to distant parts. It is malignant if it invades other tissues of the host, spreads to distant parts, or both.

Nephrotoxin   A toxin known to have deleterious effects on the kidney tissue.

Neuron   A nerve cell, the basic unit of the nervous system. It consists of a nerve cell body, dendrites, and an axon.

Neutralization   Decrease the acidity or alkalinity of a substance by adding alkaline or acid materials, respectively.

Neutron   A neutral particle (i.e., one without an electrical charge) of approximate unit mass present in all atomic nuclei, except those of ordinary (or light) hydrogen. Neutrons are used to initiate the fission an fusion process.

Neutron Chain Reaction   A process in which some of the neutrons released in one fission event cause other fissions to occur.

Noble Gas   A gas that is either completely un-reactive or reacts only to a very limited extent with other elements. The noble gases are helium argon, neon, krypton, xenon, and radon.

Noise   Commonly defined as unwanted sound, usually expressed in decibels on the A scale (dBA), the scale thought to most approximate human hearing. Noise is characterized by both frequency (pitch) and pressure (intensity).

Noise Exposure   Exposure to any unwanted sound. Overexposure to occupational noise in the United States is considered to be 90 dBA over an eight-hour time weighted average (TWA).

Noise Reduction Rating (NRR)   A measure of the effectiveness of a given hearing protector, usually expressed in decibels. Assuming a complete and perfect fit, the NRR is the difference between the sound pressure levels outside the ear and those inside the ear.

Nonionizing Radiation   Electromagnetic radiation, such as ultra-violet, laser, infrared, microwave, and radio-frequency radiation, that does not cause ionization.

Nuclear Disintegration   A process resulting in the change of a radioactive nucleus through the emission of alpha or beta particles.

Nuclear Fission   A type of nuclear transformation characterized by the splitting of a nucleus into at least two other nuclei and the release of a relatively large amount of energy.

Nuclear Fusion   The joining (coalescence) of two of more nuclei.

Nuclear Radiation Spectrum   The frequency distribution of nuclear or ionizing radiation with respect to energy.

Nucleon   Common name for a constituent particle of the nucleus. It is applied to protons, neutrons, and any other particle found to exist in the nucleus.

Nucleus   (1) *Radiation.* The atomic nucleus; the small, centrally located, positively charged region of the atom that carries essentially all the mass. Except for the nucleus of ordinary (or light) hydrogen, which is a single proton, all atomic nuclei contain both protons and neutrons. (2) *Biology.* The structure within cells that contains chromosomes and one or more nucleoli.

Nuclide   A general term referring to any nuclear species, either stable (of which there are about 270 in number) and unstable (of which there are about 500), of the chemical elements.

Nystagmus   Involuntary movement of the eyeballs often experienced by workers who continuously subject their eyes to abnormal or unaccustomed movements. The condition is often accompanied by headaches, dizziness, and fatigue. The most prevalent form of occupational nystagmus occurs in miners.

Obligate Anaerobes   microorganisms that are strictly intolerant of oxygen in their environment.

Occupational Illness   Any abnormal condition or disorder, other than resulting from and occupational injury, caused by exposure to environmental factors associated with employment. This includes any acute or chronic illnesses or diseases that may be caused by inhalation, absorption, ingestion, or direct contact.

Occupational Injury   Any injury that results from a work accident or from a single instantaneous exposure in the work environment.

Occupational Medicine   A branch of medicine dedicated to the appraisal, maintenance, restoration, and improvement of workers' health through the scientific application of preventive medicine, emergency medical care, rehabilitation, epidemiology, and environmental medicine.

Octave Band   As applied to noise, a bandwidth that has an upper band frequency that is twice its lower band frequency.

Odor   The characteristic of a substance that makes it perceptible to the sense of smell.

Odor Threshold   The minimum concentration of a substance that can be detected and identified by a majority of the exposed population.

Oncogene   A viral gene, found in some retroviruses, that may transform the host cell from normal to neoplastic. More than 30 oncogenes have been identified in humans.

Oncogenic   A substance that causes tumors, whether benign or malignant.

Open-Face Filter Cassette   A cassette holding a filter that collects airborne particulate matter (usually fibers) on removal of the entire lid and not just the small inlet plug of the cassette.

Opportunistic Infection   An infection caused by a microorganism that does not ordinarily cause disease but can become pathogenic under some circumstances.

Organ   Organized group of tissues that perform one or more definite functions in an organism.

Organic   Of, pertaining to, or derived from living organisms. Also, in chemistry, refers to substances containing carbon compounds.

Organism   Any living biological entity composed of one or more cells.

Osmosis   Diffusion that proceeds through a semi-permeable membrane separating two miscible solutions and that tends to equalize concentrations.

Oxidant   A substance that contains oxygen and that may react chemically in air to produce a new substance.

Oxidation   In chemistry, the combination of oxygen with a substance (or the removal of hydrogen from a substance) that causes a loss of electrons from the atom of a substance.

Oxygen Deficiency   A condition in which the concentration of oxygen by volume is insufficient to maintain normal respiration. It exists in atmospheres in which the percentage of oxygen by volume, which is 21 percent under normal conditions, drops below 19.5 percent.

Ozone (O_3)   A reactive chemical that is an irritant to the eyes and respiratory tissues. Ozone occurs at a concentration of about 0.01 ppm in the atmosphere. levels above 0.1 ppm are considered toxic in the work place.

Ozone Depletion   Refers to the destruction of the stratospheric ozone layer that shields the earth from the harmful effects of the sun's ultra-violet radiation.

Ozone Layer   Segments of the stratosphere (7 to 10 miles above the earth's surface) in which ozone is found in abundance and provides a natural protection against the sun's ultra-violet radiation.

Pair Production   The process whereby a gamma-ray (or x-ray) photon, with energy in excess of 1.02 MeV is converted into a positive electron and an negative electron as it passes near a nucleus of an atom. As a result, the photon ceases to exist.

Pandemic   A disease affecting the population of a widespread region, a nation, or the world (example: AIDS).

Parasite   An organism (often microbial) that lives at expense of the host in or on which the organism resides. Parasites are not necessarily harmful to their host.

Particles   Small, distinct masses of solid or liquid matter such as dust, fume, mist, or smoke.

Particulate   Fine liquid or solid particles such as dust, smoke, mist, fumes, or smog, found in air or emissions.

Pathogen   Any viral, rickettsial, bacterial, fungal, or parasitic microorganism capable of causing disease in humans, animals, or plants.

Pathogenicity   The capacity of an agent to cause disease in a susceptible host.

Pathology   The branch of medicine concerned with all aspects of disease, but having special reference to the causes and development of abnormal conditions and any structural or functional changes that occur as a result of disease.

Peak Sound Pressure Level   The maximum instantaneous level that occurs over any specified time period and is usually measured in decibels.

Penetration   A mechanism by which a substance enters through a barrier via an imperfection such as a rip or a tear.

Percentage Depth Dose   Amount of radiation delivered at a specified depth in tissue, expressed as a percentage of the amount delivered at the skin.

Permanent Gas   Term used to describe a gas that cannot be liquefied at normal ambient temperatures.

Permeable   Refers to a substance or barrier that affords passage or penetration despite the absence of any damage (i.e., tear or rip) in the barrier.

Permeation   Movement, on a molecular level, of a substance through a barrier.

Permissible Dose (radiation)   The amount of radiation that may be received by an individual within a specified period with expectation of no harmful effects.

Permissible Exposure Limit (PEL)   The specific level established by OSHA for airborne human exposure to industrial chemicals and also for exposure to noise.

Permit-Required Confined Space   An enclosed space that is large enough and so configured that a person can enter it and perform work. This space requires written authorization prior to entry and usually has one or more of the following characteristics: (1) It has a known potential to contain a hazardous atmosphere. (2) It contains a material that has the potential for engulfing an entrant. (3) Its internal configuration presents a trapping or asphyxiating hazard. (4) It contains other recognized serious safety or health hazards.

Personal Decontamination   The removal of biological, chemical, or radioactive material from the body of clothing by appropriate mechanical and/or chemical means.

Personal Monitoring   (1) *Radiation.* The determination, using standard survey meters, of the degree of radioactive contamination on individuals, and the determination, using dosimetry devices, of dosages received. (2) *Industrial Hygiene.* The practice of having an individual wear monitoring or sampling devices during their workday to measure exposures to various hazardous substances or agents.

Personal Protective Equipment (PPE)   Any of a number of devices or types of equipment (hard-hats, gloves, goggles, etc.) worn to provide protection against various hazards.

Petrochemical   An organic compound, such as gasoline, kerosene, or petroleum, that has been obtained from petroleum or natural gas.

pH   The negative logarithm of the hydrogen ion concentration [H+]. A measure of the degree to which a substance is acidic or basic. The pH values are indicated on a scale of 1–14, with those above the neutral seven considered "bases" and those below seven considered "acids."

Phagocyte   A cell that engulfs and destroys foreign particles or microorganisms by digestion.

Phosphorescence   The emission by a substance as a result of previous absorption of radiation of shorter wavelength. In contrast to fluorescence, the emission may continue for a considerable time after cessation of the exiting radiation.

Photochemical Oxidants   Air pollutants, such as aldehydes, acids, and nitrates, formed by the action of sunlight on oxides of nitrogen and hydrocarbons.

Photochemical Smog   Air pollution the action of sunlight on various pollutants emitted from numerous sources, such as hydrocarbons emitted as products of incomplete combustion.

Photographic Dosimetry   *See* **Film Badge**.

Photometry   The measurement of the intensity of light.

Photon   The quantum of energy emitted or absorbed in the form of electromagnetic radiation whose energy value is the product of its frequency and Plank's constant (E-hv).

Physiology   The science concerned with the normal vital processes of organisms, especially as to their normal functioning rather than to their anatomical structure.

Picocurie (pCi)   A measurement of radioactivity equal to one million-millionth (or trillionth) of a curie and represents about 2.2 radioactive particle disintegrations per minute.

Plank's Constant   A natural constant (v) of proportionality (h) relating the frequency of a quantum of energy to the total energy of the quantum, expressed as $h = E/v = 6.6 \times 10^{-27}$ erg-s.

Plasma   The fluid (non-cellular portion) of the circulating blood, as distinguished from the serum obtained after coagulation.

Plumbism   A chronic poisoning of humans caused by the absorption of lead or lead salts.

Pneumoconiosis   Inflammation often leading to fibrosis of the lungs and caused by the inhalation of dust associated with various occupations (e.g., mining). The disease is characterized by chest pains, cough, cyanosis, and fatigue.

Pocket Dosimeter   A direct-reading portable unit, usually shaped like a pen with a pocket clip, generally used to measure exposures to gamma- and x-radiation.

Poison   Any substance that, when administered to a living organism, causes a harmful effect. Most substance are harmful at some dose and may be harmless at very low doses.

Polymerization   A chemical reaction in which a high molecular weight material is reduced by the addition to or condensation of a simpler compound. For example, the production of polystyrene from styrene.

Positron   Particle equal in mass but opposite in charge to the electron. Also referred to a positive electron.

Potentiation   An increased toxicologic effect by one agent or another resulting in a combined effect that is greater than the simple sum of those of the individual agents.

Precipitate   A collection or deposit of solid particles that have settles out of solution.

Presbycusis   Hearing loss due to normal aging, as opposed to hearing loss due to environmental or occupational exposure to noise.

Pressure   Force applied or distributed over an area and measured as force per unit area.

Pressure-Demand Respirator   A type or respirator that provides a positive pressure during both inhalation and exhalation and delivers an air flow of about 115 L/min (4 cfm) before a negative pressure is measurable at the face mask.

Primary Electron   The electron ejected from an atom by an initial ionizing event, as caused by a photon or beta particle.

Primary Radiation   That which arises directly from the target of an x-ray tube or from a radioactive source.

Prognosis   A forecast of the probable cause and/or outcome of a disease or injury.

Prophylaxis   Preventive measure against disease such as the immunization against hepatitis B virus.

Proportional Counter   An instrument in which a gas-filled radiation detection tube or chamber receives pulses that are proportional to the number of ions formed in the gas by the primary ionizing particles.

Proton   An elementary nuclear particle with a positive electric charge equal numerically to the charge of the electron and having a rest mass of 1.007575 atomic mass units.

Protozoan   Microscopic, single functional cell units of aggregations of non-differentiated cells loosely held together and not forming tissues. Some are pathogenic in humans and other may invade tissues and cause other diseases to develop.

Psychrophiles   An instrument for the measurement of dry- and wet-bulb temperature.

Psychrotrophic   Refers to microorganisms that are cold-temperature tolerant (capable of surviving in temperature between 0° C and 20° C).

Pure tone   Sound characterized by a single frequency.

Purging   The initial step in adjusting the atmosphere of a confined space to acceptable standards either by displacing the air in the space with a fluid or vapor (inert gas, water, steam), or by carrying out forced-air ventilation.

Pyrophoric   Refers to materials that ignite spontaneously in the presence of sufficient oxygen.

Q fever   An rickettsial infection often seen among meat and livestock handlers. Can be contracted by inhalation of dust particles that are infected with the infectious organism. The organisms are found in the hides of sheep and cattle. It is contagious between humans and its symptoms include sudden hay fever, chills, headache, muscle pain, and coughing.

Quantum   The smallest quantity of energy, responding to the energy of electromagnetic radiation, that can be associated with a given phenomenon.

Quantum theory   A theory based on the concept that energy is radiated intermittently in units of definite magnitude called quanta and absorbed in like manner.

Quarantine   The application of measures to prevent contact between un-infected persons and persons suspected of being infected.

Radiant energy   The energy of electromagnetic waves from sources such as radio waves, visible light, infra-red, x-rays, and gamma rays.

Radiant heat load   Energy that is transformed into heat when it strikes an object. The human body can both emit and receive radiant energy.

Radiation   The emission and propagation of energy in the form of waves or corpuscular emissions of particles through space or through a material medium.

Radiation Absorbed Dose (RAD)   The unit of absorbed dose that is equal to 100 ergs/g. It is a measure of the energy imparted to matter by ionizing particles per unit mass of irradiated material at the point of exposure.

Radiation hazard   A situation in which persons might receive ionizing radiation in excess of the applicable maximum permissible dose or in which radiation damage might be done to materials.

Radiation Protection Guide (RPG)   The total amount of ionizing radiation dose over certain periods of time that may be permitted to persons whose occupation involves exposure to such radiation. It is equivalent to what was formerly called the Maximum Permissible Exposure (MPE).

Radiation sickness   A self-limiting syndrome characterized by nausea, vomiting, diarrhea, and psychic depression, following exposure to appreciable doses of ionizing radiation, particularly to the abdominal region.

Radioactive substances   Those substances that emit radiation.

Radioactivity   The spontaneous emission of radiation, generally alpha or beta particles often accompanied by gamma rays, from the nucleus of an unstable isotope. As a result, the radioactive isotope is converted (or "decays") into the isotope of a different element, that may or may not be radioactive. Ultimately, as a result of one or more stages of radioactive decay, a stable (non-radioactive) end product is formed.

Radioactivity Concentration Guide (RCG)   The maximum permissible amount of any specified radioisotope that may be allowed to accumulate in the body.

Radiography, industrial   The use of penetrating radiation such as x-rays, gamma rays, or neutrons, to make visual images of the insides of objects.

Radiology   The branch of medicine that deals with the diagnostic and therapeutic applications of radiant energy.

Radiometer   An instrument used to demonstrate the transformation of radiant energy into mechanical energy.

Radionuclide   Radioactive element that is characterized in terms of its atomic mass and atomic number that can be made artificially or can be naturally occurring. They have potentially mutagenic effects on the human body.

Rancid   Refers to a musty, rank taste or smell that is usually associated with fats that have undergone decomposition.

Raynaud's disease   Also known as vibration syndrome, this disease consists of vasospastic, neuromuscular, and arthritic disorders usually of the hands and upper limbs. It is associated with the operation of hand-held or manually supported or guided machines that produce intense vibration in the frequency range of 10 to 1000 Hz. Symptoms include pain in the fingers, loss of manual dexterity, stiffness in the joints, radiographically observable changes in the bones and joints, whitening and numbness in one or more fingers of either hand, skin atrophy, and occasionally gangrene.

Reactivity   A measure of the tendency to undergo a chemical change.

Reagent   A substance that produces a chemical reaction that can be used to detect, measure, or produce another substance.

Recombinant Bacterium   A type of microorganism whose genetic makeup has been deliberately altered by introduction of new genetic elements.

Recording Thermometer   A thermometer with a recording device used to obtain a permanent record of temperature.

Relative Biological Effectiveness (RBE)   The ratio of the number of rads of gamma radiation or x-radiation of a certain energy that will produce a specified biological effect to the number of rads of another radiation required to produce the same effect.

Relative Humidity   The amount of moisture in the air as compared with the maximum amount that the air could contain at the same temperature, expressed as a percentage.

Resonance Capture   An inelastic nuclear collision occurring when the nucleus exhibits a strong tendency to capture incident particles, photons, or particular energies.

Respirable Dust   Airborne particulate matter capable of passing through upper respiratory system and being deposited in the lungs. Such particles are typically less than 10 microns in diameter.

Respiration   Process that involves the oxidation of inorganic or organic molecules.

Respiratory System   The group of organs concerned with the exchange of oxygen and carbon dioxide in organisms. In higher animals, it consists successively of the air passage through the mouth, nose, and throat; the trachea; the bronchi; the bronchioles; and the alveoli of the lungs.

Reverberation   The persistence of sound after direct reception of the sound has ceased.

Rickets   A nutritional disease caused by a deficiency of Vitamin D in the diet.

Rickettsia   A class of microbial agents resembling small bacteria that multiply by simple fission, but only within a living cell.

Risk   The likelihood or possibility of hazardous consequences in terms of the potential severity for ill health or injury and the probability that such consequences will occur.

Risk Assessment   The qualitative and quantitative evaluation performed in an effort to define the risk posed to human health and/or the environment by the presence or potential presence of hazards in the work place.

Risk Management   The process of evaluating alternative regulatory and non-regulatory responses to risk and selecting among them. The selection process necessarily requires the consideration of legal, economic, and social factors.

Roentgen (R)   A unit of exposure dose of gamma- or x-radiation. It is defined as the quantity of gamma- or x-radiation that will produce (in 0.001293 gram of air [1 cc] at 0° C and 760 mm of Hg pressure) ions carrying one electrostatic unit quantity of electricity of either sign.

Roentgen Equivalent Man (or Mammal) (REM)   A unit of biological dose of radiation in which the number expressing the relative biological effectiveness of radiation is equal to the number or rads absorbed, multiplied by the RBE of the given radiation (for a specified effect). It is the quantity of ionizing radiation of any type that, when absorbed by a human (or other mammal), produces a physiological effect equivalent to that produced by the absorption of one roentgen of gamma- or x-radiation.

Safety   As used in this text, the development of systems and techniques to assure that individuals in occupational settings and their environment are relatively free from conditions that could cause death or serious physical harm.

Saline   A solution of salt (sodium chloride) in water.

Salmonella Test   *See* **Ames Test.**

Salutary   Pertaining to health or a state of well-being.

Sample   In industrial hygiene, a process consisting of the withdrawal, isolation, or concentration of a fractional part of a whole as related to environmental media such as air, water, soil, or other liquid, solid, or gas, for analysis by acceptable procedures. Samples can be either cumulative (i.e., collected over time), in which gas the result is expressed as a time-weighted average, or instantaneous.

Sampling Train   Generic term used to describe the combination of collection media, sampling pump, tubing, and a flow rate measuring device.

Sanitation   The formulation and application of measures designed to protect the public health.

Sarcoma   Malignant neoplasm composed of cells imitating the appearance of supportive and lymphatic tissues.

Scatermia   The type of toxemia wherein chemical toxins are absorbed through the intestines.

Scattered Radiation   Term used to describe radiation that, during its passage through a substance, has been deviated in direction.

Scintillation Counter   The combination of a phosphor (that converts ionizing particle energy into a light pulse), a photomultiplier (that converts the light pulse to many electric pulses), and associated circuitry for counting electric pulse. It is used primarily for the routine detection of radioactive material in various media.

Scrubber   An air pollution device that uses a spray of water or reactant, or a dry process, to trap pollutants in emissions.

Scurvy   A disease caused by a lack of ascorbic acid (Vitamin C) in the diet. It is characterized by the presence of anemia, edema, ulceration of gums, and hemorrhages into the skin and from the mucous membranes.

Secondary Radiation   That which originates as the result of absorption of other radioactivity in matter. It may be either electromagnetic or particulate in nature.

Selective Localization   As applies to radioisotopes, the accumulation of a particular isotope to a significantly greater degree in certain cells or tissues.

Self-Contained Breathing Apparatus (SCBA)   A device, worn by an individual, that provides a containerized source of high-pressure breathing air to a pressurized mask through a regulator. Normally used in atmospheres containing extremely high levels of air contaminants, unknown concentrations of hazardous substances, or, in some instances, in oxygen deficient atmospheres.

Sensitization   An allergic-like reaction. A process occurring in some individuals, over a non-specific time period, in which the induction of acquired sensitivity or of allergic symptoms is manifested without any associated infection or disease. Sensitization can occur with chemicals, drugs, or other substances of exposure. Once sensitization has occurred, the individual will display allergic reactions whenever in contact with the substance.

Sensorineural Hearing Loss   Usually an irreversible hearing loss involving the auditory nerve and the loss of hair cells in the organ of Corti in the ear, caused by exposure to excessive noise levels, viruses, congenital defects, or drug toxicity.

Sepsis   The presence of pathogenic microorganisms or their toxins in the blood or other tissue.

Severity Rate   A formula applied to OSHA-recordable injuries that result in lost time from the work place and expressed as the days of lost-time $\times$ 200,000 $\div$ number of hours of worked.

Shaver's Disease   A disease of the lungs found in workers exposed to fumes containing aluminum oxide and silica in the process of smelting bauxite in the manufacture of corundum (a form of aluminum oxide). A type of pneumoconiosis, this disease is marked by opacities on x-ray film and is sometimes accompanied by decreased lung function.

Shielding   Any material or obstruction that absorbs radiation to protect personnel or materials from the effect of nuclear radiation exposure.

Short-Term Exposure Limit (STEL)   The concentration to which workers may be exposed continuously for a short period of time (usually 15 minutes) without suffering adverse health effects.

Silicosis   A lung disease caused by long-term exposure to silica dusts and crystals. It is marked by a loss of elasticity of the lung tissue and development of silica-containing nodules in the lungs, resulting in decreased lung function, shortness of breadth, enlargement of the heart, and, often, lung cancer. Also known as miner's asthma, grinder's consumption, miner's phthisis, potter's rot, and stonemason's disease.

Skin Dose   In radiology, the dose at the center of an irradiation field on skin.

Smog   A combination of atmospheric smoke and fog composed of a complex mixture of gases, dust, and droplets that appear as atmospheric haze.

Smoke   Small gas-borne particles resulting from incomplete combustion, consisting mainly of carbon and other combustible materials.

Solvent   Substance (usually liquid) capable of dissolving or dispersing one or more other substances.

Sound   Any pressure variation in air, water, or other media that the human ear can detect. Sound is characterized by both frequency (pitch) and pressure (intensity).

Sound Intensity   The average rate at which sound energy is transmitted through a unit area normal to the direction of sound propagation, measured in joules per square meter per second (J/m2/sec) or in decibels referenced to 10^{-12} watts/m^2.

Sound Level Meter   An instrument used for measuring sound pressure levels in decibels, referenced to 0.0002 microbar. Readings can also be made in specific octave bands ranging from 75 Hz to 10,000 Hz.

Sound Power   The total sound energy radiated by the source per unit time, measured in decibels and referenced to 10^{-12} watt.

Sound   The root mean square value of the pressure changes above and below atmospheric pressure, normally described in decibels referenced to 2×10^{-5} newtons/m^2. Sound pressure level can be measured in dynes per square centimeter, microbars, or pascals.

Sound Velocity   The speed of propagation of a sound wave through a medium. The velocity of sound is 1130 ft/s in air, 4700 ft/s in water, 13,000 ft/s in wood, and 16,500 ft/s in steel.

Source   In radiation, a discrete amount of radioactive material or radiation-producing equipment. The term usually refers to radioactive material specifically packaged for scientific or industrial use.

Stable Isotope   A non-radioactive isotope of an element.

Stable Materials   Chemicals that have the ability to resist changes in their composition despite exposure to water, air, pressure, temperature extremes, or shock.

Static Pressure   The potential pressure exerted in all directions by a fluid at rest.

Sterilant   A chemical or physical agent used to destroy microorganisms.

Sterilization   The destruction of all microorganisms in or on an object using heat, steam, chemical agents, ultra-violet radiation, or a combination of these.

Streptococci   Non-motile, non-spore-forming, aerobic to facultatively anaerobic bacteria that occur regularly in the mouth and intestines of human and animals and in food and dairy products. Some species are pathogenic for humans.

Supplied-Air Respirator   A device that supplies breathing air to the wearer through hoses carrying a continuous flow of air, usually from a breathing-air compressor or from a bank air supply under pressure. A series of regulators and valves is often employed to assure proper air flow. Often used in highly contaminated atmospheres or oxygen deficient atmospheres.

Susceptible   Describes a person or animal lacking sufficient resistance to prevent disease if or when exposed to a particular pathogenic agent.

Swab Test   A surface contact method of micro-biological examination of surfaces, usually used to assess the effectiveness of cleaning and sanitizing procedures.

Swinging-Vane Anemometer   A device that measures air velocity and static pressure by utilizing the compression of a calibrated spring by the moving airstream. Also known as a velometer.

Symbiosis   A mutually beneficial cohabitation of dissimilar organisms.

Symptom   Subjective evidence of disease or illness, as indicated by a departure from the normal in structure, function, or sensation, as experience by an individual.

Syndrome   The combined set of signs and symptoms that typify a particular disease.

Synergism   The cooperative interaction of two or more chemicals or other substances or phenomena that produces a total effect greater than that of the sum of the individual effects.

Systemic   Relating to the body as a whole.

Tenosynovitis   A cumulative trauma disorder (CTD) of the tendon produced by a swelling of the synovial sheath which surrounds it. The surfaces of the tendon become irritated, rough, and bumpy, and movement is impeded. Often related to long-term exposure to repetitive motion, inappropriate posture or positioning of machinery, or exposure to vibration from equipment.

Teratogen   A substance capable of producing physical defects in a fetus. The result may include fetal or embryonic mortality or in the birth of offspring with defects.

Thermal analysis   The measurement of changes in properties (physical or chemical) of materials as a function of temperature change, usually brought about by heating or cooling at a uniform rate.

Threshold dose   The dose of a chemical below which an adverse effect does not occur.

Threshold Limit Value (TLV)   Refers to airborne concentration of a substance and represents conditions under which the American Conference of Governmental Industrial Hygienists (ACGIH) believes that nearly all workers may be repeatedly exposed day after day without adverse health effects. TLV levels are revised annually and are normally based on an assumed 8-hour work day for 40 hours per week.

Time-Weighted Average (TWA)   An exposure averaged over a given time period, often an 8-hour work day. Threshold limit values or permissible exposure limits are often based on 8-hour time-weighted averages.

Tinnitus   A ringing, buzzing, or other similar subjective or pathological sensation in the ears.

Tissue   A group of cells with similar structure and function. There are four basic tissues in the body   muscle, nerve, epithelium, and connective tissues (including blood, bone, and cartilage).

Toxic   Term used to describe a chemical that has the ability to cause harmful or fatal effects on exposure to humans, animals, or plants. The level of toxicity is normally based on a scientific evaluation of the dose/response relationship. Basically, the evaluation of the results of varying degrees of exposure to toxic agents determines the level of their respective toxicity.

Toxicant   A poisonous agent that kills or injures humans, animals or plants.

Toxicity   The state of being poisonous; the capacity of a substance to induce damage to living tissue. Toxicity can be acute (immediately harmful), chronic (harmful over time), local (effects only the site of exposure), or systemic (distributed throughout the boy).

Toxicology   The science and study of chemicals and their adverse systemic effects, including source, chemical composition, action, tests, and antidotes.

Toxin   Proteins or conjugated protein substances that are lethal to other organisms. They are produced by some higher plants, certain animals, and pathogenic bacteria.

Transmission   The direct (contact or droplet-spread) or indirect (vectorborne, vehicleborne, or airborne) transfer of an infectious agent from a source to a susceptible host.

Trigger finger   An occupational cumulative trauma disorder (CTD) in which a tendon in the index finger is compressed from exterior swelling. The condition can be produced by repeatedly pulling a trigger or by using hand tools that have sharp edges that press into the tissue or whose handles are so far apart that only the end segments of the user's hand grasp them, while the middle segments remain straight.

Turbidimeter   A device that measure the amount of suspended solids in a liquid.

Turbidity   The presence of suspended material, such as clay, silt, finely divided organic material, plankton, and other inorganic material, in water.

Tyrotoxism   Poisoning due to the presence of a toxin in milk, cheese, or other dairy products.

Ultrasonic noise   Noise characterized by frequency greater than audible noise, typically greater than 20,000 Hz.

Ultra-violet (UV) rays   Radiation from the sun or other sources that can be useful or potentially harmful. UV ray from one part of the spectrum enhance life and are useful in some medical and dental procedures. UV ray from other parts of the spectrum to which humans are exposed can cause skin cancer or other tissue damage.

Universal precautions   An approach to infection control that is based on the premise that all human blood and certain human body fluids are to be treated as if known to be infectious for HIV, HBV, and/or other bloodborne pathogens (whether or not actual infection is indeed the case).

Unstable material   A chemical substance that will vigorously polymerize, decompose, condense, or become self-reactive on exposure to air, water, shock, pressure, or temperature fluctuation.

Upper Explosive Limit (UEL)   *See* **Upper Flammability Limit (UFL).**

Upper Flammability Limit (UFL)   The concentration of a substance in air, usually expressed as a volume percent, above which combustion cannot be supported at normal room temperature because the mixture of air and fuel is too "rich" (i.e., too much fuel) and therefore, insufficient oxygen. Combustion in air of a flammable material can occur only at concentrations between the lower and upper flammability limits. This area is referred to as the flammability range.

Vaccine   A preparation containing killed or living whole or parts of microorganisms have antigenic properties that are used to induce specific immunity in the recipient against an infectious agent.

Vapor   A term used for a substance that, although present in the gaseous phase, usually exists as a liquid or solid at room temperature and pressure.

Vapor Capture System   Any combination of hoods and ventilation systems that captures or contains organic vapors in order that they may be directed to an abatement or recovery device.

Vaporization   Conversion of a solid or liquid into a vapor without change in the chemical composition of the material involved.

Vapor Pressure   The pressure exerted when a solid or liquid is in equilibrium with its own vapor. Usually expressed in millimeters of mercury or pounds per square inch, vapor pressure is a function of the substance and the temperature, and is often used as a measure of how rapidly a liquid will evaporate.

Velometer   *See* Swing-vane anemometer.

Ventilation   One of the principle methods of controlling health hazards, it can be described generally as the causing of fresh air to circulate and replace contaminated air that is simultaneously removed.

Venturi-Type Collectors   A technology used to remove dust from exhaust air by employing a high-velocity airstream that is used to break up water fed into a venturi throat.

Vesicant   An agent capable of producing small circumscribed elevations of the skin that contain fluid. Some vesicant gases (such as mustard gas) have been used as chemical warfare agents.

Vibration Syndrome   *See* Raynaud's disease.

Villi   Minute, worm-like vascular processes on certain membranes (plural: villus).

Virus   A sub-microscopic, non-cellular particle, composed of a nucleic acid core and a protein shell that reproduce only within host cells and are the cause of many diseases.

Volatile   Describes chemicals that tend to evaporate rapidly (such as ether, chloroform, and benzene).

Volatile Organic Compound (VOC)   Organic compounds that are involved in atmospheric photochemical reactions.

Volatility   The measure of the tendency of a chemical to vaporize or evaporate at ambient conditions.

Water Reactive   Describes chemicals, such as sodium or potassium, that react with water to release a gas that may be flammable and/or may present a health hazard because of the toxic properties of the end product.

Wave Length   The distance in the line of an advancing wave from any point to a like point on the next wave.

Wet Bulb Globe Temperature (WBGT) Index   An index of the heat stress in humans when work is performed in hot environments.

Wet Bulb Temperature (WBT)   The temperature obtained from the wet-bulb thermometer that considers the influence of evaporating water in air to determine humidity levels.

Wet Bulb Thermometer   A dry-bulb thermometer that has its bulb covered with a cloth saturated with water.

Work Strain   The physiological response reaction of the body to the application of work stress. Not always harmful or traumatic, it can be marked by an increased heart rate and localized strain on body members, such as the wrists.

Work Stress   In biomechanics, the action of any external force acting on a body during the performance of work. Its end product is work strain.

X-Radiation   Highly penetrating radiation similar to gamma-radiation. However, unlike gamma rays, x-rays are not generated from the nucleus of the atom but from the surrounding electrons. They are produced by electron bombardment.

X-Ray Tube   A electron tube designed for the conversion of electrical energy into x-ray energy.

Xeroderma   Dry and often rough skin.

Yellow Fever   An acute, dangerous, often fatal infectious disease of warm climates, transmitted by the bite of a mosquito of the genus Aedes. Characterized by jaundice, vomiting, hemorrhages, and, of course, fever.

Zoology   The study of animals.

Zoonosis   An infection or infectious disease transmissible under natural conditions between vertebrate animals and humans. In many cases, transmission of foodborne pathogens occurs indirectly from food animals or animal products.

Zygote   A cell that is produced by the joining of two sex or germ cells (gametes).

Bibliography

Anderson, Marion. 1983. *Occupational Lung Diseases: An Introduction.* ALA Occupational Health Committee: American Lung Association.

ANSI S1.23-1976(R1983). *Method for Designation of Sound Power Emitted by Machinery and Equipment.* New York: American National Standards Institute.

ANSI Z88.2-1980. *Practices for Respiratory Protection.* New York: American National Standards Institute.

Anthony, Catherine Parker, and Norma Jane Kolthoff. 1971. *Textbook of Anatomy and Physiology.* St. Louis: C. V. Mosby Company.

Ashford, Nicholas A., and Claudia S. Miller. 1991. *Chemical Exposures.* New York: Van Nostrand Reinhold.

Berger, E. H., and W. D. Ward, J. C. Morrill, L. H. Royster. 1986. *Noise and Hearing Conservation Manual.* Akron, OH: American Industrial Hygiene Association.

Brauer, Roger L. 1990. *Safety and Health for Engineers.* New York: Van Nostrand Reinhold.

Brief, Richard S. 1975. *Basic Industrial Hygiene Training Manual.* New York: Exxon Corporation for The American Institute of Industrial Hygiene.

Brody, Jane E. 1982. *Guide to Personal Health,* 7th Edition. New York: Times Books.

Bueche, John R. 1972. *Principles of Physics,* 2nd Edition. New York: McGraw-Hill.

Burdon, Kenneth L., and Robert P. Williams. 1968. *Microbiology,* 6th Edition. New York: The MacMillan Company.

Coluccio, Vincent, M., 1994. *Lead-Based Paint Hazards: Assessment and Management.* New York: Van Nostrand Reinhold.

Considine, Douglas M. 1995. *Van Nostrand's Scientific Encyclopedia*, Volumes 1 and 2, 8th Edition. New York: Van Nostrand Reinhold.

Danse, Ilene R. 1991. *Common Sense Toxics in the Workplace.* New York: Van Nostrand Reinhold.

Davisson, John F. What Risk Does TB Pose in the Workplace? *Occupational Hazards,* May 1994. pp. 84-87.

Dobie, Robert A., M.D. 1993. *Medical-Legal Evaluation of Hearing Loss.* New York: Van Nostrand Reinhold.

Eastman Kodak Company. 1991. *Ergonomic Design for People at Work*, Volumes 1 and 2. New York: Van Nostrand Reinhold.

Eller, P. M. (Editor) 1984. *Manual of Analytical Methods*, 3rd Edition. Cincinnati: National Institute for Occupational Safety and Health.

Feder, Harold A. 1991. *Succeeding as an Expert Witness.* New York: Van Nostrand Reinhold.

Forsberg, Krister, and S. Z. Mansdorf. 1993. *Quick Selection Guide to Chemical Protective Clothing*, 2nd Edition. New York: Van Nostrand Reinhold.

Freeman, Stanley H. 1991. *Injury and Litigation Prevention, Theory and Practice.* New York: Van Nostrand Reinhold.

Gray, Henry. 1977. *Gray's Anatomy*, 15th Edition. New York: Bounty Books.

Grenard, Steve, and Gustav J. Beck, Geroge W. Rich. 1971. *Introduction to Respiratory Therapy.* Sarasota, FL: Glenn Educational Medical Services, Inc.

Gustafson, David I. 1993. *Pesticides in Drinking Water.* New York: Van Nostrand Reinhold.

Hodgson, Ernest, R. B. Mailman, and J. E. Chambers. 1988. *Dictionary of Toxicology.* New York: Van Nostrand Reinhold.

Holum, John R. 1968. *Elements of General and Biological Chemistry*, 2nd Edition. New York: John Wiley & Sons, Inc.

Hosty, John W., and Patricia Foster. 1990. *A Practical Guide to Chemical Spill Response.* New York: Van Nostrand Reinhold.

Johnson, Johnson, and Little. "Expertise in Trial Advocacy: Some Considerations for Inquiry into its Nature and Development." *Campbell Law Review,* Vol. 7, No. 2, Fall 1984, p.125.

Khalil, Tarek M., et al. 1993. *Ergonomics in Back Pain.* New York: Van Nostrand Reinhold.

Lewis, Grace, R. 1994. *1,001 Chemicals in Everyday Products.* New York: Van Nostrand Reinhold.

Lewis, Richard J., Sr. 1993. *Hawley's Condensed Chemical Dictionary*, 12th Edition. New York: Van Nostrand Reinhold.

Lewis, Richard J., Sr. 1993. *Hazardous Chemicals Desk Reference*, 3rd Edition. New York: Van Nostrand Reinhold.

Lewis, Richard, J., Sr. 1994. *Rapid Guide to Hazardous Chemicals in the Workplace*, 3rd Edition. New York: Van Nostrand Reinhold.

Lewis, Richard J., Sr. 1992. *Sax's Dangerous Properties of Industrial Materials*, Volumes 1, 2, and 3, 8th Edition. New York: Van Nostrand Reinhold.

Lisella, Frank S. 1994. *The VNR Dictionary of Environmental Health and Safety*. New York: Van Nostrand Reinhold.

MacLeod, Dan. 1994. *The Ergonomics Edge*. New York: Van Nostrand Reinhold.

Maslansky, Carol J., and Steven P. Maslansky. 1993. *Air Monitoring Instrumentation*. New York: Van Nostrand Reinhold.

Meyer, Eugene. 1990. *Chemistry of Hazardous Materials*. Englewood Cliffs, NJ: Prentice Hall.

Ness, Shirley A. 1991. *Air Monitoring for Toxic Exposures: An Integrated Approach*. New York: Van Nostrand Reinhold

Ness, Shirley A. 1994. *Surface and Dermal Monitoring for Toxic Exposures*. New York: Van Nostrand Reinhold.

Patnaik, Pradyot. 1992. *A Comprehensive Guide to the Hazardous Properties of Chemical Substances*. New York: Van Nostrand Reinhold.

Pelmear, Peter L., William Taylor, and Donald E. Wasserman. 1992. *Hand-Arm Vibration: A Comprehensive Guide for Occupational Health Professionals*. New York: Van Nostrand Reinhold.

Plog, Barabara A. Editor. 1988. *Fundamentals of Industrial Hygiene*, 3rd Edition. Chicago: National Safety Council.

Que Hee, Shane S. 1993. *Biological Monitoring: An Introduction*. New York: Van Nostrand Reinhold.

Rhodes, Don, and Carlton Blanton. 1990. "Alice Hamilton: Occupational Safety and health Legend." *Professional Safety*, Vol. 35, No. 8, August 1990. pp. 11–15.

Roach, S. A. 1977. A Rational Basis for Air Sampling Programs. *Annual Occupational Hygiene*, 20:65-84.

Sherman, Janette D. 1988. *Chemical Exposure and Disease*. New York: Van Nostrand Reinhold.

Stepkin, Richard L., and Ralph E. Mosely. 1984. *Noise Control, A Guide for Workers and Employers*. Des Plaines, IL: American Society of Safety Engineers.

U. S. Department of Health & Human Services. 1974. *Biohazards Safety Guide.* Washington, DC: U. S. Department of Health & Human Services (formerly, the Department of Health, Education and Welfare), National Institute of Health, Centers for Disease Control.

U. S. Department of Health & Human Services. 1974. *Classification of Etiologic Agents on the Basis of Hazard.* Atlanta: U. S. Department of Health & Human Services (formerly, the Department of Health, Education and Welfare), Public Health Service, Centers for Disease Control.

U. S. Department of Health & Human Services. 1989. *Criteria for a Recommended Standard on Occupational Exposure to Hand-Arm Vibration.* Cincinnati: U. S. Department of Health & Human Services (formerly, the Department of Health, Education and Welfare), Public Health Service, Centers for Disease Control.

U. S. Department of Health & Human Services. 1988. *Guidelines for Protecting the Safety and Health of Healthcare Workers.* Atlanta: U. S. Department of Health & Human Services (formerly, the Department of Health, Education and Welfare), Public Health Service, Centers for Disease Control.

U. S. Department of Labor. July 1993. *Regulations Relating to Labor—General Industry.* Occupational Safety and Health Administration, Code of Federal Regulations, Title 29, Part 1910. Washington, DC: Office of the Federal Register, Nationa Archives and Records Administration.

Vincoli, Jeffrey W. 1993. *Basic Guide to Environmental Compliance.* New York: Van Nostrand Reinhold.

Wald, Peter. H., M.D., M.P.H., and Gregg M. Stave, M.D., J.D., M.P.H. 1994. *Physical and Biological Hazards in the Workplace.* New York: Van Nostrand Reinhold.

Willeke, Klaus, and Paul A. Baron. 1993. *Aerosol Measurement: Principles, Techniques, and Applications.* New York: Van Nostrand Reinhold.

Williams, Phillip L., and James L. Burson. 1985. *Industrial Toxicology: Safety and Health Applications in the Workplace.* New York: Van Nostrand Reinhold.

Index

9 780471 286783